Quality Assurance for Chemistry and Environmental Science

Metrology from pH Measurement to Nuclear Waste Disposal

Günther Meinrath
Petra Schneider

Quality Assurance for Chemistry and Enviromental Science

Metrology from pH Measurement
to Nuclear Waste Disposal

With 93 Figures and 39 Tables

With CD-ROM

 Springer

Dr. Günther Meinrath
RER Consultants Passau
Schießstattweg 3A
94032 Passau
Germany
e-mail: rer@panet.de

Dr. Petra Schneider
Am Tummelsgrund 27
01462 Mobschatz
Germany

Library of Congress Control Number: 2007924720

ISBN-13 978-3-540-71271-8 Springer Berlin Heidelberg New York
DOI 10.1007/978-3-540-71273-2

Springer is a part of Springer Science+Business Media

springer.com

© Springer-Verlag Berlin Heidelberg 2007

Cover design: design&production GmbH, Heidelberg
Typesetting and production: LE-TEX Jelonek, Schmidt & Vöckler GbR, Leipzig, Germany
Printed on acid-free paper   52/3180/YL - 5 4 3 2 1 0

Measurement results are information.
Without knowledge concerning its determination
and its uncertainty, however, it is just rumour.

Preface

A more systematic introduction of metrological concepts in the chemical measurement community is taking off at the beginning of this twenty-first century: the introduction of metrology in chemistry ("MiC") is proceeding.

What is now needed is a better knowledge of basic concepts of measurement (and associated terms "labeling" these concepts) for the chemical measurement community. That requires better understanding and a more systematic explanation of these concepts, in order to produce the necessary justification of metrology in chemical measurement.

Is all of that really needed? The answer is "yes". A measurement result cannot be a mere declaration of an isolated figure. Any author of a measurement result must be able to demonstrate where his/her result comes from (its metrological traceability) and locate this measurement result in a larger conceptual framework that is common to all measurement. In addition, since we communicate more intensively as well as globally *about* measurement results, and *by means of* measurement results, in a large variety of applications, we need to do so using concepts and operational procedures which are intercontinentally understood and which are described ("labeled") by means of intercontinentally agreed terms.

In communication between parties, languages are needed as vehicle for the ideas that we want to exchange. When measurement results are involved in such a vehicle language, concepts about measurement are needed that are understood in the same way by all parties concerned. Commonly, which nowadays means intercontinentally, agreed terms in one language are necessary as "labels" for these concepts.

Terms are the tools in the writings we use in our relations, especially in relations of a border-crossing and culture crossing nature. That leads to the need of a *correct translation* of such intercontinentally agreed terms into (necessarily different) terms in many other languages, perhaps 30–40 or more. All of this must first be achieved in one language, presumably English, otherwise, any translation attempt from English into other languages would be futile. We also need to talk clearly to ourselves in the first place, in order to precisely formulate our thoughts. There too, clarity is of the utmost importance. Lack of clarity in conceptual thinking about measurement and all its features only generates lack of clarity of the text we write. Lack of clarity in our writings also influences, albeit unwillingly, the clarity of our thinking. Lack of clarity in thinking always constitutes a major impediment to understanding, and therefore to any agreement in whatever field of application, whether it be in intercontinental agreements on trade, on monitoring the implementation of border-crossing environmental regulations, or of dispute settlements at the World Trade Organisation.

Clarity about a measurement result also requires knowing and understanding its limitations, i.e. evaluating the degree of doubt which must be associated with *any* measurement result. In measurements of the twenty-first century, we call that *measurement uncertainty*. That is a delicate matter, since we have a large (sometimes very large!) tendency to underestimate that uncertainty. We do want to look "good" with the "uncertainty" of our measurement result by displaying small "error bars" (obsolete wording) and do not realise that the price to pay for incorrect small "error bars" is high: many so-called significant differences in measured quantity values do not mean anything at all.

In a time where the teaching of basic and general concepts in measurement has almost vanished from the chemistry curricula in universities and technical schools throughout the world, this book is a valuable contribution to remedying this worrying deficiency.

Professor Dr. P. De Bièvre
Editor-in-Chief
'Accreditation and Quality Assurance – Journal for Quality, Comparability and Reliability in Chemical Measurement'

Contents

List of Symbols

A	area, cross-section, Debye–Hückel factor
a	activity (amount of substance), estimated intercept
a'	intercept
b	estimated slope
b'	slope
Co	Courant number
c	speed of light
D	Debye–Hückel term, diameter
$D_{m,i}$	molecular coefficient of diffusion of substance j
d	path length
$d()$	maximum deviation in Kolmogorov–Smirnov test
EDF	empirical distribution function
F	function
$F_{r,n-r,\alpha}$	Fisher's F distribution at r and n − r degrees of freedom and confidence level α
G	total number of bootstrap permutations
ΔG_R°	Gibbs free energy of reaction R at conditions $^{\circ}$
g	constant of gravity
H_l	longitudinal dispersion tensor
ΔH	enthalpy difference
h	height, suction (in m)
I	ionic strength
ILC	interlaboratory comparison
j	concentration flow vector
K	formation constant
k	permeability, metrological expansion factor
K_f	hydraulic conductivity
L	length
m	mass
m_i	molal activity of substance i
$N(\sigma, \mu)$	normal distribution
n	sample size
K_D	sorption coefficient
P	purity, porosity, model parameter

$P(x)$	Poisson distribution of x
PDF	probability density function
Pe	Peclet number
p	pressure
r	number of parameters (in a mathematical formula)
R	gas constant, retardation coefficient, Reynolds number
RND	pseudo-random number
s	sample standard deviation
S_{xx}	sum of residuals
T	absolute temperature
t_x	time at moment x
$t_{df, \alpha}$	Student's t at df degrees of freedom and confidence level α
U	expanded uncertainty
u	uncertainty
u_c	combined uncertainty
v	sample variance
V	population variance
w_i	i-th statistical weight
$x(i)$	mole fraction of substance i
z_i	ionic charge of substance i
z_t	t-th statistical residual
α	confidence level
α_l	longitudinal dispersivity
β	formation constants
χ^2	chi square distribution
Δ	difference
δ	density
ε	molar absorption coefficient
ε_i	i-th residual
η	dynamic viscosity
ν	kinematic viscosity
γ_i	activity of substance i
Φ	cumulative normal distribution
ϕ	aquifer water content
μ	statistical mean value, viscosity
Ω	amount of water flowing
ϱ	coefficient of correlation
σ	population standard deviation
$\in$	interaction coefficient
∂	partial derivative
[i]	molar concentration of substance i

Introduction

Metrology in chemistry is a rather young discipline within metrology. The development of metrology in chemistry affects all fields relying upon information derived from chemical measurement. Not only chemistry itself is affected from introducing metrological concepts in chemistry, but also food chemistry, clinical chemistry and environmental chemistry, as well as geochemistry, hydrogeology and climatic research, to name a few examples. In many situations, the people working in these fields will face the evolving metrological requirements unprepared. It is not an uncommon experience of the authors that the term "metrology" is considered to be a misprint of "meteorology". In traditional professional training of chemists, metrology in chemistry has not played any role and the concepts of metrology are at present largely unknown. Instead, traditional concepts, e.g. detection limit, repeatabilities or expert judgement, are common. Those concepts have their merits, but only metrological principles can assure comparability within different measurements of the same quantity. These principles, despite their urgent necessity, are not at present an element of chemical measurement. Taking a look to other faculties, e.g. engineering sciences (especially production engineering) quickly highlights the enormous perspectives chemical measurements can offer in future, on the condition of a metrological network properly implemented and maintained.

The first part of this treatise gives an introduction into metrology, in general focusing on metrology in chemistry. The key concepts of metrology, e.g. metrological traceability, the complete measurement uncertainty budgets and cause-and-effect analysis with application to chemical measurements, are presented on the basis of selected examples. These examples include some statistical concepts such as robust regression and computer-intensive resampling methods. The second part deals with geochemical modeling with a focus on the limitations imposed on computer calculations due to the limited accuracy and precision of input data. The third part illustrates the application of metrological principles to hydrology. Guidelines are presented to judge field measurement values, e.g. for permeabilities, on the basis of fundamental criteria. While the first two parts have been written by G. Meinrath, the third part is mainly written by P. Schneider.

The topics presented in this book are of interest to a wide audience. The analytical chemist may take this book as a primer in metrology, while the hydrogeologist may learn to judge measurement uncertainty of field data. Chemical engineers will get inside the limits of accuracy to which chemical data can be determined, which in turn form a basis to judge the reliability of their computations, simulations and predictions. Water resources managers, both on the local and national level, will find some support in judging proposals arriving on their desks. Last but not least, metrology in chemistry interferes with political issues, e.g. regulatory discharge limits for contaminants, nuclear waste disposal and mining site remediation. In the field of waste management in general, chemical measurements are essential and often lead to controversy, where the language of metrology may become an important tool to improve communication. Nuclear waste disposal is a prominent field where mutual trust is essential to master future challenges. Decision-makers on all administrative and political levels will profit from the book, as metrology provides the common language to communicate about data, their quality and limitations within the framework of international contracts and agreements.

It is our pleasure to acknowledge the suggestions and comments obtained by our valued colleagues Professor Dr. Bernd Delakowitz (Zittau/Germany) and Professor Dr. Broder Merkel (Freiberg/Germany). Further input came from Petra Spitzer at Physikalisch-Technische Bundesanstalt Braunschweig, Professor Dr. Christian Ekberg and A. Ödegaard-Jensen at Chalmers University of Technology Göteborg/Sweden and the group of Professor Dr. S. Lis at Adam Mickiewicz-University Poznań/Poland. While all persons involved in the production of this book have done their best to avoid any shortcomings, we are fully aware that the book may contain errors and misprints. We are thankful to the readers who will direct us to such cases. The CD shipped with this book holds computer programs. The programs are meant as illustrations of how the concepts outlined in this book may work in practise. However, none of the programmers is a professional. In addition, the enormous complexity and variety of modern operating systems makes it impossible to generate computer code running on all machines without problems. We ask your leniency in case of trouble with the programs.

G. Meinrath and P. Schneider
Passau & Chemnitz   March 2007

to Andrea   *to my family*

Acknowledgement to Copyright Holders

The Institute of Reference Materials and Measurements has kindly permitted reproduction of Fig. 1.48 and Fig. 2.9.

The following text is added on request of EURACHEM and CITAC: The guides "*Quantifying Uncertainty in Analytical Measurement*", "*Traceability in Chemical Measurement*" and "*Quality Assurance for Research and Development and Non-routine Analysis*" are published by kind permission of EURACHEM and CITAC. The guides can also be downloaded free of charge from their websites, www.eurachem.org and www.citac.cc, respectively.

About EURACHEM

EURACHEM is a network of organisations in Europe having the objective of establishing a system for the international traceability of chemical measurements and the promotion of good quality practices. It provides a forum for the discussion of common problems and for developing an informed and considered approach to both technical and policy issues. It provides a focus for analytical chemistry and quality related issues in Europe. Each country has its own national EURACHEM organisation, and the contact details are given on the EURACHEM website. One of the principle activities is the preparation of guides and organising of conferences by the various working groups. Anyone wishing to join one of these groups either as a full member or on a "papers for comment" basis should contact their national representative.

About CITAC

CITAC (Co-operation on International Traceability in Analytical Chemistry) has similar aims to EURACHEM, but operates on a world wide basis to improve the international comparability of chemical measurement.

1 Concepts of Metrology

"The search for truth is more precious than its possession."

(A. Einstein)

Metrology is an essential element of daily life, but goes almost unnoticed. A meter rule has become a common tool in most households. A length is measured, its value is noted, and in another place, another meter rule is used to compare the length of an item with the figures noted previously. It rarely occurs that this procedure results in significant deviations. In fact, using a reference (meter rule) to determine a value of a quantity (length) by comparison (measurement) is a basic metrological activity. The procedure described above works, because the meters involved in this comparison relate to the same reference meter. The deviations between that reference meter and the custom tools available in most households of the economically developed countries are not relevant for daily application needs – the meters are fit-for-purpose. It would, on the other hand, be considered as unreasonable by most people to use these tools for measurement in the sub-millimeter range. The metrological concepts of measurement and fitness-for-purpose have become a matter of course.

Most data cannot be obtained with arbitrary accuracy. Obtaining information by comparison in most cases includes some doubt on the accuracy of this information. This doubt needs to be communicated between the parties involved in a decision-making process. It is, for example, a basic requirement that the meter rules used in the assessment of a part's dimensions are comparable to each other on the level of the part's acceptable tolerance. Consequently the tolerance itself needs to be assessed. The fundamental framework of establishing quality, fitness-for-purpose, and metrological traceability is provided by metrology. Metrology is, in short, the science of measurement and its communication within the parties affected by the result of the measurement.

It has become a common requirement in economic and social life to give some statement on the future effects of a measure taken. Often, these statements include a certain liability for the effects of the measure, because at least two parties are involved in the activity. These parties do not usually have common interests. A simple example is the seller-client situation, where the seller is interested in his profit and the client in cheap products with long lifetime of products and services. Such situations will be referred to as "situations of conflicting interests". In situations of conflicting interest, agreement by convention is not possible. The measure may be the delivery of a part with specified dimensions and the effect may be to fit into some machine. The measure might be the remediation of a contaminated site and the effect can be the improvement of water quality in a close-by water body.

The major response to the emerging relevance of metrological concepts in the field of chemical measurement is the introduction of the "Guide to the Expression of Uncertainty in Measurement" (GUM), issued by the International Standard Organisation (ISO). The GUM was issued in conjunction with the International Organisation for Legal Metrology (OIML), the International Electrotechnical Commission (IEC), the International Union for Pure and Applied Chemistry (IUPAC), the International Union of Pure and Applied Physics (IUPAP), and the International Federation for Clinical Chemistry (IFCC). Thus, an international convention exists for expressing uncertainty in measured data that is valid in many fields of technology, science and commerce. Within this convention, metrology became an essential aspect in fields where other conventions (expert judgement, bi-lateral agreements etc.) had been acceptable before. Alternative conventions have commonly been in use with data obtained by chemical analysis. Such local agreement on the basis of mutual consensus is no longer acceptable, as soon as it will be applied outside this restricted range and, most importantly, in situations of conflicting interests.

1.1
Organisation and Framework

"In questions of science, the authority of a thousand is not worth the humble reasoning of a single individual."

(G. Galilei)

1.1.1
Metrology – An Introduction into its History and its Organisation

Metrology is a very old practice. A measurement is a comparison. To compare is a basic human activity. The written documents from the earliest civilisations give testimony of a complex social framework whose economic basis was maintained by collecting and distributing valuables, food and ground. This essential task does not only require the ability to write, to count and to calculate. A successful ruler had to install and to maintain a system of measures in his territory for length, weights, time and volumes. We may recall the need for the administration of the Egyptian pharaohs to reassess the boundaries of fields along the Nile River following each flooding period without causing major dis-satisfaction among the affected population. Structured administrations were needed to accomplish these tasks. In short, metrology is an essential element of civilisation.

The application of metrological concepts to chemical data is not a new sub-ject either. The content of precious metals in ores may be taken as a prominent example. Salt is another important chemical compound where the amount had to be assessed as accurately as possible. Due to their high purity, these amounts were measured on a by weight basis, and a constant subject of fraud and treachery.

Physical measures are accepted on a conventional basis. Each authority could (and usually did) establish its own units, which in turn were overthrown with the change of authority. So, an enormous number of measures for, e.g. length have been established, modified and abolished during the history of human civilisation. Commonly it is difficult to translate the size of a his-torical base unit into currently accepted units. We may remind to the first known determination of Earth's circumference by Eratosthenes of Alexandria in 250 BC. He estimated 250 000 stadiums. Because we do not know exactly the length of a stadium in Eratosthenes' days in meters, all we can say is that his measure corresponds to about 37 000 km assuming a conversion factor of 1 stadium = 148.5 m. The multitude of weights and measures with often very local and short-term acceptance flourished with the emerging trade and manufacturing structures in late medieval societies. In 1789, more than 200 different units of the quantity "length" were used in France, all of which were named "toise". More than 2000 different measures for weight and lengths were in use.

A fundamental task of modern metrology is to build and to safeguard trust into the values obtained from a measurement (Price 2000). This trust in the measurement procedures and measurement results becomes an essential ele-ment in situations of conflicting interests which are a basic feature, e.g. of all trade actions: the client wants to acquire an item as cheaply as possible, while the vendor wants to get a price as high as possible. In medieval times, the larger settlements in Europe, often protected by massive walls, realised that trade was a beneficial activity for their own development. Local administration started to control the transactions by establishing fundamental measurement units for all relevant trade activities of mostly local validity. This activity pro-moted trade by establishing trust. Setting and control of the measurement units became an essential element of sovereignty and power. In pre-revolutionary France, the noble class controlled the units and measurement instruments. It is known from historic records that the abuse of this control sparked consider-able dissatisfaction in the French population, finally giving rise to the French Revolution (Guedj 2000).

It is therefore not surprising that despite all chaos and bloodshed following the year 1789, a metrological commission was established. This metrological commission, whose noted members were (among others) Lavoisier, Laplace, Lagrange, Coulomb and Condorcet, created those base units which are in world-wide use today: meter, kilogram, and second. The almost universal application of these base units in international relationships owes much to the far-sighted decision by the members of the metrological commission to select references which were independent from subjective decision. Instead they are derived from nature. Delambre and Méchain spent years in determining the length of the Paris meridian by triangulation between Barcelona and Dunkerque. Its length serves as a basis for the unit of the quantity "length" for which the term "mètre" was invented.

In 1875, the Meter Convention was signed among 17 nations to secure the world-wide equivalence of the unit "meter" and the derived mass unit "kilogram". Today, a complex system of international and national metrological institutes ensures equivalence of the seven base units of the metrological system: meter, kilogram, second, Kelvin, Ampère, candela, and mole. The mole is the most recent unit in the SI, added in 1972. While the mole is defined as the number of atoms equivalent with the number of atoms in 12 g of carbon-12, the process of comparing a given sample with 12 g of carbon-12 is rather complicated. There are, for example, a wide variety of matrices in which the amount of only one substance needs to be determined. There are also an enormous number of different chemical species that need to be distinguished. In some cases, not only the chemical composition but also the specific three-dimensional arrangement of the atoms within the molecule matters. There are the often rather limited stabilities of chemical compounds, e.g. some coordination compounds in solution may "exist" only in a time-average in a dynamic equilibrium where an individual entity exists only for a very small fraction of a second. To separate a species of interest from its matrix, a number of operations such as filtration, ion exchange or distillation may become necessary. In many analytical procedures, the quantity actually determined is not the atom itself, but rather some other quantity, e.g. light intensity (UV-Vis spectroscopy) or electrical current (amperometry). In such situations, calibration is necessary where the analyst has to rely on the availability of appropriate calibration standards.

The inclusion of the quantity "amount of substance" with the unit "mole" and symbol "mol" into the Système International (SI) as the seventh base unit is less a recognition of the scientific importance of analytical chemistry than a consequence of the important role the quantification of chemical elements and compounds plays in modern life and industry. The importance of SI results from an international agreement (Meter Convention) by currently 51 independent states, denouncing their sovereign rights of defining units of measurement in favour of the seven base units of meter, second, kilogram, Ampère, Kelvin, candela and mole. To ensure that, for example, a kilogram in each member state of the Metre Convention was as close together as possible, a hierarchical structure of institutions was created with the Bureau International des Poids et Mésures (BIPM) in Paris-Sèvres at top cooperating with the national metrological institutes (NMIs) as head laboratory in each member state. The BIPM is organised in committees. The highest authority is the CIPM (Comité International des Poids et Mésures). The CCQM (Comité Consultative Quantité de Matière) is responsible for the quantity "amount of substance" with unit mole and symbol mol.

The raison-d'être of this structure is economic interests. Scientific considerations play, if at all, only a marginal role. The renunciation of sovereign rights is balanced by economic benefits. It must be kept in mind that metrology is a power game, not a scientific sand box. Those who control the measurement instruments have the advantage due to superior access to information. Waiving sovereignty in a crucial subject such as weights and measures, however, cannot be solely motivated by economic benefits. In addition (and with equal importance), it is based upon a key element of human interaction: mutual trust (Quinn 2004).

In 1977, the CIPM recognised the lack of a common metrological basis to communicate uncertainty. The BIPM formed a working group in 1980 forwarding the recommendation INC-1. This recommendation should allow the expression of uncertainty in a unified manner for all technical, commercial and scientific situations. In 1986, CIPM asked the International Standard Organisation (ISO) to work on the details. ISO should derive a guide on basis of the recommendation INC-1 to establish rules for the statement of measurement uncertainty in the field of standardisation, calibration and accreditation of laboratories and metrological services.

In 1993, ISO came forward with the "Guide to the Expression of Uncertainty in Measurement" (GUM). The purpose of this guide is to "inform complete how to derive uncertainty statements" and to "create a basis for international comparability of measurement results". Several organisations are working for adapting the requirements specified in GUM to chemical measurements, especially for the determination of values of the quantity "amount of substance".

A key element of metrology is metrological traceability. Metrological traceability is a property of a value determined by a measurement which allows one to relate this value back to an accepted reference. It is the property of metrological traceability which allows to use one meter rule at home to determine a value for the length of an item and another meter rule at another place or time to reassess this length without dissatisfaction in the result. If a value is traceable, it can be reassessed at other times and other locations with the help of measurement tools that themselves are traceable to the same reference (Hässelbarth 1998).

Since 1993, the Co-operation on International Traceability in Analytical Chemistry (CITAC) has worked to communicate metrological principles to the analytical chemistry community, as well as to create understanding within the metrological community about the specific problems in chemistry. Meanwhile, metrological traceability in analytical chemistry has developed from the

BIPM to the regional metrological and chemical organisations, and is required at a laboratory accreditation according to the ISO/IEC 17025 standard. Today, CITAC's mission is to "improve metrological traceability of the results of chemical measurements everywhere in the world" (Kuselman 2004). An important result of the CITAC activities are the guides to metrological traceability in chemical measurement (EURACHEM 2004), quantification of measurement uncertainty (EURACHEM 2002) and quality assurance in the analytical laboratory (EURACHEM 1998).

Only a few of the organisations involved in the development and distribution of metrological concepts have legal authority. Metrological concepts are not always enforced by law. These concepts are mainly enforced by convention, that is a mutual agreement between partners over the widest possible range. There are countries, with the USA as a prominent example, where in daily life the SI units "meter", "Kelvin" and the derived unit "liter" are replaced by the "yard", "Fahrenheit" and "gallon". In scientific and trade affairs, however, the SI units are enforced due to the membership of the USA in the Meter Convention and the Mutual Recognition Agreement.

The fundament of mutual agreement is trust. A practical example is the Mutual Recognition Agreement (MRA) signed in 1999 by the directors of all 51 member states to the Meter Convention and the associated member states of the Meter Convention (MRA 1999). The objectives of the MRA are to establish the degree of equivalence of measurement standards by the NMIs and to provide for the mutual recognition of calibration and measurement certificates issued by the NMIs, providing governments and other parties with a secure technical foundation for wider agreements that relate to international trade, commerce and regulatory affairs. Trust in metrological procedures is a much more powerful instrument than legislation. As long as the information serving as a basis of a decision isn't questioned by the parties affected by the decision, there is no need to discuss data quality. As soon as there is conflict of interest, a common basis is needed. The world-wide metrological network is providing this basis. With the GUM, a new element has been added to this basis: a measure for the quality of information expressed by the complete measurement uncertainty budget.

The number of methods of identifying and quantifying atoms in different aggregation states and matrices has increased dramatically, spurred considerably by the field of instrumental analysis. From insights generated by quantum chemistry and quantum physics, electronic devices were manufactured allowing convenient and rapid identification and quantification of chemical elements with an ease and economy unimaginable a decade before, and that development seems set to continue. However, it is a common experience that comparable samples sent to different laboratories will not result in comparable information on the amount of substance in the samples. Metrology in chemistry can provide a more detailed substantiation of these observations.

1.1.2
The International Framework in Metrology

International trade, globalisation, standard of living and use of natural resources are affecting modern life to an extent unthinkable almost 20 years before. Information and commodities are increasingly being exchanged on a global scale. At the same time, access to natural resources is becoming more and more restricted even in developed countries. Water is a prominent example. Food quality is another issue of global importance, as are the effects of climate change. It is important to have an internationally accepted system to overcome measurement disagreement. Such a system is the International System of Units (SI). By the use of traceable measurements, the SI provides an international infrastructure for comparable measurements. This is true for all type of measurements, including chemical measurements.

The head office of the metrological infrastructure is the Bureau International des Poids et Mésures (BIPM), established by the Meter Convention. The BIPM is

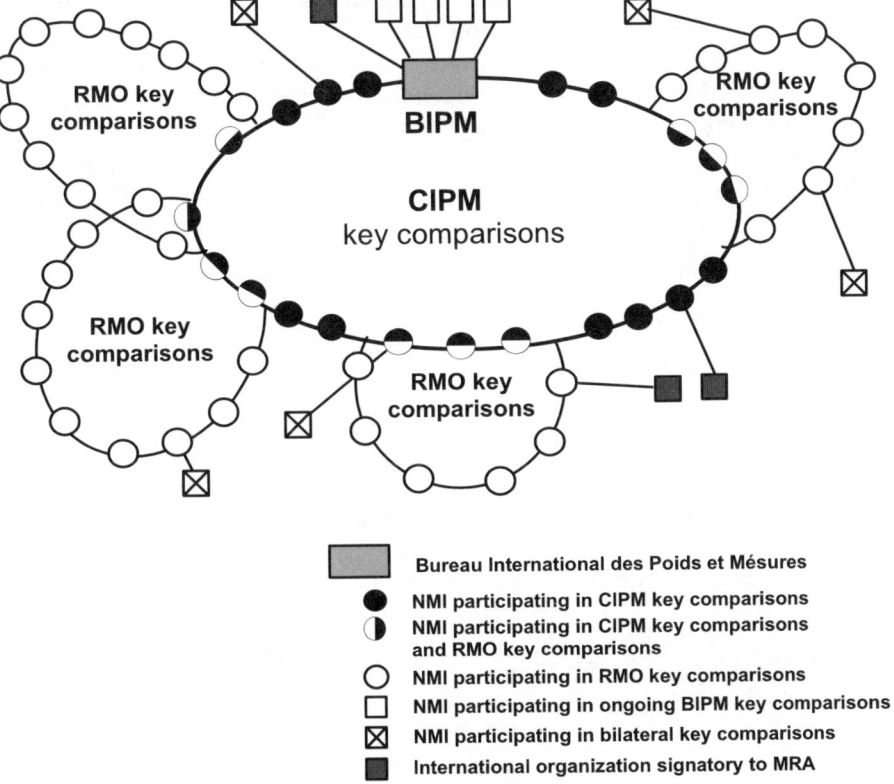

Figure 1.1. A graphical representation of the infrastructure for international comparison (according to Wielgosz 2002)

organised in consultative committees where topics on relevant issues in specific areas are discussed. In the case of metrology in chemistry, the Consultative Committee for Amount of Substance (CCQM; Comité Consultative de Quantité de Matière) is the relevant body. The members of the committees come from the respective sections of the national metrological institutions (NMIs) (Wielgosz 2002).

The NMIs are organised in regional metrological organisations (RMOs): APMP (Asia/Pacific region), EUROMET (Western Europe), COOMET (Central/Eastern Europe), SIM (Americas), SADCMET (Southern Africa) and MENAMET (Middle East/North Africa).

A central task of BIPM and the RMO is the organisation of international key comparisons. By these comparisons, the competence of the NMI and the equivalence of the results from the respective measurements are assessed. The results of the key comparisons are available to the public (KCDB 2005). These key comparisons are performed under the MRA and, most importantly, accompanied by a statement of uncertainty. These statements of uncertainty are reviewed carefully during the process laid down in the MRA and reviewed in the light of the results obtained in the key comparisons. Trust in these procedure is sufficiently high to allow the NMI to accept a measurement result obtained by another NMI. Thus, the value is not required to be repeated, thereby saving cost and time: "Measured once, accepted everywhere". Figure 1.1 gives a graphical representation of the infrastructure for international comparisons.

In Fig. 1.2, an example for the result of a key comparison organised by the BIPM is given. In the BIPM key comparison data base (KCDB 2005) further

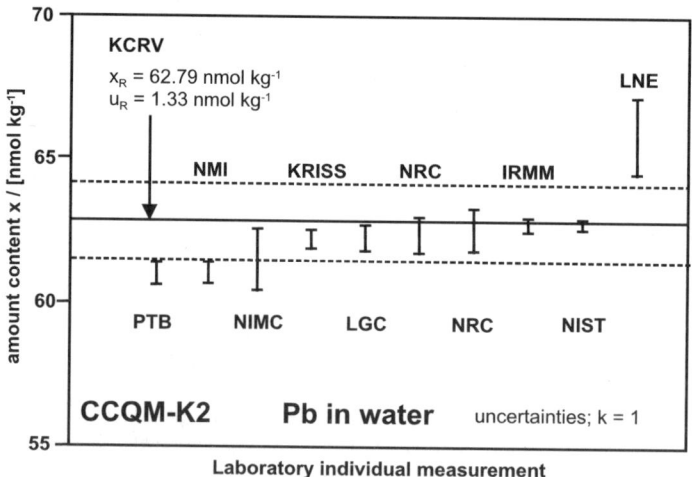

Figure 1.2. Results of key comparison "CCQM-K2" reported by different national metrological institutes (PTB: Germany; NMi: The Netherlands; NIMC: Japan; KRISS: Korea; LGC: UK; NRC: Canada, IRMM: CEC; NIST: USA; LNE: France) (according to BIPM)

examples are available for the quantity "amount of substance". Within the key comparison "CCQM-K2" concentrations of lead (Pb) in water were determined. The results of individual NMI are given, together with an measurement uncertainty budget. The amount content of Pb in the sample was determined previously by a reference method and specified together with an uncertainty budget (dashed horizontal lines). Details of this key comparison can be found at the respective website.

Such key comparisons are performed to assess the equivalence of national primary measurement procedures used to determine the amount content of lead in aqueous solutions. Even though key comparisons are performed for a wide range of chemicals, it will not be possible to investigate all materials in all matrices. This complexity of analytes and matrices is one of the challenges for metrology in chemistry (Wielgosz 2002; Clark 2003).

1.2
Convention and Definitions

"Convictions are more dangerous to truth than lies."

(F. Nietzsche)

1.2.1
A Definition of Convention

Many aspects of human life are based on conventions. Conventions are agreements between different parties. Commonly, such agreements are made in situations of mutual interests. Often conventions are abandoned if the mutual interest has disappeared. Language itself is a good example of a convention. The meaning of words is fixed in some way and the grammar rules are accepted because it allows communication. No formal treaty is necessary. We follow the rules because we enjoy the benefit of communication. There are alternatives. This is realised when we travel into a region where different language conventions are followed. We may decide to learn a foreign language because we would enjoy communicating with those people following the different convention. Somehow it would be comforting if one language convention would be valid globally. Wherever we are we could be sure to be able to communicate.

This situation is similar for the results of measurements. It would be good if the measures themselves would be globally valid. Furthermore, we want to be sure that a length we have determined here corresponds to the same length elsewhere. For international trade, such a situation is essential. Ordering parts from a foreign company requires that the dimension of the parts fit into the intended equipment. They also need to have the required quality. That is: being suitable for the intended use.

It is helpful for an understanding of the importance of metrology to consider the following definition of convention (Lewis 1969):

"A behavioural regularity R within a population P in relevant recurrent situations is a convention if (and only if):

1. There are alternatives to R.
2. Everyone conforms to R.
3. Everyone expects everyone else to conform to R.
4. Everyone prefers to conform to R, rather than any of the alternatives on condition that everyone else conforms to it.
5. The result of R is of advantage for the members of P."

An example will make this definition more evident. We are educated to the convention of stopping our cars at a red light and continuing our ride at a green light. (1) The alternative would be to stop at the green light and to drive at the red light. (2) We stop at the red light. (3) We expect everyone to stop at a red light. (4) We prefer to stop at a red light and to drive at a green light only as long as everyone does the same (some people, even though conforming as car drivers behave different as pedestrians at a red light) and (5) we enjoy the safety resulting from this convention.

It is crucial to understand that conventions are restricted to a population profiting from the consequences of adhering to the behavioural regularity R – that is in cases of mutual interest where criteria 1–5 apply. Systems of measurement units are conventional, as the name "Meter Convention" says (Price 2001). If criterion (5) is not fulfilled because there is conflict of interest, a convention does not apply. In case of conflict of interest, a measurement convention can only persist if both parties have independent means to assess the situation according to common rules and protocols. Such rules and protocols are developed under the auspices of the Meter Convention.

1.2.2
Terms of Metrology: A Measurement is a Comparison

It is important for communication that a term has the same meaning for all users. Hence the terminology of metrology needs clear definitions (de Bièvre 2004a). These definitions are summarised in the "International Vocabulary of Basic and General Terms in Metrology" (VIM) (VIM 1994). The vocabulary has been prepared simultaneously in English and French by a joint working group consisting of experts appointed by BIPM, IEC, IFCC, ISO, IUPAC, IUPAP and OIML and published in the name of these organisations.

It is outside the scope of this book to present all definitions given in the VIM. To illustrate its organisation, the following definition of a (measurable) quantity is given:

"measurable quantity":

Attribute of a phenomenon, body or substance that may be distinguished qualitatively and determined quantitatively.

Notes:

1 The term "quantity" may refer to a quantity in a general sense or to a particular quantity.

Examples:

a) Quantities in a general sense: length, time, mass, temperature, electrical resistance, amount-of-substance concentration;

b) Particular quantities: length of a given rod, electrical resistance of a given wire, amount-of-substance concentration of ethanol in a given sample of wine.

2 Quantities that can be placed in order of magnitude relative to one another are called quantities of the same kind.

3 Quantities of the same kind may be grouped together into categories of quantities, for example:
work, heat energy
thickness, circumference, wavelength.

4 Symbols for quantities are given in ISO 31.

Important for the discussion in this book are the following conventions. They are taken from VIM (VIM 1994) and, in most cases, given without examples, notes and annotations available in VIM.

Measurement procedure:

"A set of operations, described specifically, used in the performance of particular measurements according to a given method."

Measurand:

"Particular quantity subject to measurement."

Influence quantity:

"Quantity that is not the measurand but that affects the result of the measurement."

Result of a measurement:

"Value attributed to a measurand, obtained by measurement."
The definition includes the following note: "A complete statement of the result of a measurement includes information about the uncertainty of measurement".

Repeatability:
"Closeness of the agreement between the results of successive measurements of the same measurand carried out under the same conditions of measurement."

Reproducibility:
"Closeness of the agreement between the results of measurements of the same quantity carried out under changed conditions of measurement."

The following definition is central for the discussion in this book. Therefore, it will be given together with the Notes.

Uncertainty of measurement:
"A parameter, associated with the result of a measurement, that characterises the dispersion of the values that could reasonably be attributed to the measurand."

Notes:

1. *The parameter may be, for example, a standard deviation (or a given multiple of it) or the half-width of an interval having stated level of confidence.*
2. *Uncertainty of measurement comprises, in general, many components. Some of these components may be evaluated from the statistical distribution of the result of series of measurements and can be characterised by standard deviations. The other components, which can also be characterised by standard deviations, are evaluated from assumed probability distributions, based on experience or other information.*
3. *It is understood that the result of the measurement is the best estimate of the value of the measurand, and that all components of uncertainty, including those arising from systematic effects, such as components associated with corrections and reference standards, contribute to the dispersion.*

Measurement standard:
"Material measure, measuring instrument, reference material or measuring system intended to define, realise, conserve or reproduce a unit or one or more values of a quantity to serve as a reference."

(Metrological) traceability:
"Property of the result of a measurement or the value of a standard whereby it can be related to stated references, usually national or international standards, through an unbroken chain of comparisons all having stated uncertainty."

Calibration:
"The set of operations that establish, under specified conditions, the relationship between values of a quantity indicated by a measuring instrument or measuring system, or values represented by a material measure or a reference material, and the corresponding values realised by standards."

1.2.3
Concepts of Metrology

The discussion as to whether true values actually exist is very much a philosophical one. It is clear that the "true value" of the mass of the kilogram prototype at BIPM is known to be 10 000 kilogram, by definition. In most situations, however, the true value of a measurand cannot be known with arbitrary accuracy. On the other hand, experimenters implicitly assume that a measurand has a fixed value. The speed of light in vacuum, for instance, is assumed to have one value which does not change in space and time. Otherwise, the theories of Albert Einstein would not be valid, as they have been proven experimentally and theoretically. On the other hand, our knowledge of this value is affected by some doubt, which nevertheless has reduced considerably during the past 200 years. A glimpse of this discussion on true values can be obtained from de Bièvre (2000d), Meinrath (2002) and Fuentes-Arderiu (2006).

1.2.3.1
True Values and Conventional True Values

In most cases, to measure also means to approximate to a true value. By scientific measurement, information can only be obtained about reproducible phenomena. Singular events, interesting as they might be, cannot be subject to scientific experimentation. Even in a sample which undergoes chemical analysis to obtain a value for an amount concentration of an element or compound, the emphasis is on the reproducible phenomena which allow establishment of a relationship between a measurement signal and the amount of substance.

On the other hand, true values cannot be a stable basis for a measurement network, which serves the interests of the sciences only as a kind of collateral effect. A metrological framework therefore relies on artificially produced reference materials instead of mostly inaccessible "true values". To avoid confusion, the agreed values of a measurand in a reference material will be termed "conventional true value". These conventional true values always carry a measurement uncertainty. The true values, which lie in the heart of scientific investigations, do not carry a measurement uncertainty, but our knowledge of these true values necessarily carries some unavoidable doubt. Hence, while the pun is on slightly different facets, the machinery of metrology in chemistry serves both interests simultaneously.

1.2.3.2
Uncertainty and Error

The introduction of the Guide to the Expression of Uncertainty in Measurement (GUM) (ISO 1993) has stressed the importance of a reliability estimate also for chemical measurement. While uncertainty in the broad sense is no new concept in chemistry, the GUM underpins that a statement of uncertainty must be meaningful information on the accuracy of the result. Meaningful estimate implies comparable estimate. Hereby, comparability does not only mean comparability to other values obtained by the same method in the same laboratory for the same sample, but comparability over space and time. A correct interpretation of accuracy ensures that results are judged neither overly optimistically nor unduly pessimistically (Ellison et al. 1997).

Uncertainty is different from error. Error, according to VIM, is a single value: the difference between a measurement value and the true value. Hence, error can be corrected if it can be quantified. The uncertainty is an inherent component of measurement. It can be quantified but not corrected. Figure 1.3 illustrates this difference schematically. The GUM explicitly excludes gross errors of procedure from consideration within an assessment of the complete measurement uncertainty budget. Gross error of uncertainty refers, for example, to inappropriate protocols, incorrect calibration and mistakes (e.g. using the wrong calibration standard). The GUM uncertainty estimates apply only to measurement processes under statistical control. Proper quality control measures therefore are a prerequisite to apply the GUM concept.

Measurement uncertainty is a quantitative expression of doubt. Doubt is a psychological phenomenon. It cannot be the subject of a measurement process. However, it can be communicated in an objective manner. Like many other elements of human life, it is a subject of convention, based on mutual consensus and trust. There is the surprising fact that such an agreement can be reached even in situations of conflicting interest despite the differences and incompatibiities of human psychology (von Weizsäcker 1981).

The GUM specifies two types of uncertainty. These uncertainties result either from a "type A" evaluation or a "type B" evaluation. There is no fundamental difference between the both types of uncertainty. The main purpose of distinguishing both types is to ease discussion. Furthermore, the both types do not intend to replace the more familiar types "random" and "systematic".

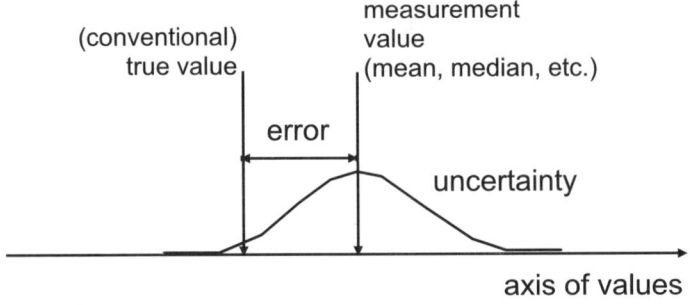

Figure 1.3. Schematic representation of the concept of error and uncertainty

Both types of evaluation of uncertainty components result from probability distributions, and are represented quantitatively by variances or standard deviations.

An uncertainty component of type A is derived from an observed probability distribution, while type B uncertainties result from other means, e.g. assumed probability distributions. The standard uncertainty of a value of a measurement is termed "combined standard uncertainty" with symbol u_c if it is obtained from several uncertainty contributions. Furthermore, an "expanded uncertainty" with symbol U is defined, which results from u_c by multiplication with an expansion factor k. The expansion factor k must be reported together with an expanded uncertainty U, e.g. U = 0.12 ($k = 2$).

Already in recommendation INC-1 (BIPM 1980), it is stated that there is not always a simple correspondence between the categories A and B and the familiar classification into "random" and "systematic". However, the term "systematic uncertainty" is likely to result in erroneous interpretation and should be avoided in a metrological discussion.

1.2.3.3
Complete Measurement Uncertainty Budget

To measure means to compare. All measurement is a comparison with a reference. The value of the measurand is given in multiples of that reference. A result of a measurement is always given as a combination of a numerical value and a name identifying the reference (Price 2001). Even Eratosthenes followed this convention when he reported the circumference of the Earth to be 250 000 stadiums. A problem is that today the reference "stadium" is known only rather imprecisely.

To perform a measurement, a measurand must be chosen and a measurement procedure applied. By comparison with a reference, a value will be obtained. Giving the measurand, the value and the reference, the result of the measurement can be communicated:

$$\text{quantity of measurand} = \text{value} \times \text{reference} . \tag{1.1}$$

After introduction of the Guide to the Expression of Uncertainty in Measurement (ISO 1993), this statement is incomplete. According to the note in the definition of the "result of a measurement", information about the uncertainty of a measurement has to be included. Thus, the result of a measurement must be communicated in a form

$$\text{quantity of measurand} = (\text{value} \pm \text{uncertainty}) \times \text{reference} . \tag{1.2}$$

The task of assigning a meaningful estimate of uncertainty to the value obtained by a measurement is a comparatively recent one. For chemists, this requirement is almost completely new. An estimate of, for example, an amount-of-substance concentration is rather rarely seen with an associate estimate of uncertainty. Here, pH may be taken as a prominent example. The quantity pH plays an important part in medical science, food science, environmental science, health science, chemistry, biology and many other fields. Thus, a considerable effort has been spent by BIPM and NMI to derive a procedure for assignment of meaningful estimates of uncertainty toward values of the quantity pH (Baucke 2002; IUPAC 2002; Spitzer and Werner 2002).

The protocols for assignment of meaningful measurement uncertainty to the value of a quantity are not quantity-specific. Hence, each experimenter is required to carefully study the applied measurement procedure. A first step towards this goal is the identification of relevant influence quantities (EURACHEM/CITAC 2002). A second step will be the quantification of the magnitude a relevant influence quantity may reasonable have on the value of the measurand. In the last step, the uncertainty components are combined into the complete measurement uncertainty budget.

Identifying influence quantities and their magnitude is an important step towards comparable measurement values. The following example may illustrate the need for a careful assessment. Table 1.1 lists weights obtained by students for pipetting 1 ml distilled water into a baker. Here, the calibrated balance is used as a reference. The pipette has been an Eppendorf pipette. The students have been given the freedom to change the tip after each sampling or to use the same tip for all ten samplings.

The tabulated value for the density δ of water at $20\,^{\circ}$C is $\delta = 0.9982\,\mathrm{kg\,dm}^{-3}$. In Fig. 1.4, the respective mean values and confidence intervals from the meas-

Table 1.1. Weights (in $\mathrm{kg\,dm}^{-3}$) obtained by pipetting 1 ml distilled water into a beaker

Group no. Sampling no.	1	2	3	4	5	6
1	1.0023	1.0124	0.9561	1.0191	1.0129	1.0125
2	1.0023	1.0117	1.0082	1.02	1.0132	1.0044
3	1.0028	1.0107	1.0017	1.0167	1.0112	1.0052
4	0.9993	1.0038	1.0108	1.0138	1.0128	1.0039
5	0.9903	1.012	1.0074	1.0112	1.0132	0.9978
6	1.0015	1.0096	1.0047	1.0171	1.0089	1.0108
7	0.9982	1.0109	1.0094	1.0141	1.018	1.0051
8	1.0139	1.0093	1.0031	1.015	1.0096	1.0123
9	1.0079	1.0085	1.0052	1.0121	1.0088	1.0023
10	1.0067	1.0103	1.0083	1.0064	1.0062	1.0048
Mean	1.0025	1.0099	1.0015	1.0146	1.0115	1.0059
SD	0.0063	0.0025	0.0162	0.004	0.0033	0.0047
0.95% CI	0.0143	0.0057	0.0366	0.009	0.0075	0.0106

SD standard deviation; *CI* confidence limit

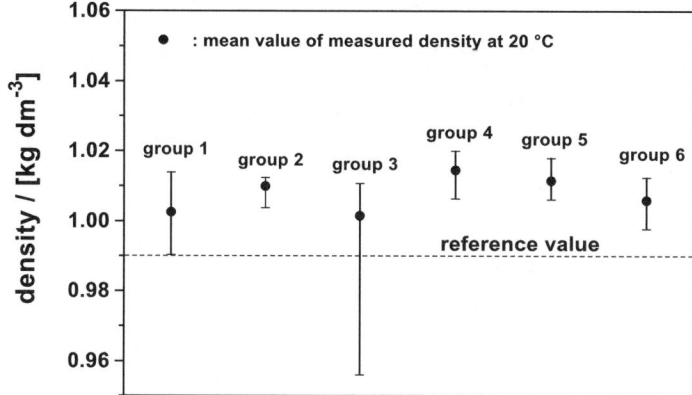

Figure 1.4. Graphical representation of the data given in Table 1.1

urement values are compared to the reference value for the density of water at 20 °C. The mean values of all groups are above the reference value. Some bias might be suspected. The data in group 3 have a 95% confidence interval with sufficient spread to cover the reference value. Its result is least precise but the most accurate.

Considering only the mean values, the value for the density of water $\delta(H_2O)$ from the six groups is $\delta(H_2O) = (1.0077 \pm 0.0052) \, \text{kg dm}^{-3}$; whereby the figures behind the "$\pm$" symbol represent the standard deviation. This value for the density considers only repeatability.

If the volume transfer is an influence quantity of a measurement, its uncertainty contribution must not be ignored. Instead of, say, 1 ml of a reagent almost 0.5–2% more would be transferred according to the data above. Because such volume operations play an important role in the preparation of stock solutions, titrants and calibration standards, the effects accumulate.

Despite the fact that only a simple operation has been performed, the different groups have obtained different mean values and the variance of results also differs. Different equipment, different calibration standards and other factors cause further variability in the result. This is schematically illustrated by Fig. 1.5. The horizontal bars represent possible results considering the complete effect of influential quantities in each step of the analytical procedure. The spread of the bars represents the distribution of the results in the respective step of the measurement process. Ignoring influential contributions, the experimenters will propagate only their respective (random) value. Eventually a result is achieved. In chemistry, the result is commonly only the mean value without a statement of uncertainty. Consequently, only the difference in the results can be acknowledged. In the case of experimenter A, the result is outside the conventional true value (say, of a standard), while experimenter B's result is within the range of the con-

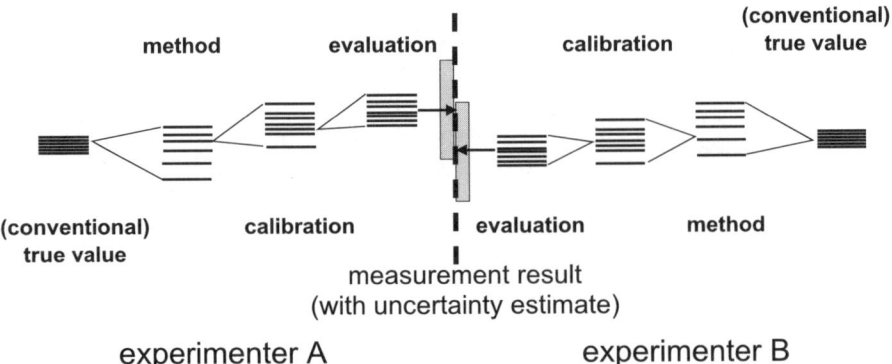

Figure 1.5. Illustration of the possible outcome from chemical analysis of a reference sample (with a conventional true value) under neglect of the uncertainty contributions from influential quantities. The forwarded mean values (*arrows*) differ. If the uncertainties are evaluated, a range (*grey bars*) is obtained. The resulting uncertainties may look wide but they overlap and they cover the range of the conventional true value (de Bièvre 2002). Note that this figure is an extension of Fig. 1.3

ventional true value range. Nevertheless, this difference is merely accidental. (It may be argued that the experimenters have just performed an operation to obtain random numbers. As with dices, where six different outcomes are possible but only one outcome is finally realised, the experiments illustrated in Fig. 1.5 have a range of possible outcomes. In the light of possible outcomes in each step of the analysis it is mere accident which value is finally realised).

If influential quantities are neglected and only the resulting mean values are used in the evaluation of the experiment, the uncertainty contributions which result from the choice of the method, from the type of calibration and from the method of evaluation are not appropriately considered. Hence, starting from the same material, both experimenters A and B will achieve differing results. Including the measurement uncertainty (grey bars), the both uncertainty ranges overlap and indicate the region where the conventional true value may be found. Note that evaluation of the measurement uncertainty neither improves the measurement process, nor does it reduce the magnitude of the uncertainty or protect against erroneous experimentation and inadequate data evaluation.

In the schematic representation of Fig. 1.5, the sample was assumed to be a well assessed reference sample with stated uncertainty. Often, however, the amount concentration of the measurand in the sample is unknown. To fully appreciate the message of Fig. 1.5, let us assume that the measurand is the amount concentration of ethanol in blood. In many countries limiting values for the amount concentration of alcohol in blood exist. Exceeding these limits may have drastic consequences for the car driver, including loss of the driver's licence with subsequent social consequences. A blood sample given to experimenter A would have exceeded that limit, a blood sample given to experimenter B would not have exceeded this limit (on the basis of the mean values): the outcome, however, is purely random under the premises of Fig. 1.5. Measurement uncertainty of breath-alcohol analysis has been discussed under metrological aspects (Gullberg 2006) underscoring the necessity of metrological traceability and comparable measurement values in daily life. Hence, assessment of the complete measurement uncertainty budgets for determination of amount concentrations is by no means a theoretical exercise. It affects many important decisions which itself affect our life. The problem of conforming to legal limits will be treated further in Chap. 1.7.

1.2.3.4
Metrological Traceability

A measurement is a comparison. Comparison is a relative process. To compare measurement values it is important that they relate to the same reference. This simple fact is the raison d'être of the SI units and their realisations. There is no use to have a definition of a reference – it must be realised and available for comparison. It is of equal importance to have these material standards all over the world. The meter rule used in a household in Northern Germany should correspond to a meter rule used in India; some granite plates ordered in India may have to fit into a building constructed in Northern Germany.

The comparability of measurements done at different places and at different times, using different methods and different personnel is of great importance (Golze 2003). This comparability is achieved by linking a measurement to an internationally accepted standard, be it a SI unit or some derived standard (Price 1996). The link between the standard and the accepted reference is the traceability chain. The definition of metrological traceability requires to have accepted standards and to have laboratories with demonstrated link between their measurements and the accepted reference (de Bièvre and Williams 2004). Metrological traceability, as a key element in metrology, is the subject of a CITAC Guide (EURACHEM/CITAC 2004).

If two parties want to compare their measurement values, they need to make the measurement values traceable to a common reference. On the international scale, the units of SI are a natural choice because these references are agreed upon by various treaties and supported by large measurement programs. Nevertheless, a traceability chain between a reference and a measurement value will look like shown in Fig. 1.6.

A common reference material has to be found, e.g., a suitable high-purity material. For the mole reference, carbon-12 has been chosen. Due to the prop-

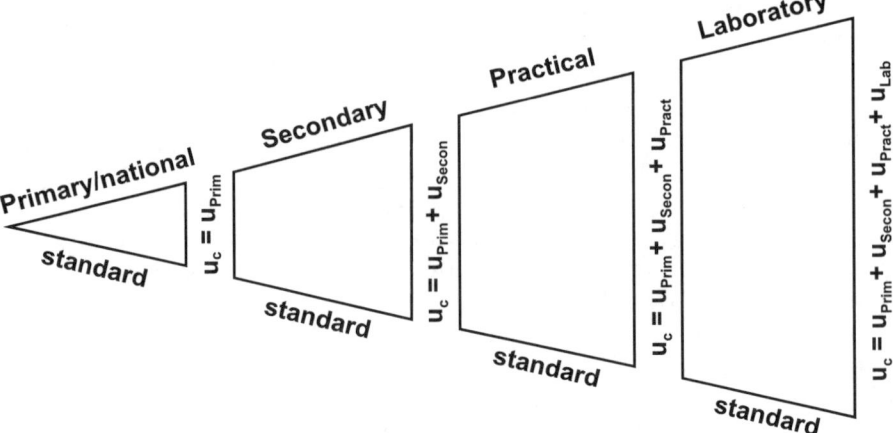

Figure 1.6. Hierarchy of measurement standards. The *leftmost tip* of the triangle is formed by the definition of the unit, e.g. the kilogram prototype which is without uncertainty. Any reference derived from this prototype has an uncertainty. In going from a primary standard to a laboratory standard will include several comparison steps, each contributing its own measurement uncertainty. Hence the complete measurement uncertainty budget, with symbol u_c, will increase with each comparison

erties discussed in connection with Fig. 1.5, the assignment of a reasonable complete measurement uncertainty budget is of considerable relevance for the comparability of measurements based upon the respective standards. The standard with the smallest uncertainty u_c is the highest in the metrological hierarchy. To achieve metrological traceability on an international level, an involvement of the national metrological institute and the interest of a nation to fund an appropriate framework is unavoidable (Richter and Güttler 2003).

The concept of metrological traceability in its details is still under discussion (de Bièvre 2000a,b,c). A meaningful statement of uncertainty is supposed to play a central role in all fields where decisions will be based upon the values obtained by measurements. A wide range of socio-economic activities are underpinned by measurements (EURACHEM/CITAC 2004). A very large number of chemical measurements every day support decisions on food safety, health, medical care, environmental protection, legislation and legal prosecution. In most cases, the parties involved into or being affected by the decisions require accurate and reliable measurement results (Zschunke 1998; Källgren et al. 2003).

In chemical measurement, traceable calibration standards are of prime interest. In almost all fields of chemical measurement, the relationship between a signal of a measurement instrument and the amount concentration of an analyte is established by calibration. The calibration standards used to establish the relationship between the measurement signal and the value assigned to the measurand are crucial elements in maintaining the traceability chain.

1.2.3.5
Cause-and-Effect Diagrams

The recommendation INC-1 (BIPM 1980) and the GUM (ISO 1993) refer to all technical, economic and scientific fields without special focus to one field. Nevertheless, the implementation of the concepts outlined in this chapter relies on a quantitative model of the measurement system, typically embodied in a mathematical equation including all relevant influential quantities (Ellison and Barwick 1998). There are challenges in applying the GUM methodology in analytical chemistry. It is, for example, not uncommon to find that the largest contributions to uncertainty result from the least predictable effects. Examples are matrix effects, recovery, filtration or components interfering with the measurement signal and sieving (Gluschke et al. 2004). It is difficult to include these influence quantities into a mathematical model of a chemical analysis. Furthermore, a reasonable estimate of their contribution to the complete measurement uncertainty budget can often only be obtained from separate experimentation. The introduction of the GUM together with the ISO 17025 (ISO 1999) has given emphasis to the metrological approach to quality measures for experimental data in chemical analysis. In several important fields of chemistry, for instance environmental chemistry, traditional practice in establishing confidence and comparability relies on the determination of overall method performance parameters, e.g. detection limits, recovery, linearity as well as repeatability and reproducibility, most of them being mainly precision measures. In evaluating these traditional measures, a part of the observed variability may include contributions from some but not all influential quantities. Combining such performance information with the GUM approach carries the risk of double-counting uncertainty contributions.

To ensure comparability of a measurement result over space and time, careful documentation of the influence quantities and their magnitude is essential. It is not uncommon to find relevant influence quantities not included into the complete measurement uncertainty budget (Holmgren et al. 2005). Thus, these results would be almost meaningless. With adequately documented influence quantities and their magnitudes, the complete measurement uncertainty budget could be corrected. The cause-and-effect approach is a powerful instrument to communicate these elements of the measurement process in a highly structured and concise way (Ellison and Barwick 1998). Cause-and-effect analysis is well known from quality management studies. The characteristic diagrams are also known as "fish-bone" or Ishikawa diagrams (ISO 1993b). Such a diagram is given in Fig. 1.7.

Figure 1.7 is derived for the evaluation of the complete measurement uncertainty budget of a copper solution serving as calibration standard. The copper standard is cleaned, weighted and dissolved in acid in a volumetric flask. In standard situations (in distinguishing from the complex situations discussed later in this text) the quantitative relationships between influential quantities

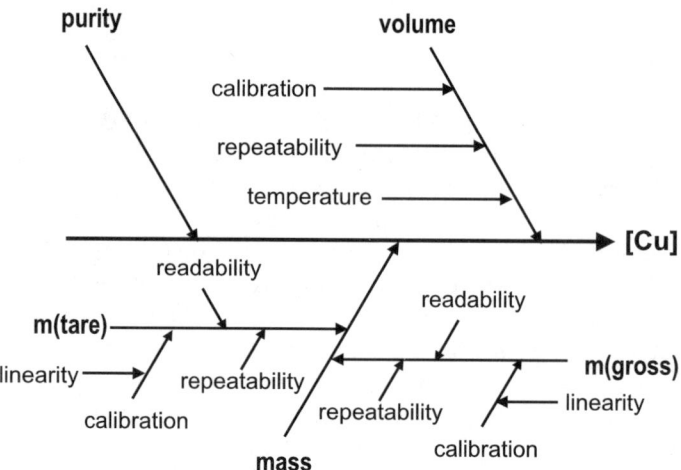

Figure 1.7. Ishikawa or cause-and-effect diagram presenting the influential quantities for the preparation of a copper calibration standard. The *main axis* of the diagram directs to the measurand, the amount concentration of copper (Cu). (According to EURACHEM/CITAC 2002)

and the measurand is represented by a mathematical equation. In the present example, this equation is given by Eq. (1.3):

$$[Cu] = \frac{1000 \cdot P \cdot m}{V} \tag{1.3}$$

where [Cu]: amount concentration of the standard in $[mg\,dm^{-3}]$; P is the purity of the metal; m is the mass of the metal in [mg]; V is the volume of the liquid in the volumetric flask in [ml]. A more detailed discussion of an equivalent example is available in EURACHEM/CITAC (2002).

A cause-and-effect diagram should be accompanied by a table giving the magnitudes of the relevant uncertainty components. The individual components are combined by use of the classical rules of error propagation. These rules are derived from the statistics of normal distributions and are practically applied almost exclusively in their first-order approximation. Its formal expression is given in Eq. (1.4):

$$u[y(x_1, x_2, \ldots, x_j)] = \sqrt{\sum_{i=1}^{j} \left[\frac{\partial y}{\partial x_i} u(x_i)\right]^2 + \sum_{\substack{i,k=1 \\ i \neq k}}^{j} \frac{\partial y}{\partial x_i}\frac{\partial y}{\partial x_k} \mathrm{cov}(x_i, x_k)} \tag{1.4}$$

The uncertainty u of the measurement result y depends on the influences $x_1, \ldots, x_j$. The first term of the root sums the variances of these influences, where the terms $\partial y/\partial x_i$ are commonly referred to as the sensitivities. The second term under the root represents the effects of covariances between the influences x_i and x_k. Often the assumption is made that the two derivatives (according to x_i and x_k) make this term negligibly small. A standard uncertainty from an influence quantity is denoted by the symbol u and related to a 68% coverage probability of a normal distribution. Hence, the measurement uncertainty is assumed to be normally distributed. From the complete assessment of uncertainty contributions from all relevant influence quantities the combined standard uncertainty u_c is obtained. The combination of uncertainty components u is done via classical progression-of-error concepts (discussed below). To increase coverage, the combined standard uncertainty can be expanded by an expansion factor k (commonly $k = 2$) to obtain the expanded standard U_c.

Thus, the concept of the complete measurement uncertainty budget relies heavily on the normal distribution with its requirement of independently and identically distributed observations. Correlation is also not considered. Since the publication of ISO's GUM, sufficient evidence has accumulated requiring a discussion of correlation effects (e.g. Hässelbarth and Bremser 2004; Ellison 2005).

There are meanwhile ample examples for the derivation and application of cause-and-effect analysis to simple situations (e.g. Ruth 2004; Hirano et al. 2005; Kolb and Hippich 2005; Osterc and Stibilj 2005), in addition to the examples given in the appendix of ISO (1993) and EURACHEM/CITAC (2002).

1.3
Statistics

1.3.1
Statistical Basics

"There are three kinds of lies: lies, damned lies and statistics"

(attributed to B. Disraeli by M. Twain)

Statistics is not a favourite among chemists (Thompson 1994). Nevertheless, the language of statistics is almost exclusively used to communicate about data sets and their properties. The successes of statistics are impressive and have contributed broadly to the development of science and technology. Appropriate statistics may strongly contribute to create trust in measurement values.

There is no intent to present an introduction to statistics in a closed form. The following pages merely intend to prepare the field for the discussion of metrological concepts in complex situations. Hence, some terms must be defined and some criteria discussed to underpin the concepts outlined in the subsequent sections. It is, nevertheless, hoped that the abbreviated discussions including examples will improve an understanding of the fundamental concepts because the almost "archetypal" facts derived from the normal distribution will be contrasted with approaches from non-parametric distributions focusing on the empirical probability distribution.

"The prime requirement of any statistical theory intended for scientific use is that it reassures oneself and others that the data have been interpreted fairly"

(B. Efron 1986)

1.3.2
Distributions

A large amount of varied information is difficult to communicate. We may think about the inhabitants of a larger city, their age and their numbers of siblings, to take a practical example. Furthermore, different types of data exist. Age is numerical but discrete, while sex is categorical. Income is almost continuous with a discrete value at zero. In the same way, measurement results can be of various types: continuous, discrete, numerical, categorical, etc. In most cases it is possible to communicate the information in the form: (value, frequency). Under certain circumstances it might be helpful to bin data [e.g. ages (in years): $0-5, 5-10, \ldots, 90-95, \ldots$]. Instead of tabulating thousands of items, a list of values can be given together with the frequency a certain value occurs. Such a presentation is a distribution.

Measurement results are almost unpredictable, in other words, measurement results are subject to random variation. Distributions are also helpful in communicating random variations. In fact, random variation of certain variables is a very common observation in the field of measurement. Comparisons in the real world are affected by a large number of influence quantities from the environment around the measurement location. Thus, the sum of influences cannot be controlled to an arbitrary level. The resulting lack of predictability of a measurement result is one reason for random variation of the measurement result.

1.3.3
Normal Distribution

The prototype of a distribution is the normal distribution. The normal distribution refers to a continuous variable. A continuous variable is a variable that can take any value in an interval $[a, b]$ (in which a can be $-\infty$ and b can be $+\infty$). The normal distribution is used commonly to express and interpret measurement results that its fundamental properties and the resulting limitations are commonly unaware or ignored. The GUM often refers to the properties of the normal distribution, e.g. the mean value, the standard deviation and the variance. Despite its abundant use, it is important to note that the concepts outlined in the GUM are not restricted to normal distribution of data and variables. Triangular or uniform distributions are simple examples for distributions other than the normal distribution.

The distribution of a random variable X for which the probability function $f(X)$ is given by Eq. (1.5)

$$f(X) = \frac{1}{\sigma\sqrt{2\pi}} \exp(\frac{(X-\mu)^2}{2\sigma^2}) , \quad -\infty < X < +\infty , \qquad (1.5)$$

is called the normal distribution of X. Because the normal distribution is completely described by the parameters σ and μ, it is often abbreviated as

$$f(X) = \mathrm{N}(\sigma, \mu) . \qquad (1.6)$$

The graph of $f(X) = \mathrm{N}(\sigma, \mu)$ is the well-known bell-shaped curve given in Fig. 1.8 for $\sigma = 0.5$ and $\mu = 40$.

The normal distribution $\mathrm{N}(\sigma, \mu)$ gives a probability density. The area under the curve gives the probability to observe a value within that area. The probability to observe a value in the interval $[-\infty, +\infty]$ is, by definition, 1. In the special case $f(X) = \mathrm{N}(0.5, 40)$ shown in Fig. 1.8, the area below $f(X)$ in the interval $(39.5, 40.5)$ is filled. The area in the interval is 0.682. This observation

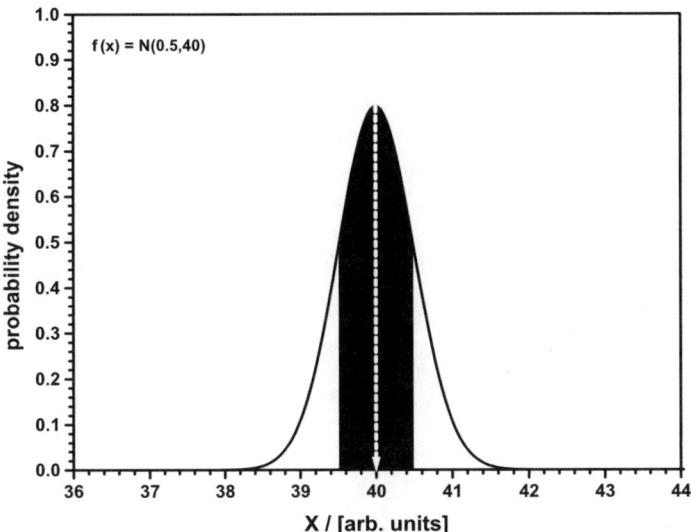

Figure 1.8. Graphical representation of N(0.5,40). The *white central down arrow* represents $\mu = 40$. The *black area* covers the *area below the curve* between 39.5 $(\mu - \sigma)$ and 40.5 $(\mu + \sigma)$. The area below N(0.5, 40) = 1. The area over the abscissa interval $(\mu - \sigma)$ and $(\mu + \sigma)$ is 0.68

can be generalised: For a normally distributed variable X, the probability to have an observation in the range $(\mu - \sigma, \mu + \sigma)$ is 0.682. In other words, in about 68% of all observations the value of the variable X will be found within this interval.

This property of the normal distribution devices a way to get the parameters σ and μ for a given set of data: From a large set of observations choose the smallest range having 68% of all data. The range is 2σ and the central value is μ. This procedure, however, is tedious and needs a large number of observations. From the definition of the normal distribution (cf. Eq. (1.5)), closed formulae can be derived to evaluate the parameters μ and σ by analytical expressions from n observations:

$$\mu = \frac{1}{n} \sum_{i=1}^{n} x_i , \qquad (1.7)$$

and

$$s = \sqrt{\frac{\sum_{i=1}^{n} (x_i - \mu)^2}{n - 1}} . \qquad (1.8)$$

By means of Eqs. (1.7) and (1.8), it is rather simple to evaluate the parameters μ and σ from experimental observations. The parameter μ (Eq. (1.7)) is known as the mean value, while σ is known as the standard deviation, introduced by Pearson in 1893 as the square root of the variance v:

$$s = \sqrt{v} . \qquad (1.9)$$

Equations (1.7)–(1.9) are well known. Calculation of mean values and standard deviations is a common task for experimental scientists. Performing these calculations daily, it is often it is forgotten that the observations x_i of the variable X must be normally distributed. Commonly there is no test performed to control this important requirement for the use of Eqs. (1.7)–(1.9). There is no a priori knowledge on the likely distribution of a set of observations $(x_1, \ldots, x_n)$.

It should be remembered that there are other parametric distributions. As an example, the Poisson distribution of a variable X is given in Eq. (1.10):

$$P(X = r) = \frac{e^{-\lambda} \lambda^r}{r!} . \qquad (1.10)$$

The Poisson distribution gives the probability to have r (non-negative, integer) observations if the probability to have such an observations is p after n attempts (to see one side of a tossed coin is $p = 1/2$, to see one face of a dice is $p = 1/6$). Then, $\lambda = np$. The mean value $\mu_P = \lambda$ and $\sigma_P = \sqrt{\lambda}$. The normal distribution and the Poisson distribution are special cases of the binomial distribution. Thus, distributions are a common way to discuss probabilities.

The "average" scientist's familiarity with the normal distribution and its parameters μ and σ is to some part due to the convenience to obtain these parameters from some experimental observations. There is, however, a further reason. This reason is the "central limit theorem".

1.3.4
Central Limit Theorem

The central limit theorem has been proposed by Laplace (who was a member of the metrological commission in Paris in 1790). There are several versions of the central limit theorem. The shortest implies: Averaging almost always leads to a bell-shaped distribution.

The following three statements are given:

1. The mean of the sampling distribution of means is equal to the mean of the populations from which the samples were drawn.
2. The variance of the sampling distribution of means is equal to the variance of the population from which the samples were drawn divided by the sample size.
3. If the original population is distributed normally, the sampling distribution of means will also be normal (this statement is exact). If the original population is not normally distributed, the sampling distribution of means will increasingly approximate a normal distribution as sample size increases.

By measuring an effect, it is implicitly assumed that the magnitude of the effect has a defined value. If speed of light is measured, it is assumed that the speed of light is a constant and variability of the observed values of the measurand is caused by random influences resulting from influence quantities during the measurement process. Thus measurement results are samples of mean values (the measurand) from random distributions. Most important is statement c. Even if the original populations are not normal, the sampling distribution of means will be normal with increasing sampling size (number of experimental observations). A computer simulation demonstrating the central limit theorem is given elsewhere (Meinrath and Kalin 2005).

The normal distribution is derived from two central assumptions. The first assumption is that the samples lie within $[-\infty, +\infty]$, the second is the independence of the random observations. No correlation may exist between independent samplings (measurements) (Williams 1978).

There are, however, exceptions known from the central limiting law. A prominent example is the Cauchy distribution. The Cauchy distribution has very wide tails. Increasing the number of draws from this distribution does not tend to reduce the error bar about the mean by a $\sqrt{N}$ rule (Sivia 1996). aaaaaaa

1.3.5
Cumulative Normal Distribution

The normal distribution of Eq. (1.5) is a probability density function. The probability P to observe a value between x_1 and x_2 for a given probability density function $f(X)$ is

$$P(x_1 < X < x_2) = \int_{x_1}^{x_2} f(X) . \tag{1.11}$$

In Fig. 1.8, the filled area beyond the curve is the integral of the normal distribution $N(0.5, 40)$ between $x_1 = 39.5$ and $x_2 = 40.5$ or

$$P(39.5 < X < 40.5) = \int_{39.5}^{40.5} \frac{1}{0.5\sqrt{2\pi}} \exp - \frac{(X - 40)^2}{0.25} . \tag{1.12}$$

This integral, however, cannot be solved analytically. Either tables with the values have to be used (in some cases requiring standardisation), or approximations need to be applied. In the special situation, where the limits of integration correspond to the standard deviation of the distribution, the area is known to be 0.68. The cumulative normal distribution (Eq. (1.13)) is given in simplified form by

$$\Phi(X < x_1) = \int_{-\infty}^{x_1} N(\sigma, \mu) . \tag{1.13}$$

Hence, the cumulative normal distribution answers the question: what is the probability to observe a value smaller than x_1. If the area in the interval $[x_1, x_2]$ is of interest, the probability $F(X < x_1)$ has to be subtracted from the probability $F(X < x_2)$. Figure 1.9 gives the cumulative probability distribution of the normal distribution shown in Fig. 1.8. The mean value μ, in the center of a symmetric distribution, is at $\Phi(\mu) = 0.5$. The filled area A between $\mu - \sigma$ and $\mu + \sigma$ in Fig. 1.8 is obtained from the cumulative distribution via $A = \Phi(\mu + \sigma) - \Phi(\mu - \sigma) = 0.84 - 0.16 = 0.68$. The range $[\mu - \sigma, \mu + \sigma]$ is therefore sometimes referred to as the 0.68 confidence interval.

By the same reasoning, the confidence interval $[\mu - 2\sigma, \mu + 2\sigma]$ can be obtained from Fig. 1.9 with $\Phi(\mu - 2\sigma) = 0.023$ and $\Phi(\mu + 2\sigma) = 0.977$ and $\Phi(4\sigma) = 0.954$, corresponding to a 95% confidence region. The confidence interval $[\mu - 3\sigma, \mu + 3\sigma]$ covers 99.7% of all observations, implying that from 1000 observations only three observations should be outside the 3σ confidence region. In practice, however, observations outside this range occur much more often than predicted by the normal distribution and giving rise to the field of outlier analysis (Barnett 1978; Beckman and Cook 1983; Meinrath et al. 2000),

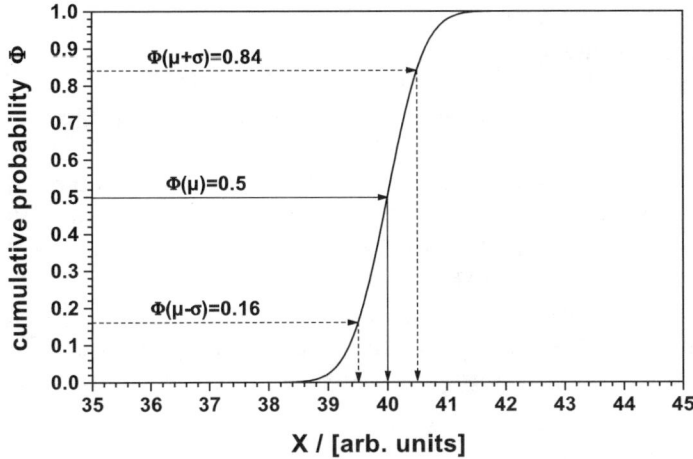

Figure 1.9. Cumulative probability distribution of the function $f(x) = N(0.5, 40)$ shown in Fig. 1.8. The mean value ($\mu = 40$) is in the center of the symmetric distribution, and hence $\Phi(\mu) = 0.5$. With $\Phi(\mu + \sigma) = 0.16$ and $\Phi(\mu - \sigma) = 0.84$, a difference $\Phi(2\sigma) = 0.68$ results

an important field which, however, lies outside the focus of this book. It must be emphasised that whatever is done to check a set of data for extraneous or influential observations, and whatever decision is taken to account for them (weighing, removing, ignoring) in the evaluation of parameters, has a considerable influence on the reported measurement values.

1.3.6
Empirical Distribution Function (EDF)

These more theoretical considerations are relevant to understand the concept of the normal distribution. The predominance of the normal distribution in data analysis is mainly a consequence of the central limit theorem. A further benefit results from the availability of closed formulas for the calculation of mean value (Eq. (1.7)) and standard deviation (Eq. (1.8)) from a set of numerical experimental observations, i.e. measurement results. These parameters characterise a normal distribution completely.

Alternatives to the normal distribution exist. The empirical distribution function (EDF) does not need assumptions on its distribution, but gives the distribution of observations directly. The median is obtained by ordering a set of n observations of a variable. The median of this ordered set is the $(0.5 \times n) + 1^{st}$ observation in the ordered list if n is odd, and the mean of the $(0.5 \times n)^{th} + (0.5 \times n) + 1)^{th}$ observation, if n is even.

In Table 1.2 the data from Table 1.1 (group 1) are given in original and sorted sequence. The number n of observations is even. The $0.5 \times n^{th}$ entry in the sorted column is 1.0023, the $0.5 \times n + 1^{th}$ entry is also 1.0023. Hence, the median is $0.5 \times (1.0023 + 1.0023) = 1.0023$. In this example, the median is rather close to the mean value.

The example shows an important deficit of the median. There is no closed formula to derive the parameter with ease comparable to the mean value. To derive a median, sorting is necessary. For ten observations, sorting by hand is feasible. However, to derive the median of all 60 observations in Table 1.1, a calculator is a welcome support. From the 60 entries in Table 1.1, the median is the mean of the summed 30th and 31st sorted entry, resulting in a median of $1.0091 \, \text{kg dm}^{-3}$.

Like the cumulative normal distribution, the empirical distribution function returns the probability to observe a measurement value smaller than a certain value x_i. In a formal expression, the empirical distribution function $EDF(x)$ is defined as

$$EDF(x) = \frac{1}{n} \sum_{i=1}^{n} H(x - x_i) , \qquad (1.14)$$

where $H(u)$ is the unit step function jumping from 0 to 1 at $u = 0$; n gives the total number of observations. The ordinate values, hence, are fixed $(0, 1/n, 2/n, ..., n/n)$. The shape of the function results from the ordered set of observations.

The observations from Table 1.2 are given in Fig. 1.10 (black squares), while the empirical distribution function $EDF(x)$ is represented by the connecting lines. $EDF(x) = 0$ for $x < 0.9903 \, \text{g}$ because there is no smaller observation. At $x = 0.9903 \, \text{g}$, the EDF jumps to 1/10. At $x = 0.9982$, the EDF jumps to 0.2 and

Table 1.2. Observations of group 1 in units of kg dm^{-3} (cf. Table 1.1)

no.	Group 1 Original sequence kg dm^{-3}	Sorted sequence kg dm^{-3}
1	1.0023	0.9903
2	1.0023	0.9982
3	1.0028	0.9993
4	0.9993	1.0015
5	0.9903	1.0023
6	1.0015	1.0023
7	0.9982	1.0028
8	1.0139	1.0067
9	1.0079	1.0079
10	1.0067	1.0139
	Mean: 1.0025	Median: 1.0023
	SD ±0.0063	SD (0.0041, 0.0056)
	95% CI ±0.0143	

SD standard deviation; *CI* confidence interval

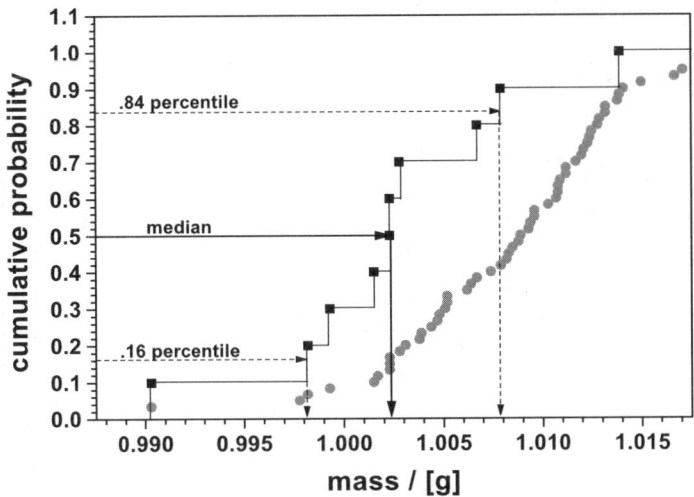

Figure 1.10. Empirical cumulative distribution function of the data given in Table 1.2. The *grey circles* show the EDF of all 60 observations from Table 1.1 for comparison

so on. It is also possible to get an estimate of uncertainty in analogy to the cumulative normal distribution in Fig. 1.9 by using the values, where the EDF has a cumulative probability of 0.16 and 0.84, respectively. These estimates are not symmetric about the median; therefore both values are given in Table 1.2.

1.3.7
Comparing EDF and Normal Distribution: Kolmogorov–Smirnov Test

An EDF becomes the smoother the more observations are included as is shown in Fig. 1.10 for the EDF of all 60 observations from Table 1.1. The question arises, how to express the similarity/dissimilarity between the two EDFs, and to what extend the both EDFs are "normal"?

The Kolmogorov–Smirnov distribution offers a quantitative estimate by comparing an EDF with a given normal distribution by returning a probability that the EDF is a random sample from the given cumulative normal distribution. The Kolmogorov–Smirnov test searches for the largest difference d(max) between the EDF and the cumulative normal distribution. The probability that this difference might occur by chance is evaluated (Massey 1951).

It is thus possible for a given data set to identify the cumulative normal distribution with the minimum d(max). Note that this is not a least-squares fit (where "least-squares" stands short for "least sum-of-squared-residuals"). From the result of this procedure, an estimate is obtained that the experimental data is derived from this normal distribution with the deviations being accidental. The result of such a procedure is shown in Fig. 1.11 for the data of Table 1.2.

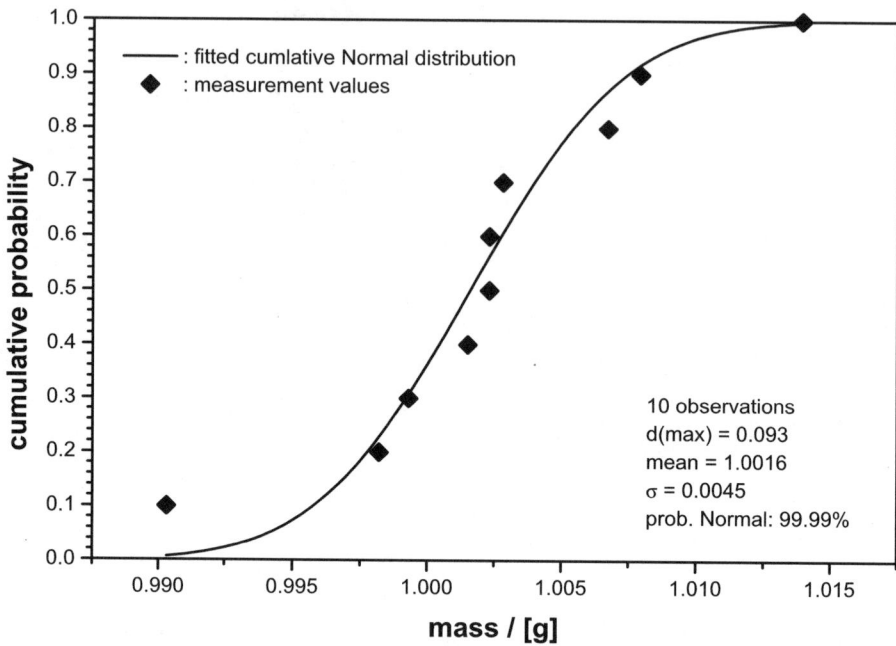

Figure 1.11. Data from Table 1.2 interpreted by the cumulative normal distribution which minimises the largest difference between experimental data and parametric distribution

The minimum distance $d(\max)$ is 0.093, evidently resulting from the smallest observation. It is possible to find normal distributions reducing this distance, however for the price of increasing the difference between normal curve and observations for another observation. Nevertheless, the Kolmogorov–Smirnov statistics gives a probability of >99.9% that the observations are normally distributed.

The Kolmogorov–Smirnov statistics can also be applied to two different EDFs, and return a probability that both EDFs result from the same specific distribution with observed differences being random. Thus, these statistics form an important link between the most important parametric distribution and empirical distribution functions and offers a convenient way to analyse and communicate properties of empirical distribution functions.

1.3.8
Linear Regression

The linear regression is the simplest of all statistical regression models. The model states that the random variable Y is related to the variable X by

$$Y = a' + b'x + \varepsilon \tag{1.15}$$

where the parameters a' and b' correspond to the intercept and slope of the line, respectively and ε denotes random disturbance. It is common to refer to ε as the "errors". In the metrological framework, the word "error" (cf. Fig. 1.3) may cause misinterpretation. The disturbances should be seen as measurement uncertainty if the observations are obtained from measurement. Residuals (unexplained variance) from a curve-fitting procedure are often the only available estimates for the disturbances.

With n observations (x_1, y_1), (x_2, y_2), ..., (x_n, y_n), it is assumed that the disturbances are random and independent observations from a normal distribution with mean 0 and standard deviation σ.

The estimates a and b of the parameters a' and b' can be obtained by closed formulae.

$$a = \bar{y} - b\bar{x} \tag{1.16}$$

and

$$b = \frac{S_{xy}}{S_{xx}} \tag{1.17}$$

where $\bar{x}$ is the mean of all x_i and $\bar{y}$ is the mean of all y_i. The sums S_{xx} and S_{xy} are defined as follows:

$$S_{xx} = \sum_{i=1}^{n} x_i^2 - n\bar{x}^2 \tag{1.18}$$

$$S_{xy} = \sum_{i=1}^{n} x_i y_i - n\bar{x}\bar{y} . \tag{1.19}$$

The variance estimate s_e^2 is obtained from Eq. (1.19):

$$s_e^2 = \frac{S_{xx}S_{yy} - S_{xy}}{(n-2)S_{xx}} \tag{1.20}$$

where S_{yy} is formed analogous to S_{xx}.

Linear regression is a convenient way to obtain estimates of a trend in data. It is included in most calculators and does not need iteration to find the slope and intercept parameters. Equations (1.15)–(1.19) are derived from the least-squares approach, assuming that a and b are the best linear unbiased estimators (BLUE), provided the assumption on which this result is based are fulfilled. There are seven requirements (Bates and Watts 1988):

List 1.3.1: Requirements for obtaining BLUE regression parameters

  a) The expectation function (Eq. (1.15)) is correct.
  b) The response is expectation plus disturbance.
  c) The disturbance is independent of the expectation function.
  d) Each disturbance has a normal distribution.

e) Each disturbance has zero mean.
f) The disturbances have equal variances.
g) The disturbances are independently distributed.

It is common to use the linear regression model without testing these assumptions. In part, they may be difficult to verify. There exist a wide range of diagnostic means for analysis which are, however, commonly not applied in chemistry publications. It is important to realise that there are also closed formulas to evaluate uncertainty limits for the parameters estimates a and b.

A $(1 - \alpha)$ confidence interval for slope b can be obtained from

$$b \pm t_{n-2,\alpha}\sqrt{\frac{s^2}{S_{xx}}} \tag{1.21}$$

while the upper and lower confidence limits s_y^o and s_y^u for an estimated Y value at $x = x_0$ is given by Eq.(1.22) for n observations:

$$\begin{matrix} s_y^o \\ s_y^u \end{matrix} = \pm t_{n-2,\alpha}\sqrt{s^2\left\{1 + \frac{1}{n} + \frac{(x_0 - \bar{x})^2}{S_{xx}}\right\}} . \tag{1.22}$$

Hence, it is possible within the assumptions of linear regression to derive uncertainty estimates for the parameters derived by linear regression. Keeping in mind that linear regression is the most common way to obtain linear calibration curves in instrumental analysis, this feature allows the evaluation of measurement uncertainty estimates for calibration data and an approximate quantification of the influence quantity "calibration".

1.3.9
Coverage and Confidence Regions

The confidence limits derived by Eqs. (1.21) and (1.22) are marginal confidence limits at a confidence level $100(1 - \alpha)$%. This confidence level implies that the mean of the sampling distribution (the unknown "true" value) will be found with a probability $1 - \alpha$ within the given limits. For small sample sets $[(n - 2) < 30]$ the confidence region must be adjusted. This adjustment is an effect of the normal distribution assumption which is an intrinsic part of linear least squares regression. However, the normal distribution is only a large sample approximation to the incomplete beta distribution. Thus, without appropriate correction, the confidence would be overestimated, especially at small sample sizes. In calibration it is not uncommon to use just three to five calibration standards. The smaller the data set (or more correctly, the degrees of freedom), the higher the probability of overestimating the precision of the result.

With the help of a computer simulation, the overestimation of the confidence in a mean value from experimental data can be demonstrated. Figure 1.12 shows (solid line) a normal distribution with mean = 10 and $\sigma = 1$, N(1, 10). The mean value of this distribution represents the "true value". This value is usu-

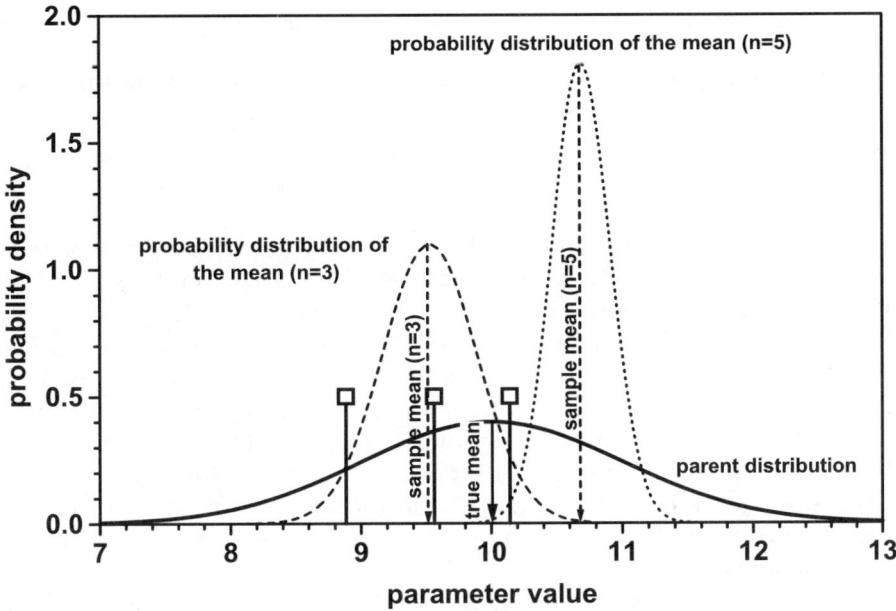

Figure 1.12. An example of an experimental test of the coverage probabilities using a sampling distribution N(1, 10) and n draws from this distribution for estimation of the mean and a confidence region for the mean by application of Eqs. (1.7) and (1.8). The $n = 3$ distribution is derived from the three "observations" represented by *lines with an open square top*. The five individual observations from which the $n = 5$ distribution results are not shown for sake of clarity. This simulation does not intend to estimate the spread of the parent distribution but only the probable location of the mean value

ally unknown in a real measurement, but known here because the fundamental process investigated by measurement (sampling from this distribution) is simulated. This distribution represents the mean value of an observable and the normally distributed disturbances (which represent the measurement uncertainty). Random samples from N(1, 10) are obtained n times (representing n measurements). The mean value of these samples is obtained by Eq. (1.7) and a confidence interval to find the true value (10) inside the confidence region is obtained by repeated sampling. According to Fig. 1.4, the "true value" should be found in 68% of all samplings within an interval of 2σ about the sampling mean and in 95% of all cases within a 4σ interval around the sampling mean. Figure 1.12 shows two examples of the sampling distributions obtained from testing the "true" mean, for $n = 3$ and $n = 5$. Form the distribution obtained for $n = 3$, the "true" mean is not within the 2σ interval but within the 4σ interval. For the $n = 5$ distribution, the true mean is not within the 2σ and the 4σ interval. It should be kept in mind that the both sampling distributions are just illustrating examples. To obtain estimates of the coverage, many repetitions of this analysis must be done.

An outcome of this procedure, where the coverage probabilities have been obtained from 5000 repetitions for different numbers of n is shown in Table 1.3. The samples were drawn, and the sample mean and its standard deviation were evaluated. The result was tested whether the true mean (10) was inside a certain confidence region of the estimated distribution. Each simulation was repeated 6 times. Table 1.3 gives the percentage of successes which can be compared to the expected value of 90%.

However, the coverage is generally found lower than 90% for $n = 3, 5, 10$ and 30. The discrepancy between expected and observed coverage is the larger the smaller the sample size. The six repetitions show that the discrepancies are not a random, accidental effect but occur systematically.

This discrepancy has been observed already much earlier. In 1906, W. Gosset derived a correcting distribution, which he had to publish under the pseudonym "Student". This distribution is known as the Student's t distribution and is tabled in almost all textbooks of statistics (Student 1908). By increasing the width of the estimated distribution of the mean using the respective values of the t distribution leads to coverages very close to 90% in each case and each run.

This simulation experiment practically shows the appropriateness of the Student t correction. The t distribution is a parametric distribution depending on more parameters than just two in the case on the normal distribution. It is tabled for the desired confidence level $(1 - \alpha)$ and the degrees of freedom df. The df are obtained from the sample size n minus the estimated parameters. In the present case, only one parameter (a mean value) is estimated, so $df = n - 1$. Included into Table 1.3 are the respective values for $t_{d.f., \alpha}$ together with the resulting coverage after expanding the confidence region by the factor $t_{d.f., \alpha}$. This experiment shows that the neglect of an appropriate correction for small sample size may easily result in some 10–15% of neglected uncertainty. Whether this magnitude is of relevance within a complete measurement uncertainty budget must be decided case by case.

Table 1.3. Comparison of coverage probability for a mean value on basis of $n = 3, 5, 10$ and 30 samples drawn from the parent distribution ($\mu = 10$, $o = 1$). Expected coverage is 90% (5000 simulations per run)

$t_{d.f., .10}$	$n = 3$	$df = 2$ 2.92	$n = 5$	$df = 4$ 2.132	$n = 10$	$df = 9$ 1.833	$n = 30$	$df = 29$ 1.697
Run 1	74.90%	89.70%	82.80%	90.50%	86.60%	89.90%	88.70%	89.5
Run 2	76.10%	89.40%	83.20%	90.20%	85.60%	89.20%	89.40%	90.2
Run 3	76.60%	90.00%	82.30%	90.00%	87.30%	90.30%	88.50%	89.3
Run 4	76.30%	89.60%	82.60%	89.90%	86.10%	90.50%	89.10%	90.0
Run 5	76.70%	90.60%	82.50%	90.60%	86.60%	90.00%	88.70%	89.6
Run 6	74.90%	89.80%	82.90%	90.60%	86.60%	90.00%	89.30%	90.9

1.3.10
Correlation

Linear regression is also applied in the evaluation of experimental data, e.g. in the standard-addition method, in Gran analysis etc. The slope of a fitted line, e.g. in a solvent extraction study, can be compared to a value from stoichiometry. The value of blank solutions may be related to the intercept of a regression line. However, the intercept and the slope of a regression line are not independent of each other. In contrary, they are strongly correlated (Mandel and Linnig 1957). The confidence intervals of a slope (Eq. (1.21)) and intercept (Eq. (1.22) for $x_0 = 0$) enclose a square area in a diagram with the variables intercept a and slope b forming the axes. These are so-called marginal confidence regions if corrected by multiplication of an appropriate $t_{d.f., \alpha}$ to give a confidence interval.

The joint confidence region (which results independently from sampling theory, likelihood inference and Bayesian inference (Bates and Watts 1988)) is an ellipse in (a, b) space and given by

$$2F_{2, n-2, \alpha} s_e^2 = n(a - a)^2 + 2 \left(\sum x_i \right) (b - b)(a - a) + \left(\sum x_i^2 \right) (b - b)^2$$

$$(1.23)$$

where $F_{2, n-2, \alpha}$ is Fisher's F at 2 and $n - 2$ degrees of freedom with confidence level $(1 - \alpha)$ and s_e^2 is the standard deviation of the regression line defined as

$$s_e^2 = \frac{1}{n - 2} \left(\sum y_i^2 - a \sum y_i - b \sum x_i y_i \right) .$$

$$(1.24)$$

The discrepancies between marginal and joint confidence region can be considerable, as is shown in Fig. 1.13. In fact, the areas of the both regions are also rather different. The use of marginal confidence regions may overestimate the uncertainty considerably. Furthermore, it should be noted that almost 20% of the joint confidence region are outside the marginal confidence region. Therefore, the correlation between regression parameters a and b introduces risks for both overestimation and underestimation of the regression parameter values!

The importance of these effects for assessing comparable measurement uncertainty budgets is not yet investigated in detail. Correlation does not only affect linear calibration curves but is relevant for all parameters being simultaneously evaluated by curve fitting from a data set. A practical example may be found for spectroscopic data (Meinrath 2000).

1.3.11
Progression of Error

Metrology is applying statistical concepts partly because the concept of a normal distribution is suitable to communicate uncertainty in a concise and con-

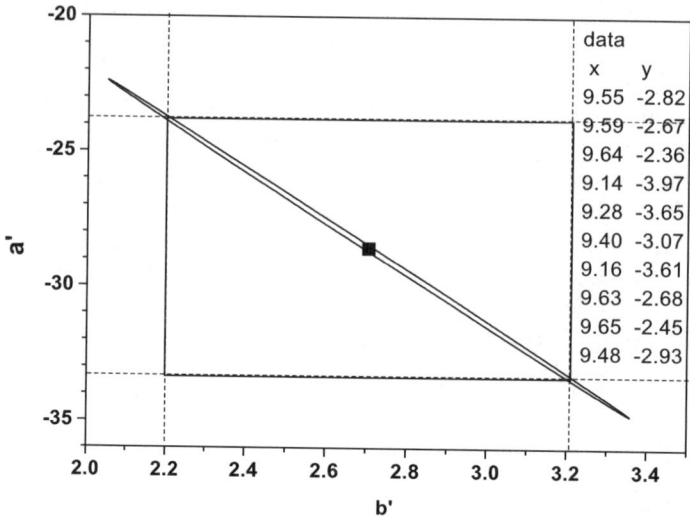

Figure 1.13. Square marginal and elliptical joint confidence regions ($\alpha = 0.05$) for slope b and intercept a of the data given in the figure

venient way. The complete measurement uncertainty budget is an attempt to characterise the underlying distribution of a measurand under the assumption that the spread of this distribution is caused by identifiable influences. The central limit theorem gives this assumption a theoretical basis. In fact, comparison of empirical distributions with a normal distribution by application of the Kolmogorov–Smirnov statistics supports this assumption as long as the correlation between disturbances and parameters of the model function are negligible. The presentation of a measurement value y of a measurand m including its measurement uncertainty u_c in the common form $m = y \pm u_c$ corresponds to the abbreviated notion of a normal distribution $N(u_c, y)$. The mean value y is the best estimate of the measurand value, while u_c gives the spread of the underlying distribution corresponding to a symmetric 0.68 confidence interval about the mean. By the same rationale U_c ($k = 2$) corresponds to a 0.90 confidence region.

The complete measurement uncertainty budget is obtained by combining the uncertainty contributions of the influence quantities. Since it is assumed that all influence quantities result from normal distributions, the complete measurement uncertainty budget is obtained by progression of standard deviations. Statistics devices the following relationships for the progression of standard deviations:

a) addition and subtraction: $s = x + y - z$

$$u_c = \sqrt{\sigma_x^2 + \sigma_y^2 + \sigma_z^2} \tag{1.25}$$

b) multiplication and division: $s = \frac{xy}{z}$

$$\frac{u_c}{s} = \sqrt{\left(\frac{\sigma_x}{x}\right)^2 + \left(\frac{\sigma_y}{y}\right)^2 + \left(\frac{\sigma_z}{z}\right)^2} \; . \tag{1.26}$$

These relationships immediately indicate that the identification of influence quantities not need to be done to the extreme. If an influence quantity has a contribution of 10% and two others contribute 3% and 1%, respectively, then the resulting value for the combined standard uncertainty u_c is

$$u_c = (10^2 + 3^2 + 1^2) = 10.5 \; . \tag{1.27}$$

Thus, the largest uncertainty contribution is commonly determining the complete measurement uncertainty budget. On the other hand, a detailed analysis of the magnitudes of different uncertainty components can will be essential in optimising measurement processes because an overall improvement can only be obtained if the predominant influence quantities are identified (de Bièvre 2004b).

1.4
Metrology in Standard Situations

"Doubt is uncomfortable, certainty is ridiculous:"

  (Voltaire)

1.4.1
Assessing Measurement Uncertainty

At present, there are three independent approaches for an evaluation of the complete measurement uncertainty budget in use in metrology in chemistry. These are termed "top-down", "bottom-up", and the Nordtest approach (Magnusson et al. 2004). The bottom-up approach will be discussed in some detail, while only an abbreviated presentation of the remaining two approaches will be given.

1.4.2
Top-down Approach

The top-down approach relies on interlaboratory comparisons. A sample is prepared, where the measurand is quantified by a laboratory with the necessary competence (commonly a national metrological institute or an accredited laboratory) by a reference method (ideally a primary method) and distributed among other competent laboratories. The complete measurement uncertainty budget results from the variability of the values returned by the laboratories together with the measurement uncertainty budgets. This approach does not need a detailed analysis of the measurement process. This approach is most suitable for samples that can be prepared with necessary accuracy, where a suitable primary method for analysis is available. The requirement of suitability includes stability during transportation and up to the measurement. Furthermore, a sufficient number of competent laboratories for this measurand must be available for participation. The logistic and economic burden can be considerable. Commonly, only larger organisations, e.g. NMIs and national accreditation institutions, can afford the expenses. However, a large number of proficiency tests are performed in this way on a routine basis (e.g. CAEAL 2004; EPTIS 2005).

1.4.3
Nordtest Approach

The Nordtest approach was developed for environmental testing laboratories in the Nordic countries (Norway, Finland, Sweden and Denmark). It uses quality control and validation data. Such data are available to the target laboratories as a component of the European accreditation guideline (EA 2002). The Nordtest measurement uncertainty model is based upon the GUM and uses the fishbone diagram approach. However, the quantities considered are not the individual influence quantities but general terms like reproducibility in the laboratory, method and laboratory bias and reproducibility between laboratories. Thus, the Nordtest approach also requires information from proficiency tests or other interlaboratory comparisons. The measurement uncertainty model from the Nordtest approach is shown in Fig. 1.14. The fishbone diagram covers the

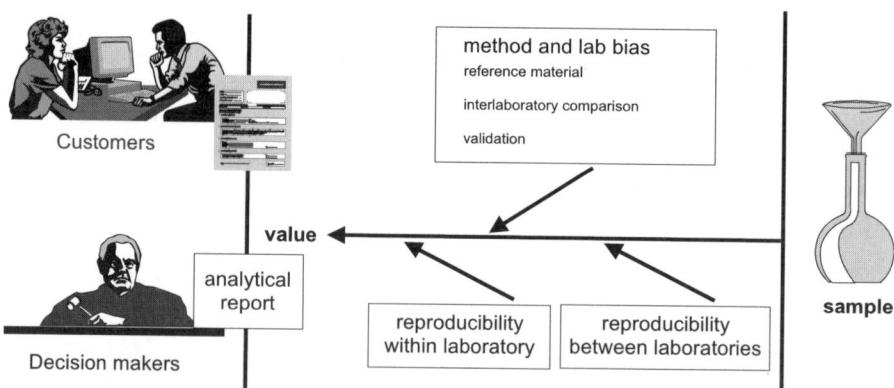

Figure 1.14. Nordtest approach measurement uncertainty model. Reproducibility within-laboratory is combined with estimates of the method and laboratory bias (from Magnusson et al. 2004)

analytical process from sample arrival in the laboratory to the report forwarded to the client/decision-maker.

For routine measurement, laboratories often use various kind of control charts which allow a control of the analytical performance over time. Thus various effects on the result, e.g. differences between sample replicates and trends over time, can be assessed. The use of such charts is required by various accreditation bodies and good laboratory practice guide lines for quality assurance/quality control.

The Nordtest approach requires information from routine analysis of an analyte performed in the laboratory over a longer period of time. Furthermore, the analyte must also be subjected to proficiency testing schemes in order to get access to between-laboratory performance information. Only in those cases is the Nordtest approach applicable. Clearly, for field measurements, which play a crucial role in environmental analysis (e.g. hydrogeological studies (Meinrath et al. 1999)), such control charts are not available. Furthermore, due to the fact that the analyte, e.g. uranium, occurs in a matrix specific for a certain location, proficiency tests are unlikely to become available.

1.4.4
Bottom-up Approach: Measurement of pH in a Low-Ionic Strength Water

The bottom-up approach has been established mainly by the Co-operation in International Traceability in Analytical Chemistry (CITAC) and EURACHEM, an association of European analytical laboratories. Its principles are laid down in two essential guides: "Quantifying Uncertainty in Chemical Measurement" (EURACHEM/CITAC 2002) and "Traceability in Chemical Measurement" (EURACHEM/CITAC 2004). Both guides give examples for the application of the bottom-up approach to non-routine measurements for which the Nordtest approach cannot be applied. The bottom-up approach is the answer to the enormous variability of analytical tasks a laboratory may face due to the millions of known chemical compounds in three aggregation states and an almost infinite number of possible matrices with concentration ranges covering 18 (and more) orders of magnitudes in amount concentrations (Jenks 2004). The measurand is often rather specific, especially in combination with the concentration range, matrix and interfering substances.

As a specific example, the estimation of a complete measurement uncertainty budget for the quantity pH will be given to illustrate the application of the GUM implementation by EURACHEM/CITAC. The quantity pH plays an important role in many fields, also outside chemistry, e.g. food safety, health science and medical care, hydrogeology and environmental protection/restoration etc. (Kalin et al. 2005). Hence pH is measured routinely, however reported almost exclusively without statement of uncertainty (Spitzer and Meinrath 2002). The link between the value of pH in an unknown sample and the measured potential of a glass combination electrode is commonly obtained from a calibration using two standard buffer solutions of known pH. This procedure has been criticised and a multi-point calibration procedure has been proposed (Meinrath and Spitzer 2000; Naumann et al. 2002).

The measurand pH is defined as:

$$pH = -\lg a_H = -\lg \frac{m_H \gamma_H}{m^0} \tag{1.28}$$

where a is relative activity, m is molality, γ is the activity coefficient and m^0 is standard molality ($1\,mol\,kg^{-1}$) (Baucke 2002).

Given a sample with unknown pH, pH(X), and two standard solutions, S_1 and S_2, with known pH, pH(S_1) and pH(S_2), the $pH(X)$ is obtained from the straight line equation:

$$\frac{pH(X) - pH(S_1)}{pH(S_2) - pH(S_1)} = \frac{E(X) - E(S_1)}{E(S_2) - E(S_1)} \tag{1.29}$$

where $E(S_1)$, $E(S_2)$ and $E(X)$ give the experimental potential of the glass combination electrode in the pH standards, and the unknown sample X, respectively. From Eq. (1.29) the mathematical expression for the measurand, pH(X), is obtained:

$$pH(X) = pH(S_1) + pH(S_2) - pH(S_1)\frac{(E(X) - E(S_1))}{(E(S_2) - E(S_1))} \ . \tag{1.30}$$

In the second step, the influential quantities must be identified. These influential quantities come from (examples):

a The environment: (temperature).
b The equipment: (resolution of the pH meter, stability of the electrode potential).
c The standards: (measurement uncertainty of the pH(S_i) and the potentials $E(S_i)$).
d The sample: (measurement uncertainty of the measured potential, ionic strength).
e The method: (stirring effects, calibration buffer selection).
f The operator: (electrode pretreatment, rinsing, attitude to comply with norms).

In the third step, a cause-and-effect diagram is developed (Fig. 1.15). The main branches to the measurand are formed by the quantities in the mathematical expression in Eq. (1.30). The influence quantities for these main influences must be identified from experience. It is essential that no major influence quantities must be ignored.

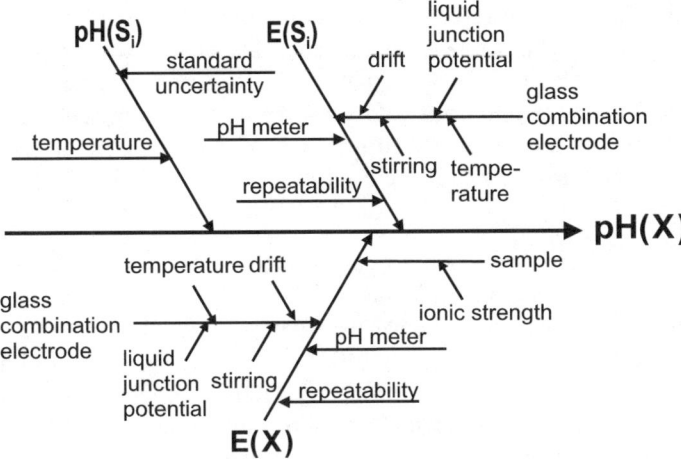

Figure 1.15. Cause-and-effect diagram for a pH measurement of unknown sample X by the two-point calibration approach (Spitzer and Werner 2002; IUPAC 2002)

The following influence quantities are included into the cause-and-effect diagram:

- For all values obtained from measured quantities, e.g. $E(S_1)$ and $E(S_2)$, repeatability must be assessed.
- Temperature is affecting almost all quantities in various ways. Because these various ways must be assessed independently for each influence quantity, "temperature" cannot be combined into a single branch as is suggested in EURACHEM/CITAC (2002).
- The influence quantities affecting the glass combination electrode are valid for all measured potentials. The ionic strength of the standards S are controlled, while the ionic strength of the sample is commonly unknown. Ionic strength affects the activity coefficient γ_H of the H^+ and thus (cf. Eq. (1.29)) has a direct influence on the value obtained for the quantity pH.
- The pH meter contributes to the uncertainty due to the reading value. Modern digital pH meters are rather robust equipment. The number of decimal positions on the display may vary and contribute a small amount of uncertainty, commonly taken into account by $u_\Delta = \Delta x / \sqrt{3}$, where Δx is 0.1 for one, 0.01 for two, 0.001 for three and so forth, available significant decimal positions on the display.
- Liquid junction potentials are a major reason for the difficulties in determination of pH by glass combination electrodes. Liquid junction potentials are consequences of diffusion of ions over liquid–liquid boundaries within the electrochemical cells. These potentials may become considerable (Baucke et al. 1993; Naumann et al. 1994).

- Stirring affects the distribution of ions in the vicinity of the liquid junctions, and hence the potential of the complete electrochemical cell. A possible way to minimise the stirring effect is to stop stirring before the measurement and wait for a stabilisation of the potential before recording the potential value.
- Drift is a time trend in the potential shown by an electrochemical cell. Modern high-quality combination glass electrodes should not exhibit a significant drift. Otherwise, the electrode should be replaced.

In a third step, influence quantities have to be quantified. In the present example a linear relationship is assumed to exist between the electrode potential in the sample and its pH, expressed by the Nernst relationship. The theoretical slope at 25 °C would be $k = 2.303\,\mathrm{dpH/dE} = 0.059\,\mathrm{V}^{-1}$. The experimental slope k', however, is commonly less (Meinrath and Spitzer 2000). This relationship is commonly quantified by linear calibration. In the present situation, the relationship has to be obtained from only two points from which two parameters (slope and intercept) have to be determined. Hence, there are zero degrees of freedom. It is not possible to assess the contribution of measurement uncertainty from the calibration process because two points always define a straight line.

In such a case, sensitivity coefficients can be used. A sensitivity coefficient, c, can be obtained by simple derivation from the mathematical model in Eq. (1.30):

$$\frac{\partial \mathrm{pH}(X)}{\partial \mathrm{pH}(S_1)} = c(\mathrm{pH}(S_1)) = 1 - \frac{E(X) - E(S_1)}{E(S_2) - E(S_1)} \tag{1.31}$$

$$\frac{\partial \mathrm{pH}(X)}{\partial \mathrm{pH}(S_2)} = c(\mathrm{pH}(S_2)) = \frac{E(X) - E(S_1)}{E(S_2) - E(S_1)} \tag{1.32}$$

$$\frac{\partial pH(X)}{\partial E(S_1)} = c(E(S_1)) = \frac{\mathrm{pH}(S_1) + \mathrm{pH}(S_2)}{E(S_1) - E(S_2)} \tag{1.33}$$

$$+ \{\mathrm{pH}(S_2) - \mathrm{pH}(S_1)\} \frac{E(X) - E(S_1)}{(E(S_2) - E(S_1))^2} \tag{1.34}$$

$$\frac{\partial \mathrm{pH}(X)}{\partial E(S_2)} = c(E(S_2)) = \mathrm{pH}(S_1) - \mathrm{pH}(S_2)\frac{E(X) - E(S_1)}{(E(S_2) - E(S_1))^2} \tag{1.35}$$

$$\frac{\partial \mathrm{pH}(X)}{\partial E(X)} = c(E(X)) = \frac{-\mathrm{pH}(S_1) + \mathrm{pH}(S_2)}{-E(S_1) + E(S_2)}\ . \tag{1.36}$$

Through these five equations, the quantitative information on the uncertainties in the influence quantities can be transformed into numerical information about the complete measurement uncertainty budget of the measurand, $\mathrm{pH}(X)$.

Example data from a two-point calibration is used to see the machinery work. Table 1.4 gives data obtained from field measurement of a low-ionic

Table 1.4. Experimental data and uncertainty contributions from influence quantities for a pH measurement by two-point calibration

Calibration data		pH	E
	S_1	4.01	169.8 mV
	S_2	7	1.8 mV
	$E(X)$		41.5 mV
Branch	Influence quantity x		$u(x)$
pH(S)	Temperature		0.2 K
	Standard uncertainty		0.01
E(S)	Glass electrode	Drift	0.2 mV
		Temperature	0.3 K
		Liquid junction potential	1 mV
		Stirring	0 mV
	pH meter		0.06 mV
	Repeatability		0.05 mV
E(X)	Glass electrode	Drift	0.014 mV
		Temperature	0.02 mV
		Liquid junction potential	1 mV
		Stirring	0 mV
	Repeatability		0.5 mV
	pH meter		0.06 mV
	Sample	Ionic strength	0.2 mV

strength water from a drinking water well, measured after thermal equilibration to 18°C outside air temperature. A portable voltmeter is used, giving potentials with one decimal position. The sample was very low in ionic strength. A stable reading could be obtained only after almost 12 min equilibration time while drift was not significant with the limited resolution of the meter. Stirring effects have been tested but found to be insignificant.

Due to the very limited number of calibration points, no statistical approach can be applied to obtain information on most uncertainties. These are mostly only ISO type B uncertainties. The repeatability of E(X) has been assessed by measuring different samples of the unknown solution. The repeatability was determined by evaluating mean value and a 68% confidence limit. The latter is given in Table 1.4. The magnitude of these uncertainties must come from separate experimentation, or experience. Table 1.4 summarises the available information.

The expectation value of the measurand is evaluated from the mathematical model in Eq. (1.30):

$$pH(X) = 4.01 + 7.00 - 4.01 \left(\frac{41.5 \, mV - 169.8 \, mV}{1.8 \, mV - 169.8 \, mV} \right) = 7.95 \ .$$

It is now necessary to evaluate the measurement uncertainty for each branch. This is done by additive progression-of-error evaluation according Eq. (1.24).

a) $pH(S_1)$, $pH(S_2)$

The certificate of the pH standard solution indicates that the pH changes for 0.004 units K^{-1}. The $0.15 K$ uncertainty attributed to the influence of varying temperatures during measurement contribute $u_T = 6 \cdot 10^{-4}$, while the standard uncertainty of each solution is 0.01. Thus

$$u(pH(S)) = \sqrt{(6 \cdot 10^{-4})^2 + (10^{-2})^2} = 0.010017 = 0.01 \qquad (1.37)$$

thus being a practical demonstration of Eq. (1.26).

b) $E(S_1)$, $E(S_2)$

The temperature effect on the potential is about 60 times higher than on pH and amounts to about $u_T = 1.5 \cdot 10^{-3}\,mV$. The drift of the electrode was assessed separately by following the potential in a pH standard solution (equilibrated with the surrounding) over some time. From this observation, the value given is a reasonable estimate of doubt. Thus, the combined standard uncertainty, u_{GCE}, of the glass combination electrode is

$$u_{GCE} = \sqrt{(1.5 \cdot 10^{-3})^2 + (0.2)^2 + (1)^2} = 1.02\,mV = 1.0\,mV \qquad (1.38)$$

It is obvious that only uncertainty effects of similar magnitude need to be considered. Thus, the liquid junction potential defines the uncertainty contributed to the potentials measured in the pH standards and the standard uncertainty is

$$u_E = 1\,mV$$

c) $E(X)$

By the same reasoning, the liquid junction potential defines the magnitude of the combined standard uncertainty contributed by the glass electrode to $E(X)$.

$$u_{GCE} = 1\,mV$$

The other influences, repeatability, pH meter and sample are clearly smaller. For the measurement uncertainty of the influence quantity $E(X)$, u_X

$$u_X = \sqrt{(0.5)^2 + (0.06)^2 + (0.2)^2 + (1.0)^2} = 1.14\,mV = 1.1\,mV\,,$$

where the uncertainty u_{pHM} of the pH meter reading results from $u_{pHM} = 0.1/\sqrt{3}$.

From this information, the complete measurement uncertainty budget of this measurement can be evaluated. The necessary steps are performed in Table 1.5.

Table 1.5. Evaluation of the complete measurement uncertainty budget of a field pH measurement

Quantity	Value x_i	Standard uncertainty u_i	Sensitivity coefficient c_i	Uncertainty contribution $u_{i,\,pH(x)}^2 = u_i^2 \times c_i^2$	Percent uncertainty contribution ui, pH(x) %
pH(S1)	4.01	0.01	0.236	$5.57 \cdot 10^6$	1
pH(S2)	7	0.01	0.763	$5.82 \cdot 10^5$	10
E(S1)	169.8 mV	1 mV	0.004	$1.60 \cdot 10^5$	2
E(S2)	1.8 mV	1 mV	0.014	$1.96 \cdot 10^4$	33
E(X)	41.5 mV	1.1 mV	−0.018	$3.24 \cdot 10^4$	54
$\Sigma[u_i,\,pH(X)^2]$				$6.0 \cdot 10^4$	100
Combined standard uncertainty $u_c = \sqrt{\Sigma\left[u_{i,\,pH(x)}^2\right]}$				0.025	
Expanded combined standard uncertainty U_c ($k = 2$)				0.05	

The right column of percentile uncertainty contributions shows that potential measurements, especially in the solutions S_2 and X, have the highest contribution to the measurement uncertainty budget of a low ionic strength sample. There is little effect in reducing the uncertainty contributions of the pH standard solutions. Since E(X) and E(S_2) have the highest contribution to the complete measurement uncertainty budget (87%) and the uncertainty mainly results from liquid junction potential effects, the complete measurement uncertainty budget of the value of the unknown sample is due to liquid junction effects. These effects are, however, difficult to assess experimentally. Thus, a measurement of the quantity pH by a two-point calibration depends considerably on the subjective estimation of the magnitude of liquid junction potentials.

The water sample thus has a pH = 7.95 ± 0.03. The second decimal position is uncertain, hence the result should be given as 7.9(5) ± 0.03. A value for pH of a water sample from the same source at a later time, e.g. another sampling campaign, should be considered significantly different if the both expanded measurement uncertainty budgets do not overlap.

The number of calculations may seem daunting on first view. However, pH measurements are typical routine measurements. The evaluation procedure can readily be implemented in a spread sheet or a small computer program. For the assessment of the individual uncertainty components an intensive discussion will be necessary giving rise to appropriate research to substantiate subjective judgement. Methods for measuring pH involving more Type A uncertainties will avoid subjective judgements and inclusion of possibly inadequate information from separate experimentation.

1.5
Metrology in Complex Situations

"We know accurately only when we know little; with knowledge doubt increases."

(J.W. von Goethe)

1.5.1
Calibration by Linear Relationships

In some measurement tasks, it is possible to write down a mathematical equation that relates the measurand with its contributing effects. An example of this approach, evaluated to a large part by EURACHEM/CITAC and with applications documented in numerous examples in the metrological literature (cf. http://www.measurementuncertainty.org), has been given in the previous sections for one of the most important chemical measurands, pH.

In a large number of situations, a mathematical formula is not available for the complete measurement process leading to value for a measurand. Prime examples are all analytical methods in which two and more parameters are derived from one set of measurements. In such situations, it is sometimes possible to write down a separate mathematical equation for each measurand, but important information, e.g. on correlation of the measurands, is not included. Further examples are those methods where the measurand(s) must be evaluated by iterative procedures. Non-linear equations are practical examples.

Those situations will be termed "complex situations" in the following. Complex situations also include those analytical procedures, where a closed mathematical formula could be applied, but where alternative methods give more satisfactory results, especially in terms of building confidence into the derived value of the measurand. Complex situations are found, e.g. in linear and non-linear calibration, in the evaluation of thermodynamic data from a variety of common techniques or in situations where theoretical parameters need to be obtained from measured data by iterative curve-fitting procedures.

The GUM gives some examples that in most cases refer directly or indirectly to the normal distribution. It is important to notice that the assumption of normally distributed measurement values is not required from the GUM. Furthermore, the GUM states explicitly that "if all quantities on which the result of a measurement depends are varied, its uncertainty can be evaluated by statistical means" (ISO 1993). It is always preferential to build the confidence in a measurement value from the information obtained during experimentation. Relying on parametric distributions and often unprovable assumptions on the magnitude of certain influence quantities always carries the risk that the data was affected by outlying data (Dean and Dixon 1951; Beckman and Cook 1983), correlation (Box and Draper 1986; Beran 1992) or non-normal distribution of parameters and/or disturbances (Alper and Gelb 1991; Meinrath 2000b).

1.5.2
Calibration by Ordinary Linear Regression (OLS)

1.5.2.1
Example: Measurement of pH by Multi-Point Calibration

There is some criticism of the two-point calibration approach to pH measurement. A crucial point is the extremely small sample size obtained by just two data points in order to evaluate a complete measurement uncertainty budget. The discussion in the section "Coverage and confidence regions", especially in connection with Table 1.3, has shown that a small sample size drastically reduces the probability that the true value (that is, the measurand), is actually within the confidence limits (with a probability of approximately 68% for the combined standard uncertainty u_c and approximately 95% for the expanded uncertainty U_c). By using two data points to establish the straight line relationship between the measurement signal and the measurand, the degree of freedom is zero and valid statistical reasoning is not possible. Statistical experience shows, however, that the uncertainty margins may vary widely in such situations. In addition, other supporting data analysis, e.g. analysis for extraneous data, analysis of linearity and analysis of disturbance distribution, cannot be performed. Thus, under the aspect of building confidence into a measurement result, the two-point calibration is a procedure attracting some criticism.

An important alternative is the so-called multi-point calibration, where the calibration line is built by using several standard solutions with known pH. The 2002 IUPAC recommendation on the measurement of pH (IUPAC 2002) lists 13 suitable solution compositions together with the pH of the resulting solutions. There is no need to rely on just two calibration points. The calibration line can be found by linear regression. Confidence limits on the regression line can be obtained from regression statistics, Eq. (1.21) and Eq. (1.22). An estimate of the unknown pH can then be derived from Eq. (1.39):

$$s_y = t_{n-2,\alpha/2} s_e \sqrt{1 + \frac{1}{n} + \frac{(x_0 - \bar{x})^2}{\sum\limits_{i=1}^{n} (x_i - \bar{x})^2}} \tag{1.39}$$

where s_e is given by Eq. (1.20) and n is the number of data points from which the regression line is derived. The standard deviation s_y is transferred into a confidence level $(1 - \alpha)$ by multiplication with the appropriate Student $t_{n-2,\alpha/2}$. An application of this methodology can be found in Meinrath (1997). Equation (1.39) describes a bone-shaped confidence band about the regression line with a minimum at pH $= \bar{x}$. The uncertainty increases with increasing difference $(x_0 - \bar{x})$, where x_0 is the measured pH. Multi-point calibration performed in such a way may yield more reliable estimates for the standard uncertainty u_c of the value derived for quantity pH.

There are, however, at least two weak points in this procedure. First, linear regression requires that the abscissa values are known without uncertainty. Standard solutions with known pH always carry an uncertainty. Fortunately, this uncertainty is commonly small compared to other uncertainty contributions (Meinrath and Spitzer 2000). Second, calibration is a two-step procedure (Osborne 1991). First, a regression line is evaluated under the assumption that the abscissa values are free of uncertainty. Then, an ordinate value (the potential in the unknown sample) is determined and a pH is read from the calibration line.

An alternative is to evaluate the regression line including the measurement uncertainty from Eq. (1.39). In the second stage of the procedure, the measurement uncertainty of the value of pH(X) in the unknown sample X is read from the intercepts of the measured potential $E(X) \pm u_c$ with the confidence band derived for the regression line. This procedure is shown in Fig. 1.16.

The necessary mathematical expression may be obtained by evaluating the intercept between E(X) with the regression line and the ordinate values $E(X)+u_c$ and $E(X) - u_c$ with the lower and upper confidence bands, respectively. The resulting expressions are given in Eq. (1.40):

$$
\begin{aligned}
u(pH)_o \\
u(pH)_u
\end{aligned}
= (pH(X) - pH_m)g \pm \frac{\dfrac{t\,s_e}{b}\left\{ \dfrac{(pH(X) - pH_m)^2}{\sum_i (pH(i) - pH_m)^2} + \dfrac{1 - g}{n} \right\}^{\frac{1}{2}}}{1 - g}
\qquad (1.40)
$$

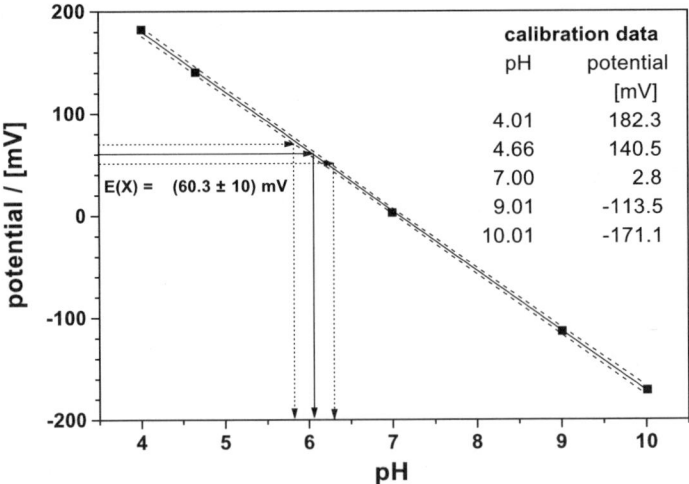

Figure 1.16. Calibration of a glass combination electrode by a multi-point calibration procedure. For an unknown solution with a potential E(X) = 60.3 mV and an (exaggerated) standard uncertainty $u_c(E(X)) = 10\,mV$, a pH $6.0_5 \pm 0.09$ is evaluated

where

$$g = \frac{t^2}{\left\{ b / \sqrt{\dfrac{s_e^2}{\sum_i (\mathrm{pH(i)} - \mathrm{pH}_m)^2}} \right\}^2} \cdot \qquad (1.41)$$

and pH_m gives the mean value of pH values of the calibration pH standards. The Student t correction for the sample size is represented by t, where t is abbreviated for $t_{n-2,\,\alpha/2}$. Formally, Eq. (1.40) is correct. However, the numerical result may differ considerably with the number of significant decimal points. One reason is the dependence of the resulting standard uncertainty u_c on g (Eq. (1.41)). The parameter g is obtained from several quotients of differences. These quotients of differences are rather susceptible to rounding effects. The following numerical example will illustrate the point.

About 30% of the calculated standard uncertainty (Table 1.6) varies due to differences in rounding by two significant decimal positions and by four significant decimal positions. Nevertheless, numerical effects like rounding are not yet considered in the normative guides on the evaluation of the complete measurement uncertainty budget.

Linear regression is based on the assumption of uncertainty-free abscissa values. In case of pH measurement it has been shown that this assumption is acceptable for pH standards with standard uncertainties in the order of $u_c = 0.1$. Under the usual conditions of a pH measurement with well-maintained equipment in low ionic strength media, the ordinate uncertainties are much larger than the abscissa uncertainties. In such cases, the abscissa uncertainties can be neglected (Meinrath and Spitzer 2000).

An often-heard argument against multi-point calibration is the increased demand of time and economic resources due to the need of additional calibration standards. It must be understood that data quality cannot be obtained by some mathematical or statistical manipulations, e.g. by using two-point calibration where the confidence limits cannot be tested due to zero degrees of freedom. Under aspects of quality control and quality assurance, such procedures may draw criticism.

Table 1.6. Value of the combined standard uncertainty of a value of pH from a multi-point calibration procedure using five pH standard solutions (with calibration data given in Fig. 1.16). For $u_c(\mathrm{E}(X))$ an estimated standard uncertainty of $u_c = 3.5\,\mathrm{mV}$ is assumed

Decimal positions	g	$u_c(\mathrm{pH}(X))$
2	$6.9916 \cdot 10^{-4}$	0.06
3	$4.1707 \cdot 10^{-4}$	0.05
4	$3.9596 \cdot 10^{-4}$	0.04
5	$3.9600 \cdot 10^{-4}$	0.04

Linear regression is a standard application to obtain a calibration line in analytical chemistry. It is common, for instance in calibration of AAS or ICP-OES instruments, to use only two or three calibration standards. The use of inadequate statistics, omission of relevant uncertainty contributions, e.g. Student's t corrections, may result in considerable overestimation of the reliability of a measurement result. The methodology outlined in this section provides a viable way towards reliable calibration. It is applicable under those circumstances where calibration standards with controlled standard uncertainty are available.

1.5.2.2
Pitfalls of Linear Regression

Ordinary linear regression, as discussed in the previous section, is bound to several requirements (cf. Chap. 1.3). If one or several of these requirements are not fulfilled, the estimates for slope and intercept are not any longer best linear unbiased estimators. Alternative schemes to identify these parameters can be applied. The discussion of linear trends, however, has focused mainly on finding the most appropriate straight line. Several methods have been discussed. For these methods, appropriate estimators for confidence limits are commonly not available.

For calibration, standard solutions with known content must be available. These solutions should have amount concentrations of the desired measurand homogeneously distributed over the calibration range. Hence, outlying data, high leverage points and other, often severe, pitfalls of linear regression should not occur (Chatterjee and Hadi 1986). There are, however, other applications in environmental analysis on basis of chemical data where the methodology outline in the preceding section would result in meaningless data.

A regular misinterpretation is the use of the Pearson coefficient of correlation as a measure of linearity (Thompson 1994; Hibbert 2005). The Pearson correlation coefficient r is obtained from Eq. (1.42):

$$r = \frac{n \sum x_i y_i - \left(\sum x_i\right)\left(\sum y_i\right)}{\sqrt{\left[n \sum x_i^2 - \left(\sum x_i\right)^2\right]\left[n \sum y_i^2 - \left(\sum y_i\right)^2\right]}} . \tag{1.42}$$

The coefficient of correlation has been a helpful figure before the advent of computing machines with graphical output. It indicates the connection between the ordinate and abscissa values and the least-squares regression line. Nowadays, the coefficient of correlation is easily calculated from Eq. (1.42) and routinely presented in the output of many linear least-squares regression codes. It has been amply outlined (e.g. Huber 2004; Hibbert 2005) that there is not much information in r in case of linear regression.

1.5.3
Alternative Methods for Obtaining Straight Line Parameters

Linear regression must be applied only in such situations where the uncertainties in the abscissa data can be assumed to be negligible. For other situations, alternative models exist to find a linear relationship between abscissa values x and responses y. Among these, the following methods will be briefly outlined:

    a orthogonal least squares regression
    b weighted least squares regression
    c bivariate least squares regression
    d robust regression techniques

1.5.3.1
Orthogonal Least Squares Regression

While in linear least squares regression the optimisation criterion is to reduce the sum of squares disturbances between the regression line and the y data points, orthogonal least squares regression minimises the distances between a data point and the regression line. This condition is met if the vector pointing to a data point is orthogonal to the regression line. This difference in assessing the disturbances is shown schematically in Fig. 1.17.

A closed formula is available to calculate an estimate of the slope b' from the data:

$$b = \frac{OR}{2} + sg(r)\sqrt{\left(\frac{OR}{2}\right)^2 + 1} \qquad (1.43)$$

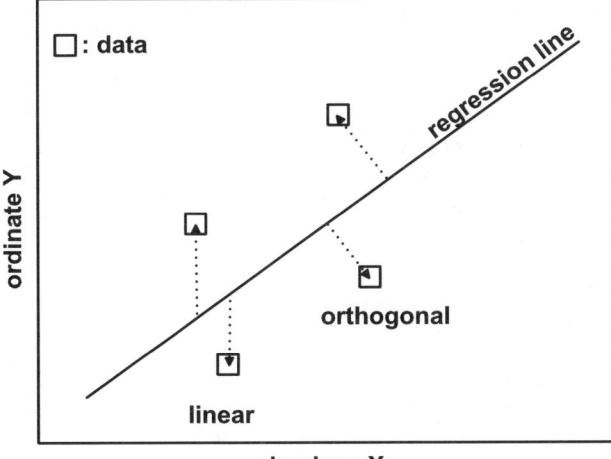

Figure 1.17. Schematic representation of the difference in calculating the disturbances in linear regression and orthogonal regression

where $sg(r)$ is the sign of the Pearson correlation coefficient r (Eq. (1.42)) and OR is obtained via Eq. (1.43):

$$OR = \frac{\sum (y - \bar{y})^2 - \sum (x - \bar{x})^2}{\sum (x - \bar{x})(y - \bar{y})} \tag{1.44}$$

where $\bar{x}$ and $\bar{y}$ are the arithmetic means of the x and y data. As soon as b is available, the intercept a is obtained from Eq. (1.16).

1.5.3.2
Weighted Least Squares Regression

Weighted least squares linear regression is a regression method which accounts for disturbances which seem to result from different distributions. This behaviour is termed heteroscedasticity. Heteroscedastic data result, for instance, if the uncertainty in a signal increases with measurand concentration. Weighted linear regression takes into account heteroscedasticity, but still requires that the uncertainties in the abscissa values are negligible. While the linear regression line results from minimisation of the sum of squared disturbances ε,

$$\sum \varepsilon_i^2 \rightarrow \min ! \tag{1.45}$$

in weighted regression a weight is assigned to each disturbance ε satisfying the condition

$$\sum \left(\frac{\varepsilon_i^2}{w_i}\right) \rightarrow \min ! \tag{1.46}$$

The weights w_i are reasonably chosen as the variances σ_i of each ε_i:

$$w_i = \frac{1}{\sigma_i^2} \tag{1.47}$$

The final weights are obtained from a normalisation:

$$W_i = w_i \left(\frac{n}{\sum\limits_{i=1}^{n} w_i}\right) \tag{1.48}$$

The estimator for the slope, b_w, is obtained from Eq. (1.16) as

$$b_w = \frac{\sum W_i x_i \sum W_i y_i - n \sum W_i x_i y_i}{(\sum W_i x_i)^2 - n \sum w_i x_i^2} \tag{1.49a}$$

and

$$a_w = \frac{\sum W_i y_i \sum W_i x_i^2 - \sum W_i x_i y_i \sum W_i x_i}{n \sum W_i x_i^2 - (\sum W_i x_i)^2} \tag{1.49b}$$

Again, the estimate of a_w is available from Eq. (1.14). With the definition of the disturbance ε by Eq. (1.13) rewritten to include the weights, an estimate for the weighted residual variance s_w^2 (Eq. (1.50)) is obtained

$$s_w^2 = \frac{\sum\limits_{i=1}^{n} W_i(y_i - a_w - b_w x_i)^2}{(n-2)} . \tag{1.50}$$

The confidence region is available from Eq. (1.51):

$$2F_{2,\,n-2,\,\alpha} s_w^2 = n(\alpha_w - a_w)^2 + 2\left(\sum W_i x_i\right)(b_w - b_w)(a_w - a_w)$$
$$+ \left(\sum W_i x_i^2\right)(b_w - b_w)^2 . \tag{1.51}$$

A confidence band $s_{w,y}$ for the predicted y at an abscissa value x can be derived from Eq. (1.52):

$$\genfrac{}{}{0pt}{}{s_{w,y}^o}{s_{w,y}^u} = \pm t_{n-2,\,\alpha} \sqrt{s_w^2 \left\{ 1 + \frac{1}{n} + \frac{(x_0 - \bar{x})^2}{\sum\limits_{i=1}^{n} W_i x_i^2 - n\bar{x}_w} \right\}} . \tag{1.52}$$

The intercept between the ordinate value $y + u_c$ and $y - u_c$ and the confidence bands are obtained from equations analogous to Eqs. (1.40) and (1.41):

$$\genfrac{}{}{0pt}{}{u(x)_o}{u(x)_u} = (x - \bar{x}_w)g \pm \frac{\dfrac{t_{n-2,\,\alpha/2}\, s_w}{b_w}\left\{ \dfrac{(x - \bar{x}_w)^2}{\sum\limits_{i} W_i(x_i - \bar{x}_w)^2} + \dfrac{1-g}{n} \right\}^{\frac{1}{2}}}{1-g} \tag{1.53}$$

with

$$g = \frac{t_{n-2,\,\alpha/2}^2}{\left\{ b_w / \sqrt{\dfrac{s_w^2}{\sum\limits_{i} W_i(x_i - \bar{x}_w)^2}} \right\}^2} . \tag{1.54}$$

The confidence limits obtained by weighted regression are usually smaller compared with those obtained for the same data set by linear regression due to the fact that the higher variances for high concentration values are down-weighted.

Weighted linear regression is an appropriate method for instrument calibration where the uncertainty increases with the absolute amount concentration of the measurand. There is little reason to be put off by Eqs. (1.47)–(1.53). These equations are easily implemented into a few lines of computer code. There is more concern about the appropriate determination of the weights. The procedure includes repeated determinations to obtain the standard deviations σ_i (Eq. (1.47)). It is easily possible to "generate" nice standard uncertainties by applying "suitable" weights. The forwarded results in such a case would be, however, neither comparable nor reasonable.

1.5.3.3
Bivariate Regression

Bivariate linear regression is recommended if the uncertainties in the abscissa data are not negligible. There are several techniques for accounting for uncertainty in both axes. The technique discussed here provides the variance–covariance matrix and is therefore helpful in assessing prediction uncertainties (Lisy et al. 1990; Riu and Rius 1996). The model function differs from Eq. (1.55):

$$y + \varepsilon_y = a' + b'(x + \varepsilon_x) , \tag{1.55}$$

whereby σ_y and σ_x are the standard deviations of the disturbances ε_y and ε_x. The standard deviations may be different for each data pair ($\varepsilon_{x,i}$, $\varepsilon_{y,i}$). The parameter estimates a and b can be obtained by a weighted regression scheme where the ith disturbance is defined as

$$R_i = y_i - f(x_i, a, b) \tag{1.56}$$

$$S = \sum w_i R_i^2 \tag{1.57}$$

$$\sigma_i^2 = w_i = s_{y,i}^2 + b^2 s_{x,i}^2 - 2b \, \text{cov}(x_i, y_i) \tag{1.58}$$

where $s_{y,i}$ and $s_{x,i}$ stand for the variances of each ($\varepsilon_{x,i}$, $\varepsilon_{y,i}$) and $\text{cov}(x_i, y_i,)$ accounts for covariance effects between x and y.

It is helpful to derive the derivatives which minimise Eq. (1.57). The resulting equations will provide matrices from which confidence bands can be obtained. The sum S (Eq. (1.57)) is minimised with the equations

$$\frac{\partial S}{\partial a} = 0 = \sum_{i=1}^{n} \left[\frac{R_i}{\sigma_i^2} \right]^2 \frac{\partial \sigma_i^2}{\partial a} \tag{1.59a}$$

$$\frac{\partial S}{\partial b} = 0 = \sum_{i=1}^{n} \left[\frac{R_i}{\sigma_i^2} \right]^2 \frac{\partial \sigma_i^2}{\partial b} . \tag{1.59b}$$

The partial derivatives of R_i^2 are

$$\frac{\partial R_i^2}{\partial a} = 2(a + b x_i) - 2y_i \tag{1.60a}$$

$$\frac{\partial R_i^2}{\partial b} = 2 \left(a x_i + b x_i^2 \right) - 2x_i y_i . \tag{1.60b}$$

Substituting Eq. (1.60b) into Eq. (1.59b) results in

$$a \sum \frac{1}{\sigma_i^2} + b \sum \frac{x_i}{\sigma_i^2} = \sum \left\{ \frac{1}{2} \left[\frac{R_i^2}{\sigma_i^2} \right]^2 \frac{\partial \sigma_i^2}{\partial a} + \frac{y_i}{\sigma_i^2} \right\} \tag{1.61a}$$

$$a \sum \frac{x_i}{\sigma_i^2} + b \sum \frac{x_i^2}{\sigma_i^2} = \sum \left\{ \frac{1}{2} \left[\frac{R_i^2}{\sigma_i^2} \right]^2 \frac{\partial \sigma_i^2}{\partial b} + \frac{x_i y_i}{\sigma_i^2} \right\} . \tag{1.61b}$$

Equation (1.61b) look rather clumsy, but can be conveniently written as matrices

$$\boldsymbol{Ra} = \boldsymbol{g} \tag{1.62}$$

with

$$\boldsymbol{R} = \begin{bmatrix} \sum \dfrac{1}{\sigma_i^2} & \sum \dfrac{x_i}{\sigma_i^2} \\[2ex] \sum \dfrac{x_i}{\sigma_i^2} & \sum \dfrac{x_i^2}{\sigma_i^2} \end{bmatrix} \tag{1.63}$$

$$\boldsymbol{a} = \begin{pmatrix} a \\ b \end{pmatrix} \tag{1.64}$$

$$\boldsymbol{g} = \begin{bmatrix} \sum \left\{ \dfrac{y_i}{\sigma_i^2} + \dfrac{1}{2} \left[\dfrac{R_i}{\sigma_i^2} \right]^2 \dfrac{\partial \sigma_i^2}{\partial a} \right\} \\[3ex] \sum \left\{ \dfrac{x_i y_i}{\sigma_i^2} + \dfrac{1}{2} \left[\dfrac{R_i}{\sigma_i^2} \right]^2 \dfrac{\partial \sigma_i^2}{\partial b} \right\} \end{bmatrix} \tag{1.65}$$

$$\frac{\partial \sigma_i^2}{\partial a} = 0 \tag{1.66}$$

and

$$\frac{\partial \sigma_i^2}{\partial b} = 2b\, s_{x,i}^2 - 2\mathrm{cov}(x_i\, y_i) \,, \tag{1.67}$$

where in most practical situations few information on the covariance $\mathrm{cov}(x_i\, y_i)$ is available. Eq. (1.62) is solved by Eq. (1.68)

$$\boldsymbol{a} = \boldsymbol{R}^{-1} \boldsymbol{g} \,. \tag{1.68}$$

Hence, the parameters a and b are obtained for given σ_i^2 (Eqs. (1.58), (1.67)). Thus, this method is iterative.

Multiplying matrix R^{-1} by the experimental sum of weighted sum of disturbances, wSOR

$$\text{wSOR} = \frac{1}{n-2} \sum \frac{1}{w_i}(y_i - a - b\,x_i)^2 \tag{1.69}$$

gives the variance-covariance matrix X

$$X = \text{wSOR}\, R^{-1} \tag{1.70}$$

A $(1 - \alpha)$ marginal confidence limit is obtained from Eq. (1.71b)

$$a \pm t_{n-2,\,\alpha/2}\sqrt{\text{wSOR}\, R_{11}^{-1}} \tag{1.71a}$$

$$b \pm t_{n-2,\,\alpha/2}\sqrt{\text{wSOR}\, R_{22}^{-1}} \tag{1.71b}$$

where R_{ii}^{-1} represent the ith diagonal element of R^{-1}.

The joint confidence ellipse is given by Eq. (1.72) (cf. Eq. (1.22)):

$$2\text{wSORF}_{2,\,n-2,\,1-\alpha} = \sum \frac{1}{w_i}(a - a)^2 + 2 \sum \frac{x_i}{w_i}(a - a)(b - b)$$

$$+ \sum \frac{x_i^2}{w_i}(b - b)^2 . \tag{1.72}$$

The $(1-\alpha)$ confidence bands about the regression line for an expected response at x_o are obtained from

$$\begin{matrix} s_{y,\,o} \\ s_{y,\,u} \end{matrix} = \pm t_{n-2,\,\alpha/2}\sqrt{\text{wSOR} \begin{pmatrix} 1 & x_0 \end{pmatrix} R^{-1} \begin{pmatrix} 1 \\ x_0 \end{pmatrix}} \tag{1.73}$$

1.5.3.4
Robust Regression

L_1 **regression.** The familiar least sum of squared disturbances (SOR) criterion for finding the "best-fit" parameters of a straight regression line (which is also common in non-linear regression) is valid only if seven requirements (cf. List 1.3.8) are valid. If the data are non-normal, if heteroscedasticity is present or if the observations are not independent of each other, then the preference of the SOR criterion is lost. The disadvantages, e.g. the high influence of data points with large distances on the position of the straight line (high leverage) remain. Thus, a valid strategy to obtain some (without any preference over straight lines obtained by minimising other criteria) straight line parameters is to modify the optimisation criterion. Within the maximum likelihood theory, the common least squares criterion is termed L_2 criterion. The L_1 criterion minimises

$$\text{SOR} = \sum |y_i - \bar{y}_i| . \tag{1.74}$$

The overemphasis on data pairs with large deviations (a result of squaring the disturbances) is reduced. On the other hand, a closed formula for the least SOR parameters, a and b, is not available and iterative search algorithms must be applied.

Least median of squares (LMS). Among the pitfalls of all linear regression are its rather poor break-down characteristics. Due to the minimisation criterion (sum of squared disturbances), the disturbance with the largest deviation from the least-squares estimate will have the highest influence on the position of the straight line and, hence, the values of a and b. Given a straight line, a single extraneous observation may shift this line considerably. Such data points are therefore termed "influential data". Influential data points do not necessarily qualify as incorrect data or "outliers". The normal distribution does not exclude the occasional occurrence of values far from the mean. There are ample techniques to analyse data on influential data points, high leverage points and outlying data (Dixon 1950; Dean and Dixon 1957; Draper and Smith 1981; Chatterjee and Hadi 1986; Rohrabacher 1991). The fact remains that an ordinary least squares regression line will shift its position if a single extraneous observation occurs, where an extraneous observation is defined as an observation not belonging to the distribution of the observed quantity. Thus, breakdown of least-squares regression is zero. L_1 regression is a method reducing the influence of extraneous data points. Nevertheless, its break-down is also zero. Break-down describes how many extraneous data may be in a data set before the regression parameters shift. For linear regression, no extraneous data must be in the data set. There are methods, however, where almost 50% of the data set may be outliers without influencing the parameter estimates for slope and intercept.

An example is robust regression with a breakdown of almost 50%. This implies that the regression line will not shift even if almost 50% of the data points are extraneous observations (Rousseeuw 1993) A robust regression technique is "least median of squared disturbances" (LMS). In LMS, the position of the regression line is not defined by the minimum sum of squared disturbances between measured ordinate values and fitted values, but by the minimum median value of the squared disturbances.

LMS requires that a straight line is fitted, the disturbances calculated and sorted (cf. Table 1.2 and associated text). The median value is recorded and another line is tested with the goal to find a line with a smaller median residual. To obtain such a line, only a random walk search algorithms is available (Rousseeuw and van Zomeren 1990). No "best estimator" criterion is available for LMS. On the other hand, LMS will in most situations very rapidly identify outlying observations by comparing an OLS straight line with its LMS analogue. On the basis of such a comparison, further investigation on the structure of a data set may be performed.

1.5.4
Application Examples

1.5.4.1
Comparing Inference Regions

Calibration in analytical chemistry may either relate instrument signals with analyte concentrations or concentration estimates from two different methods. Under certain circumstances, time and cost considerations require to apply an analytical method that will not recover the complete amount of the measurand, for instance in wet digestion of silicate sediments. Possibly methods are available which recover the total amount of the analyte but which are not available on a routine basis. Depending on the circumstances, the latter method can be applied to calibrate the former. Under these circumstances, ordinary linear regression (OLS) is not the first choice because some of its assumptions are violated. Some example data are given in Table 1.7. To determine meaningful weights, the required number of analyses and samples multiplies because the estimates of the weights need to mirror as close as possible the variances of the

Table 1.7. Calibration data for two analytical methods of uranium determination in sediment. Method 1 is a non-destructive neutron activation method, while method 2 includes digestion of the sample

	Method 1 $(\mu g\,dm^{-3})$	$s_x^2\ (= u_c(x_i)^2)$	Method 2 $(\mu g\,dm^{-3})$	$s_y^2\ (= u_c(y_i)^2)$
1	1.1	2.72	10	2.482
2	7	1.7	10	2.482
3	9	2.108	5	1.77547
4	10	2.108	25	4.38
5	62.4	8.16	21	3.94201
6	79	8.84	5	1.679
7	105	10.88	36	5.694
8	116	11.56	34	5.475
9	119	12.24	41	6.278
10	188	17.68	89	11.68
11	228	20.4	127	15.33
12	639	48.28	400	38.106
13	753	55.08	438	40.88
14	972	69.36	550	49.64
15	1130	81.6	710	62.05
16	1140	78.2	670	59.13
17	1140	78.88	600	53.29
18	1400	93.84	780	67.16
19	1590	104.72	849	72.27
20	1690	110.16	930	77.38
21	2190	138.04	1330	105.12
22	2980	180.88	1830	153.3
23	3250	197.2	2100	182.5

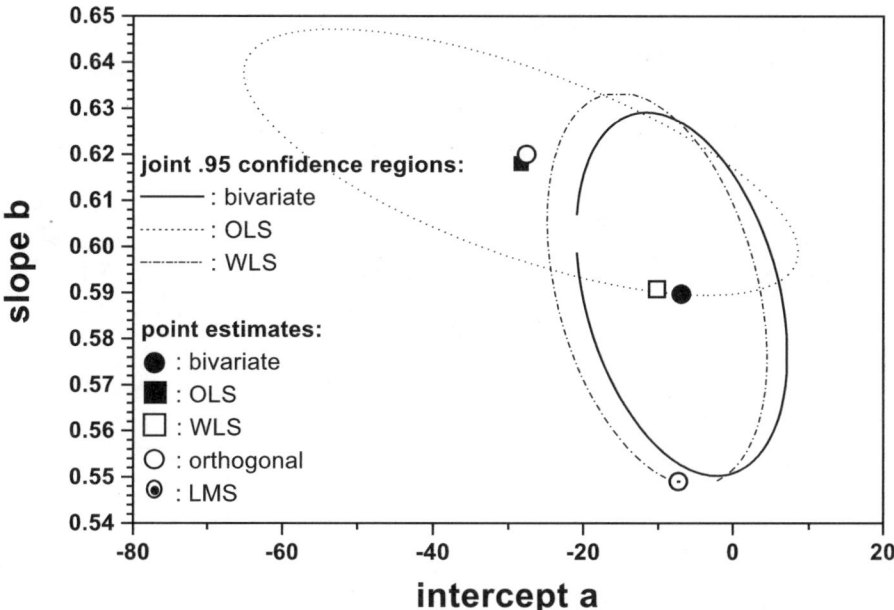

Figure 1.18. Comparison of joint 0.95 confidence regions (OLS, WLS and bivariate regression) and location estimates for slope b and intercept a for the data of Table 1.7

parent distributions. Method comparison/calibration must necessarily based upon the standard uncertainties. Using merely repeatability estimates would bias the resulting calibration curve and the associated confidence estimates of the calibration line. From the standard uncertainties, weights are obtained from Eq. (1.47), where the variances correspond to the squared standard uncertainties u_c for each data point. As in case of WLS, bivariate regression requires these standard uncertainties.

The joint 0.95 confidence regions from OLS, WLS and bivariate regression are given in Fig. 1.18, together with the expectation values for slope b and intercept a for OLS, WLS, bivariate regression, orthogonal regression and least median of squared disturbances.

OLS, LMS and orthogonal regression do not make use of the additional information available in the variances $s_{x,i}^2$ and $s_{y,i}^2$. Weighted least squares uses the variance information for the ordinate values, while only bivariate regression considers the full information. In the present situation, the positions of the joint confidence regions and location estimates of slope b and intercept a of WLS and bivariate regression are close together. The respective location estimates of the other estimation methods are either outside or at the border of the joint 0.95 confidence regions of WLS and bivariate regression. Any of these estimates may claim to be superior over the others. The LMS result as well as the sensitivity of the confidence regions towards weighing of data should trigger a more detailed investigation whether the distributional assumptions underlying OLS and orthogonal regression are fulfilled.

1.5.4.2
Establishing Confidence Bands of a Calibration Line

It is evident that a calibration relationship may differ depending on the method used for establishing the position. Furthermore, the number of data points considerably influences the confidence bands. Establishing a calibration line

Table 1.8. As(V) in 30 natural waters determined by two methods. Results in $\mu g\,dm^{-3}$, $s_{x,i}$ and $s_{y,i}$ are the standard errors of the mean used to calculate weights by Eq. (1.47) (from Ripley and Thompson 1987)

	Method 1 ($\mu g\,dm^{-3}$)	$s_{x,i}$	Method 2 ($\mu g\,dm^{-3}$)	$s_{y,i}$
1	8.71	1.92	7.35	2.07
2	7.01	1.56	7.92	2.23
3	3.28	0.76	3.4	0.96
4	5.6	1.26	5.44	1.53
5	1.55	0.39	2.07	0.59
6	1.75	0.43	2.29	0.65
7	0.73	0.22	0.66	0.19
8	3.66	0.84	3.43	0.97
9	0.9	0.25	1.25	0.36
10	9.39	2.07	6.58	1.85
11	4.39	1	3.31	0.93
12	3.69	0.84	2.72	0.77
13	0.34	0.13	2.32	0.66
14	1.94	0.47	1.5	0.43
15	2.07	0	3.5	0.99
16	1.38	0.36	1.17	0.33
17	1.81	0.45	2.31	0.66
18	1.27	0.33	1.88	0.54
19	0.82	0.23	0.44	0.13
20	1.88	0.46	1.37	0.4
21	5.66	1.27	7.04	1.98
22	0	0.06	0	0.01
23	0	0.06	0.49	0.15
24	0.4	0.15	1.29	0.37
25	0	0.06	0.37	0.12
26	1.98	0.48	2.16	0.62
27	10.21	2.24	12.53	3.51
28	4.64	1.05	3.9	1.1
29	5.66	1.27	4.66	1.31
30	19.25	4.18	15.86	4.45

with a minimum of only three data pairs requires the confidence estimates to be corrected for small sample sizes by $F_{2, 1, 0.95} = 200$ while for four calibration points $F_{2, 2, 0.95} = 19$ and five calibration points require $F_{2, 3, 0.95} = 9.55$. Thus, the neglect of an appropriate correction will reduce coverage (cf. Table 1.3) drastically and will underestimate the influence of the calibration step on the complete measurement uncertainty budget drastically.

The following example illustrates that a straight calibration line cannot be selected only on basis of the location estimates of slope and intercept. Table 1.8 presents 30 pairs of determinations of As(V) in natural river waters (Anderson et al. 1986, cited in Ripley and Thompson 1987). Method 1 applies selective reduction in combination with atomic absorption spectroscopy. Method 2 collects As by cold trapping and optical emission spectroscopy. These data

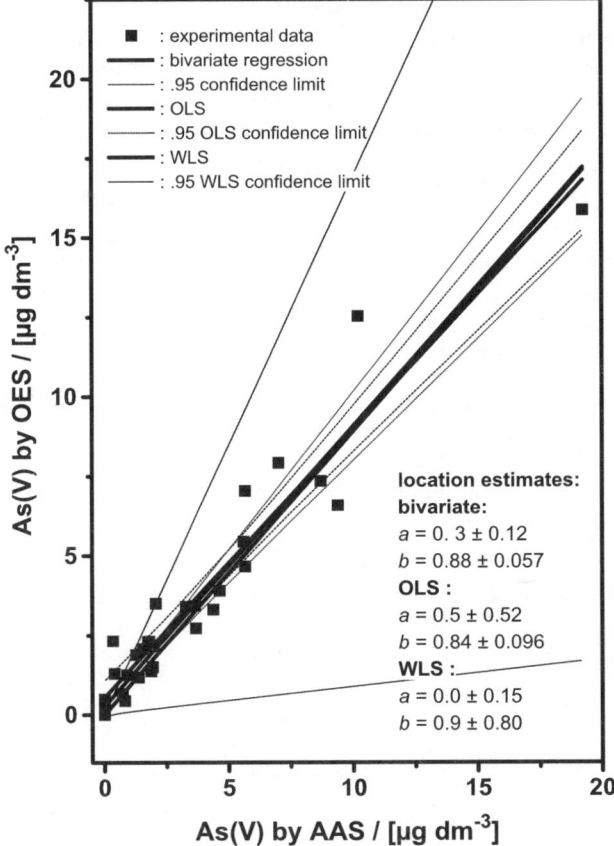

Figure 1.19. Linear fits to 30 data points of concentration of As(V) in river water (Table 1.8). The figure gives mean values of intercept a and slope b together with 0.95 confidence bands for bivariate regression, linear regression (OLS) and weighted regression (WLS)

are analysed by bivariate regression, ordinary linear regression and weighted regression. The results are given graphically in Fig. 1.19.

The mean values of the location estimates a and b are identical with those given by Ripley and Thompson (1987) for OLS and WLS within rounding errors. The confidence bands however indicate that the higher concentration values are overly down-weighted by WLS causing wide confidence intervals for data points estimated from the WLS straight line. The data sets Table 1.8 is an example set resulting in similar mean values from different methods, but widely varying confidence regions.

1.5.4.3
Importance of Robust Regression Techniques

Regression is not foolproof. This fact has been shown in the preceding sections. Having obtained calibration parameters alone is no criterion for quality. In some situations, it is a single data point which dominates the data evaluation procedure. If a calibration line is built on such data the results are meaningless. There is no subsequent amelioration possible. In most situations, the erroneous calibration will not even be detected. Such a case will be discussed in the following section to demonstrate high break down characteristics of LMS.

It is a common laboratory situation that a calibration line is determined in a certain amount concentration region of a measurand that needs to be extended later. Often, for instance for cost and time reasons, a single calibration

Table 1.9. Synthetic data illustrating the influence of high leverage and outlying data on the position of a least squares regression line

	Amount concentration ($\mu g\,dm^{-3}$)	Signal
1	0	0
2	0.4	0.49
3	0.73	0.66
4	0.82	0.44
5	0.9	1.25
6	1.27	1.88
7	1.38	1.17
8	1.55	2.07
9	1.75	2.29
10	1.81	2.31
11	1.88	1.37
12	1.94	1.5
Extraneous data D1	10	10
Extraneous data D2	10	4

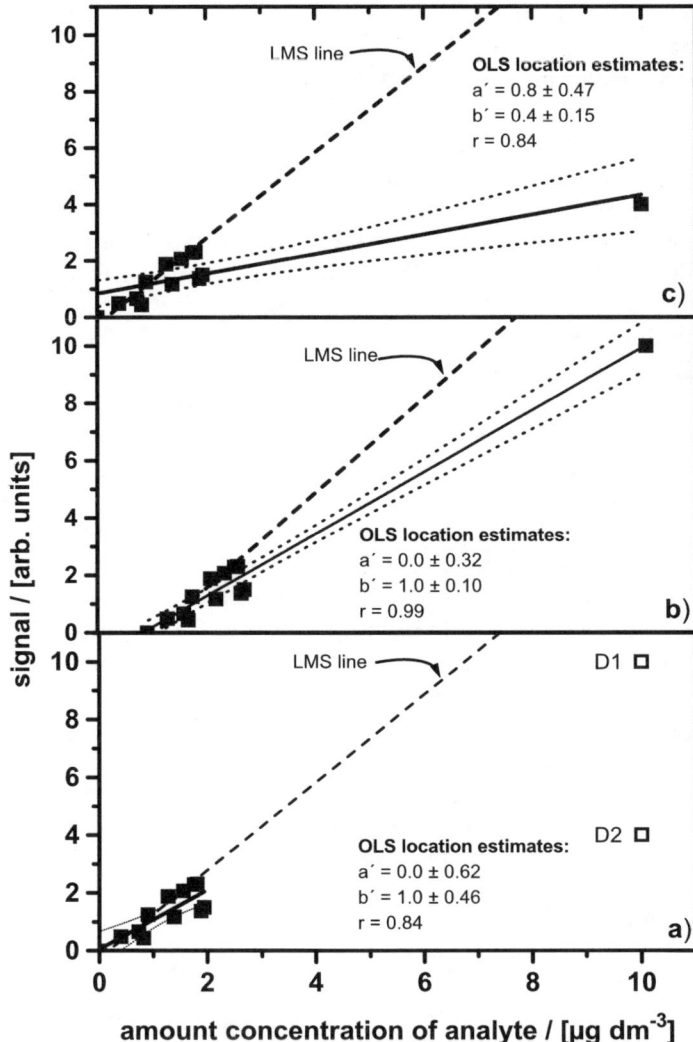

Figure 1.20. Sensitivity of calibration lines obtained from OLS on the presence of extraneous data points *D1* and *D2*, respectively. *Graph a* shows the calibration line together with the 0.95 confidence bands without both *D1* and *D2*. In *graph b*, data point *D1* is included into the regression analysis, and in *graph c D2* is included. A comparison of the resulting calibration lines shows that *D1* and *D2* determine the slope of the regression line. The 12 data points in the lower amount concentration region are almost without effect. A least median of squared disturbances (*LMS*) regression line is insensitive towards presence/absence of the extraneous data points

point is added to extend the calibration line. Table 1.9 presents some synthetic calibration data with 12 data points. The extraneous data points are D1 and D2. These both data are not necessarily outlying data but they are at a considerable distance from the point cloud of the original data.

Three different calibration lines can be obtained: the line without D1 and D2, the line including D1 and the line including D2. The calibration data by the 12 data points will compared to calibration where either the extraneous data D1 (Fig. 1.20b) or D2 (Fig. 1.20c) is included. The OLS calibration lines are given in Fig. 1.20a–c together with respective 0.95 confidence intervals. Inclusion of each of the extraneous data points into an OLS regression yields two completely different calibration lines. Obviously, the original 12 data points in the lower concentration range do not have any significant influence on the position of the calibration line while the single extraneous data point alone determines the slope. The calibration line always passes very close to either D1 and D2. Hence, both D1 and D2 are highly influential points.

The location estimates resulting from each choice of calibration data set are given together with the graphs. Adding an extraneous data point considerably extends the abscissa range and causes lower marginal 0.95 confidence limits for the slope. Including data point D1 also raises the Pearson coefficient r (cf. Eq. (1.42)). This value of r close to unity has any relevance for the reliability of the regression line. There is no reason to claim that the regression line Fig. 1.20b is more reliable than the line Fig. 1.20c.

A robust regression by the least median of squares (LMS) criterion, however, provides the same regression line independent of the presence or absence of extraneous data points. These lines are shown in all three graphs of Fig. 1.20. LMS analysis indicates the need for more detailed data analysis due to the rather large discrepancy between the LMS lines and the OLS lines.

1.5.5
Analysis of Linear Calibration Curves: A Summary

The examples given in Chap. 1.5 refer to the numerical establishment of a linear relationship between two parameters, including a confidence estimate for the parameters and values estimated from the calibration line. Calibration is a fundamental activity in chemical analysis, where linear relationships are commonly applied to relate electrical signals (generated by equipment of instrumental analysis) to analyte concentrations. A main focus of the discussion has been on methods also providing an estimate on the uncertainty inherent in the assessment of a calibration line. In practice, ordinary linear regression (OLS), weighted regression (WLS) and bivariate regression are the major calibration methods in use. For these three methods, closed formulae to derive confidence estimates for the regression parameters can be given.

Examples for other relevant methods for establishing a straight line relationship between two observables are orthogonal regression and least median of squared disturbances (LMS) regression. For these methods, no closed formulae exist to assess confidence regions for the parameters of a calibration line. Here, alternative approaches need to be considered, e.g. the iterative build-up of information via response surface curvature close to the minimum of the response surface (e.g. Spendley 1969; Box and Draper 1987; Brumby 1989; Ellison et al. 2001), Monte Carlo techniques (Rubinstein 1981; Alper and Gelb 1991) or computer-intensive statistics (Efron 1979; Efron 1981; Stine 1990). Some of these techniques will be introduced in the later sections.

The introduction of the GUM into metrology has put emphasis on estimates of uncertainty. OLS, WLS and bivariate least squares are able to provide meaningful confidence estimates together with the location estimators. However, they do this within a theory implicitly relying upon certain assumptions about the distribution of random effects. These assumptions are commonly related to the normal distribution of random influences and lack of correlation of observations and disturbances. These are required to be identically and independently distributed (i.i.d.). The application examples show that the three approaches may react sensitively to violation of the fundamental assumptions (cf. List 1.3.8). There is no a priori "best" method for establishing a calibration line.

Due to the importance of instrumental methods of chemical analysis, where the relationship between an analyte concentration and a signal is commonly established by a linear model function, calibration is a central element in the complete measurement uncertainty budget. The common practice to use only two or three standards to establish a relationship between amount concentration and signal, together with the neglect of appropriate corrections for the low degrees of freedom, suggests that here a potential for further improvement of comparability may be found. At present, the results of proficiency tests, e.g. obtained from the International Measurement Evaluation Programme (IMEP), organised regularly by IRMM, suggest that some relevant uncertainty contributions are not appropriately included by all participants (e.g. van Nevel et al. 1998; Aregbe et al. 2004; Papadakis et al. 2004; Visser 2004). It is an essential element of proficiency testing schemes, interlaboratory comparisons and round-robin studies to detect differences in the evaluation of data (Papadakis and Taylor 2004).

The standard uncertainty is not measurable. It is assessed on a subjective basis following some guidelines. It should be kept in mind that uncertainty expresses doubt. Doubt is a psychological phenomenon. A survey of literature on the determination of thermodynamic data for aqueous metal ion species revealed that words such as "verified" and "validated" are commonly applied to communicate the impression of "quality". The meaning of such words, however, is not defined, and it is impossible to extract what criteria are used to qualify certain measurement values. This example shows that the metrological concept of complete measurement uncertainty budgets is a considerable achievement carrying the potential for further improvement with growing experience.

1.6
Metrology in Complex Situations: Non-Normality, Correlation and Non-Linearity

"If a man will begin with certainties, he shall end in doubt; but if he will be content to begin with doubts, he shall end in certainties."

(F. Bacon)

1.6.1
Resampling Methods – An Introduction

Classical statistics relies considerably on the assumption of normal distribution of random influences. This assumption is backed by the Central Limit Theorem (cf. Chap. 1.3) if the random influences are identically and independently distributed (note the existence of exceptions!). On basis of these assumptions, randomness can be treated analytically by mathematical means, giving rise to a large part of statistical science being developed during the past about 250 years – and giving statistics a slightly appalling appeal (Salsburg 1985; Thompson 1994). Statistics tries to express the degree of knowledge to be obtained or available. Most statistics applied in the sciences, engineering and humanics deals with frequencies of observations; thereby replacing the measurand and the limited knowledge about its values (= doubt) by the frequency of observing a certain value of the measurand in repeated experiments. Frequency of observation is, however, only a crutch and does have its limitations. Its major advantage is its applicability during eras of scientific investigations where the major tool for the application of probability theory, the computer, was not available.

Therefore, a large part of the decisions which may affect our lives is influenced by statistical frequentist considerations. Metrology needs to quantify randomness in an objective manner. The incorporation of basic statistical concepts, e.g. mean value and variance, therefore is a natural choice.

Sometimes, the data gathered from a scientific investigation may lead to complicated statistical models where it may not be possible to implement "standard" statistical techniques to summarise the data. The estimators of the parameters may have a complicated expression, or the distribution of the parameters is unknown. A major point in chemical analysis is that complex experiments do not allow for unlimited repetition. All conclusions on the likely random processes causing variation of the measurands have to be made from a limited set of experimental data. In some cases, e.g. for parameters evaluated by neural network methods, clear formal dependencies of the parameters and measurands are not even available (Dathe and Otto 2000). Resampling methods, especially the "bootstrap" and related methods, are powerful alternatives (Efron 1979). Introductions to the bootstrap are available (Efron and Tibshirani 1986; Stine 1990; Meinrath et al. 2000; Wehrens et al. 2000).

Resampling methods rely on the information available: the experimental data. These data carry with them all information on the fundamental process of which they are a result. Thus, in the resampling methods, the observed data take the role of an underlying population. The statistics, e.g. variances, distributions and confidence intervals are obtained by drawing samples from the sample. The importance of resampling methods comes from the experience that standard errors and confidence intervals are commonly more reliable than alternatives that rely on untested assumptions (Stine 1990). Instead of concentrating on the mathematical tractability of a problem, the analyst can concentrate on the data and the information of interest.

Resampling methods are typically demanding in the computational burden. Thus, they replace mathematical computation by brute computing power. At a time where one computer represents the total annual computing power of the world in the early 1970s and predominantly being used to create computer game graphics, this burden is an easy yoke. A typical resampling plan runs as follows:

List 1.6.1: Random resampling scheme

    a obtain a sample of k empirical observations
    b create a large number f of sub-samples of size k by sampling randomly (with replacement) from the original observations
    c calculate the statistics t of interest for each of the f sub-samples
    d calculate the mean of the statistics t, $\bar{t}$.

    e obtain the standard error for t by $s_e(t) = \sqrt{\dfrac{\sum\limits_{i=1}^{b}(t_i-\bar{t})^2}{f-1}}$ (1.75)

Eq. (1.75) is rather close to the common standard error estimator for the sample mean, Eq. (1.75). A neat formula for the standard error is only available for the sample mean. Standard errors for other statistics are difficult to evaluate. A motivation for the introduction of resampling methods is an extenuation of Eq. (1.75) to estimators other than the sample mean.

An important point of the scheme given in List 1.6.1 is the magnitude of f. The resampling scheme replaces difficult mathematics with an increase of several orders of magnitude in the computing needed for a statistical analysis. While the classical approach may require calculation of some regression coefficients, resampling schemes may require several thousands of sub-samples (Efron 1979). These resampling methods are therefore also named "computer-intensive resampling methods". Upon their introduction in 1979, Efron (1979) intended to name these methods "shotgun" methods, because they are able to "shoot the head off of any problem if you can live with the mess it creates". Instead, these methods have been termed with the less martial name "bootstrap methods". The name refers to a story by KFH Freiherr von Münchhausen, where he escaped from a peat bog by lifting himself and his horse on his own hair. In the English translation, he does so by tearing his bootstraps. The name seems to have been introduced by Finifter (1972). In a similar way, bootstrap methods allow the analyst to solve his statistical problem by using the means he has at his hands: the experimental data.

Probability distributions play an important role in bootstrap methods. The cumulative distribution F with $F(x)$ giving the probability $Pr(x_i < x)$ has been introduced in Chap. 1.3. In this notation, each x_i is a random observation having the cumulative distribution F. In classical statistics, the population distribution is assumed to be normal (cf. Eq. (1.5)). The sample-to-sample variation of the sample average is well known. If the sample mean $\bar{x}$ is given by Eq. (1.7), then its variance v at n observations is

$$v(\bar{x}) = \frac{\sigma^2}{n} \tag{1.76}$$

If σ is not known it is replaced by the sample standard error s (Eq. (1.8)), resulting in

$$v(\bar{x}) = \frac{s^2}{n} . \tag{1.77}$$

If F is not the cumulative normal distribution, then v will perform poorly as an estimator of V, the population variance (Hampel et al. 1986).

The idea behind the computer-intensive resampling methods is to replace the unknown function F, which describes a population which cannot be resampled, with an estimator F, which describes a population that can be sampled repeatedly (Stine 1990). The theory of the resampling methods shows that the cumulative empirical distribution function of the statistic t obtained according the resampling scheme given in List 1.6.1 is the optimal estimator for F.

A practical example (Table 1.10) will illustrate resampling procedure using the data of group 1 from Table 1.1. Seven data sets resampled from the original data set are given. Of course, $f = 7$ is much to small for a reliable resampling estimate but this example is only given to illustrate the procedures.

The resampled data sets contain some observations more than once, while others are not included at all. In the other data sets, a previously omitted datum can be included while others are omitted. In fact, the probability that a datum is not included into a data set is

$$P(x_i \notin X_{\text{boot}}) = \left(1 - \frac{1}{n}\right)^n \approx 0.36 \tag{1.78}$$

Table 1.10. A set of seven resampled data sets from the original data set (taken from group 1 in Table 1.1) together with the statistics of interest (here the statistics of interest is the mean)

Original set	Resampled sets (sorted) (kg dm^{-3})						
	1	2	3	4	5	6	7
0.9903	1.0015	0.9993	0.9903	0.9903	0.9982	0.9903	0.9903
0.9982	1.0023	1.0015	0.9982	1.0015	0.9982	0.9903	0.9903
0.9993	1.0023	1.0015	1.0023	1.0023	0.9982	0.9903	0.9982
1.0015	1.0023	1.0015	1.0023	1.0028	0.9993	0.9903	1.0015
1.0023	1.0023	1.0023	1.0023	1.0028	0.9993	0.9982	1.0023
1.0023	1.0023	1.0023	1.0067	1.0028	0.9993	1.0015	1.0023
1.0028	1.0028	1.0023	1.0067	1.0028	1.0067	1.0067	1.0023
1.0067	1.0028	1.0028	1.0079	1.0067	1.0067	1.0067	1.0023
1.0079	1.0079	1.0139	1.0139	1.0067	1.0067	1.0079	1.0079
1.0139	1.0139	1.0139	1.0139	1.0139	1.0139	1.0079	1.0079
Mean	1.0043	1.0041	1.0045	1.0033	1.0027	0.999	1.0005

if n are large. The total number of different combinations G with resampling (without considering ordering within the combinations) is

$$G_n^n = \frac{(2n-1)!}{n!\,(n-1)!} \,. \tag{1.79}$$

Table 1.11 shows G for $n = 2-15$. Bootstrapping can be performed with sample sizes as small as seven observations. As a rule of thumb, the size b of a typical bootstrap sample B should be in the range $b = 200-500$ for an estimate of the

Table 1.11. Number of resampled data set variation for given n

n	G_n^n
2	3
3	10
4	35
5	126
6	462
7	1716
8	6435
9	24 310
10	92 378
11	352 716
12	1 352 078
13	5 200 300
14	20 058 300
15	77 558 760

standard error of a mean and in the range $b = 1000 - 2000$ for the estimate of a confidence region.

The mean $\bar{t}$ of all means obtained from the resampled data sets is:

$$\bar{t} = (1.0043 + 1.0041 + 1.0045 + 1.0033 + 1.0027 + 0.9990 + 1.0005)/7$$

$$= 1.0026\,\text{kg}\,\text{dm}^{-3}\,.$$

The estimate for the standard error $s_e(t)$ is obtained from Eq. (1.75):

$$s_e(t) = \sqrt{\frac{0.0017^2 + 0.0015^2 + 0.0019^2 + 0.0007^2 + 0.0001^2 + 0.0027^2 + 0.0021^2}{6}}$$

$$= 0.0018\,\text{kg}\,\text{dm}^{-3}\,.$$

The necessary numerical operations are rather tiresome even for a small data set of ten samples and seven resamplings. For larger data sets and adequate numbers of resamplings b, these calculations need to be performed by a computer. In Fig. 1.21, the results for resampling the original data set in Table 1.10 with $b = 25$ and $b = 1000$ are given as cumulative empirical distribution functions. The resampling set with $b = 1000$ is compared to its closest fitting normal distribution by Kolmogorov–Smirnov criterion.

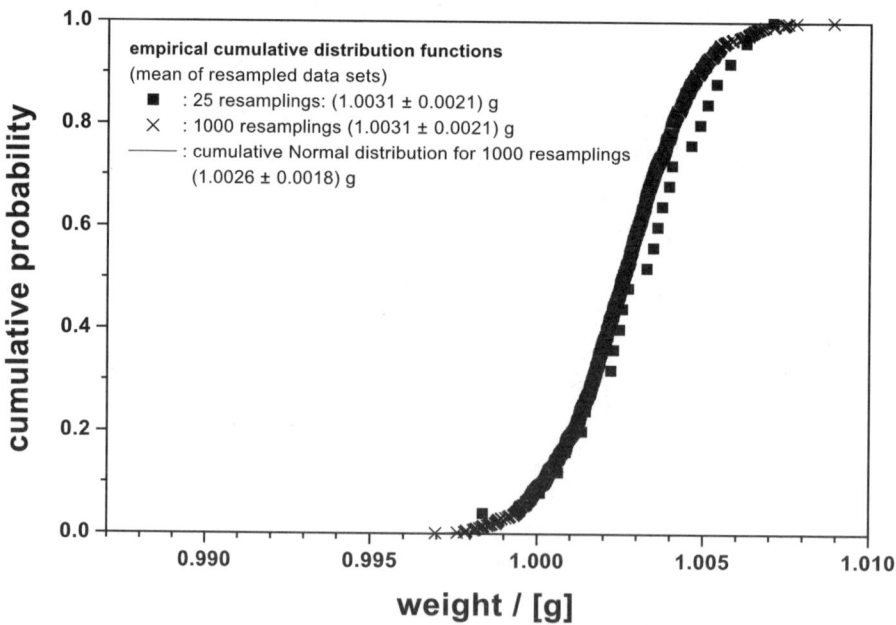

Figure 1.21. Comparison of resampling results obtained by 25 and 1000 resamplings from the original data set in Table 1.10. The resampled set with $b = 1000$ is compared to its closest fitting cumulative normal distribution. For both resampled data sets, the probability that the underlying distribution is normal is $> 99\%$

1.6.2
Bootstrapping Regression Schemes

1.6.2.1
Random Regressors

A connection between computer-intensive resampling methods and parametric statistics from Chap. 1.5 can be made by resampling regression schemes. It is possible to use computer-intensive resampling from an original data set, e.g. Table 1.7, to evaluate the bootstrap standard error of slope and intercept. To obtain the experimental data, random samples have been taken from the field. Sampling randomly from the original data therefore preserves the random structure of the experimental design.

A result of 5000 bootstrap pairs from the data in Table 1.7 evaluated by using the OLS procedure is shown in Fig. 1.22. The somewhat overly large sample has been chosen to obtain a clear picture of the distribution of the data points which describe an own confidence region. This confidence region covers almost all regions established previously by OLS, WLS and bivariate regression.

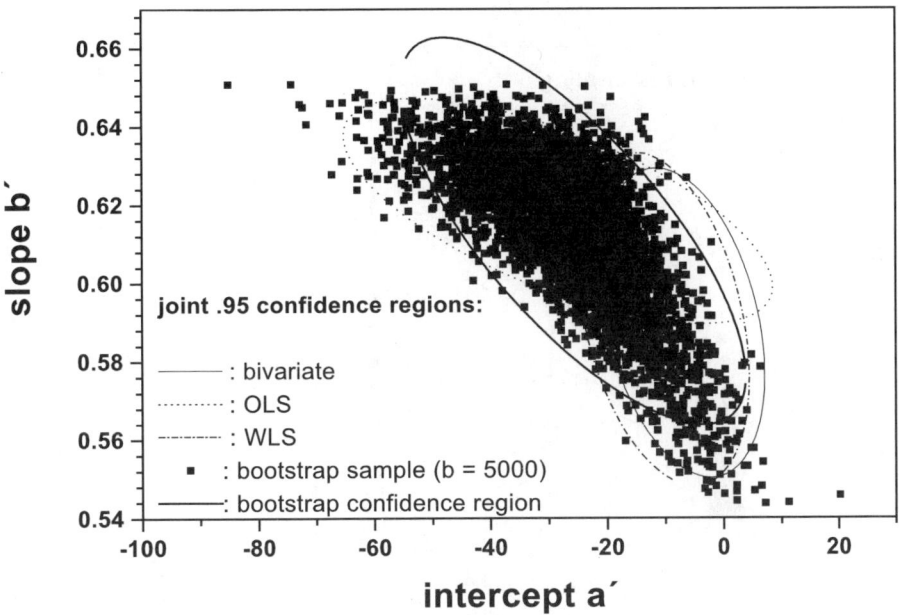

Figure 1.22. A comparison of confidence regions (cf. Fig. 1.18) for slope a and intercept b with results from 5000 bootstrap samples randomly drawn from the data in Table 1.7 and its 0.95 confidence region. The bootstrap estimates are: intercept $a = -26 \pm 11$, slope $b = 0.61 \pm 0.02$

Interpretation of a data set by classical regression schemes assumes implicitly that there exists an internal relationship between the parameters of the model and the disturbances: The optimum parameters are characterised by a minimum of the sum of disturbances. Because the disturbances are generally not accessible (being true quantities), the regression scheme replaces the disturbances by the residuals. This is a situation comparable to the bootstrap where the unknown cumulative distribution function Φ is replaced by an estimator F. There is, however, no requirement in the bootstrap for a relationship between the optimum estimate and the residual structure. The bootstrap accepts any criterion to evaluate a statistics from a data set.

The spatial distribution of a and b from the 5000 resampled bootstrap sets describes an own confidence space which is not nicely elliptically, but cudgel-shaped, a not uncommon observation from correlated distributions (Meinrath 2000b). The fact that the 5000 estimates from bootstrapped samples generate a rather homogeneous distribution shows that the assumption of independently and identically distributed disturbances is only approximately fulfilled, but not severely violated.

It is important to note that random resampling from a data set following the bootstrap scheme does not create independently distributed disturbances with constant variance. If this assumption is violated in the original data

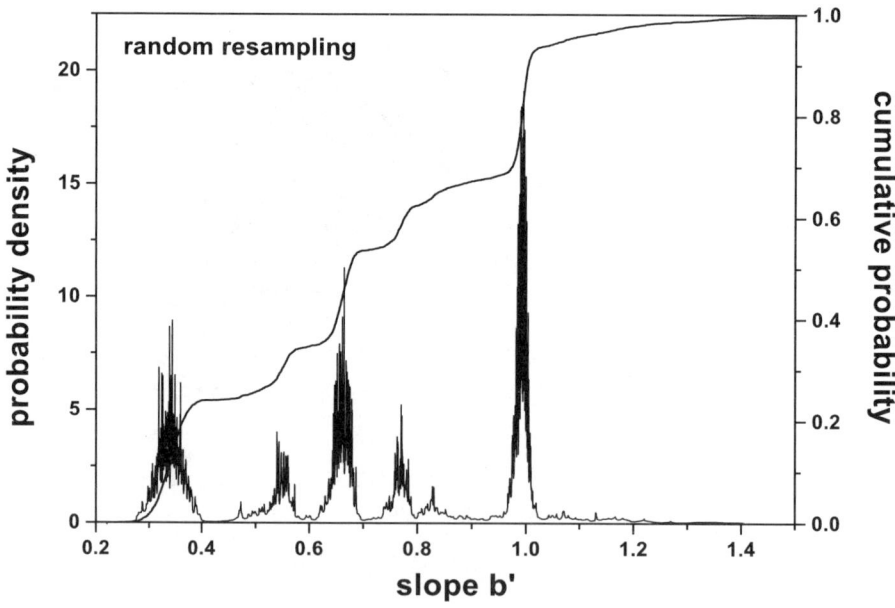

Figure 1.23. Empirical distribution function (*right axis*) and probability density (*left axis*) of slope b for synthetic data set in Table 1.9 obtained from randomly resampling data sets. The distribution is notably non-normal and multi-modal with distinct modes indicating highly

set, the OLS regression scheme has a problem of misinterpreting the data due to its intrinsic assumptions. Computer-intensive resampling is indicating such a nuisance situation as shown in Fig. 1.23 for the synthetic data set Table 1.9. The right axis gives the cumulative distribution function, while the left axis relates to its derivative, the probability density. Both distributions have the same relationship as the cumulative normal distribution and the normal distribution. The distributions in Fig. 1.23, however, are obviously non-normal but non-symmetric and multi-modal, that is, the probability density has several maxima. The bootstrap does not cure the problem of presence of extraneous data but it highlights it.

1.6.2.2
Fixed Regressors

A resampling scheme with fixed regressors is the alternative to random re-sampling. In controlled situations, for instance instrument calibration, the regressors are carefully chosen. In multi-point calibration for measurement of pH, the calibration samples are commonly chosen to homogeneously cover the expected experimental range of observations. A resampling scheme randomly modifying these often carefully selected experimental designs would be inadequate. In such situations, the experimental design must be maintained by ensuring that information is included in the resampled data set at the fixed positions. The strategy answering to this request is to resample from the disturbances.

List 1.6.2: Fixed regressors resampling scheme.

a Obtain the disturbance for each regressor and the expected values (commonly the ordinate values calculated from the best-fit parameters) by appropriate numerical interpretation of the experimental data.
b Create a large number f of sub-samples by adding a randomly selected (with replacement) disturbance to the expected value for each regressor.
c Calculate the statistics t of interest for each of the f sub-samples.
d Calculate the mean of the statistic, $\bar{t}$.

e Obtain the standard error for t by $s_e(t) = \sqrt{\dfrac{\sum\limits_{i=1}^{b}(t_i - \bar{t})^2}{f - 1}}$.

The disturbances are, as has been outlined previously, a product of the model interpreting the data. The OLS assumptions (cf. List 1.3.8) to a large part rely on a world ruled by the normal distribution. The bootstrap has the property to "smooth" the disturbances over the experimental design; an extraneous disturbance will be combined randomly with all regressors. The heteroscedasticity of the disturbances will thus result in homoscedastic bootstrap samples.

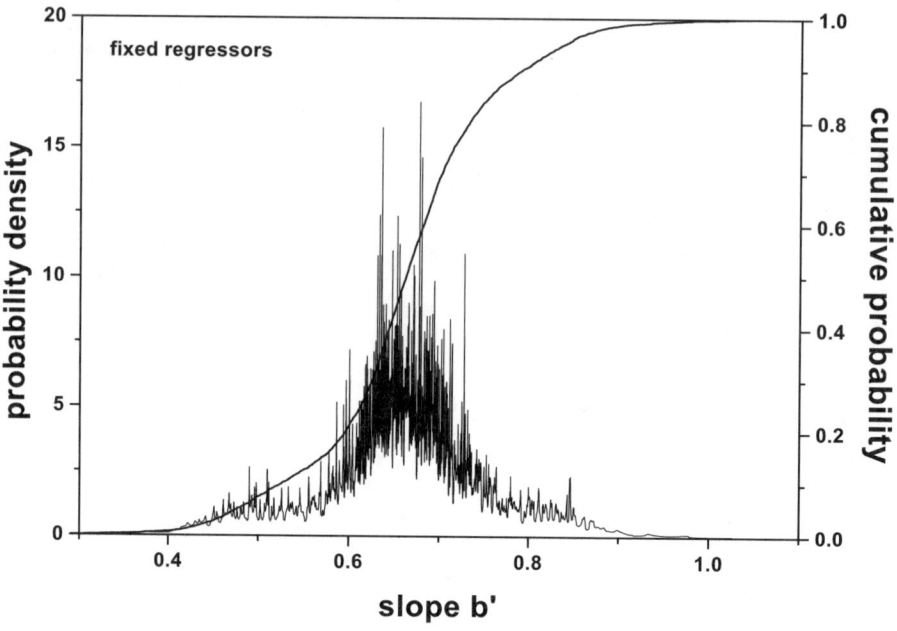

Figure 1.24. Probability density (*left axis*) and cumulative probability (*right axis*) obtained for slope *b* interpreting data set Table 1.9 by a fixed regressors scheme. Because the disturbances of the extraneous data points are smoothed over the complete experimental design, a mono-modal smoothed curve results

This behaviour is illustrated by Fig. 1.24, where the empirical distribution function (left axis) and probability density (right axis) are shown for fixed regressors resampling for the data set Table 1.9. The heteroscedasticity due to the extraneous data is smoothed into a nicely normal curve.

1.6.3
Bootstrap Confidence Bands and Inference Regions

Even though the strength of computer-intensive resampling methods is not focused on linear regression schemes, it offers some clear advantages:

- No unproven assumption of fundamental distributions.
- Not limited to L2 optimisation criterion; may also include robust regression.
- Provides an estimate of the empirical distribution function underlying the sample.
- Offers several resampling schemes (e.g. fixed and random regressors, respectively) to model the experimental design.
- Provides uncertainty estimates for statistics other than the mean.

The quantification of linear trends is among the most important tasks in metrology in chemistry that the bootstrap analogue of ordinary linear regression is given in the following. Matrix formulation will be used for sake of brevity.

1) From a given sample of k observations (x_i, y_i) draw a random sub-sample of size k with replacement.
2) From the k sub-sampled observations $(^f x_i, ^f y_i)$ form the vector X and Y by

$$X = \begin{bmatrix} 1 & ^b x_1 \\ 1 & ^b x_2 \\ 1 & ^b x_3 \\ \vdots & \vdots \\ 1 & ^b x_k \end{bmatrix} \qquad Y = \begin{bmatrix} ^b y_1 \\ ^b y_1 \\ ^b y_1 \\ \vdots \\ ^b y_1 \end{bmatrix}$$

(superscript f refers to an observation in a sub-sample)
3) Calculate the least squares estimates of the parameters a and b from Eq. (1.80):

$$\begin{bmatrix} ^f a \\ ^f b \end{bmatrix} = \left(X^T X \right)^{-1} X Y \right) . \tag{1.80}$$

4) Repeat steps 1–3 to obtain f bootstrap samples of f values for a and f values for b.
5) Calculate the mean of a and b from the b bootstrap samples
6) Calculate the vector Δ of differences between the means $\bar{a}$ and $\bar{b}$ of all $^f a$ and $^f b$, respectively, and the f individual bootstrap values with elements

$$\delta a_i = {}^f a - \bar{a} \quad \delta b_i = {}^f b - \bar{b} \quad \Delta = \begin{bmatrix} \delta a_1 & \delta b_1 \\ \delta a_2 & \delta b_2 \\ \delta a_3 & \delta b_3 \\ \vdots & \vdots \\ \delta a_b & \delta b_b \end{bmatrix}$$

7) Obtain the variance-covariance matrix V for a and b from Eq. (1.81):

$$V = DD^T/(b-1) . \tag{1.81}$$

The marginal standard deviations s of parameters result from the square roots of diagonal elements of V:

$$s(\bar{a}) = \sqrt{V_{11}} \quad s(\bar{b}) = \sqrt{V_{22}} .$$

The confidence band at a confidence level α is obtained from Eq. (1.82):

$$\begin{matrix} s_{y,o} \\ s_{y,u} \end{matrix} = \pm t_{n-1,\alpha} \sqrt{V_{11} + 2V_{12} x_0 + V_{22} x_0^2} \tag{1.82}$$

where V_{ij} is again the element of V in the i-th row and j-th column of the variance-covariance matrix V (Eq. (1.81)).

The confidence ellipse at a confidence level α is obtained from Eq. (1.83):

$$2F_{2,k-2,\alpha} = (a - \overline{a})^2 \zeta_{11} + 2(a - \overline{a})(b - \overline{b})\zeta_{12} + (b - \overline{b})^2 \zeta_{22} \qquad (1.83)$$

where elements ζ_{ij} are taken from the inverse of the variance-covariance matrix V (Eq. (1.81)).

Figure 1.25 compares the results from OLS, bootstrap regression and robust LMS for the data from Table 1.7. There is no other method for evaluating LMS confidence limits. Figure 1.25 shows that for well-behaved data sets, the results of common OLS and bootstrap regression are comparable.

Bootstrap methods do have a wide application in linear models (Davison and Hinkley 1997). The previous chapter briefly introduced bootstrap strategies to univariate linear models mainly to gain some familiarity with these methods and related concepts like empirical distribution functions and their derivative, the probability densities. In metrological situations, especially calibration, the differences between OLS and bootstrap regression are small because the situations are controlled and, almost more importantly, the number of calibration samples is small. Bootstrap methods may become important

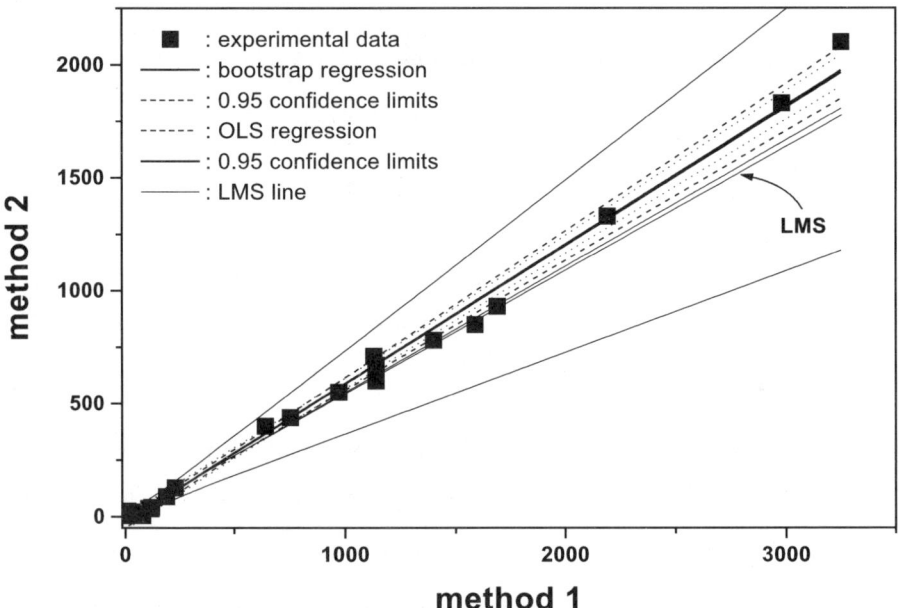

Figure 1.25. Comparison of mean straight line and respective 95% confidence bands obtained by *OLS*, *bootstrap regression* and robust *LMS* from the data in Table 1.7. For LMS, only a mean line can be obtained. Thee is no method to derive a confidence estimate

in method comparison and for robust regression. As a method of robust regression some focus was put onto the least median of squares criterion (LMS). LMS belongs to a group of trimmed squares methods that also includes, for instance, trimmed median absolute deviation regression (de Angelis et al. 1993). For such methods, there are no other confidence estimates with similar theoretical foundation.

The major importance of computer-intensive resampling method is in multi-parameter models, where the correlation between the parameters, the relationship between parameters and disturbances, the difficulty for efficient analysis of extraneous data and experimental limitations do not provide a "standard" method for the quantification of uncertainty limits.

1.7
Metrology in Complex Situations: Examples

"He who evaluates carefully the two digits after the decimal point can estimate those before."

(an unknown fellow reviewer)

1.7.1
Importance of Complex Situations in Environmental Sciences and Geochemical Modeling

If a model function after differentiation by a model parameter P still depends on other parameters, it is non-linear. Hence, Eq. (1.84) shows a linear equation, while Eq. (1.85) gives a non-linear equation

$$f(p_1, p_2, p_3) = p_1 x^2 + \sqrt{2p_2} \log(x) - p_3 \tag{1.84}$$

$$f(p) = \frac{K p}{1 - K p} \tag{1.85}$$

Finding the optimum parameter and appropriate confidence limits for p can be challenging, or even impossible. The task of deriving meaningful parameter estimates and their confidence limits is an old one, and fitting curves to data is an important field of statistics and numerical mathematics. The well-known Newton–Raphson algorithm is one iterative least-squares fitting method, the Powell–McDonald algorithm another (Powell and McDonald 1972). The Simplex algorithm (Nelder and Mead 1965; Caceci and Cacheris 1984) is an efficient algorithm that does not need derivatives. The rapid development of computing machines in the late 1960s gave rise to a larger number of concepts for the evaluation of parameters for large data sets and complex model functions. While before experimental designs had to be followed where the parameters could

be extracted after suitable linearisations by graphical methods, computers did
so with model functions of almost arbitrary complexity.

In these days, everything became possible if it could transferred into computer code, while the presence of measurement uncertainty became almost
completely ignored (Chalmers 1993). Criteria were missing to which degree
a data set could be interpreted and where the addition of an additional parameter would interpret mere noise. In chemistry, addition of a parameter could also
mean to have identified a new species. Consequently, an enormous collection
of chemical species came into being which still today fill the thermodynamic
data bases of geochemical modeling codes (Grauer 1997).

Searching for objective criteria for limiting the numerical interpretability
of chemical data is an essential task of metrology in chemistry. It is the logical
next step in the development of chemical data interpretation. Geochemical
application of chemical data often give rise to complex models. Individual informations, e.g. obtained from a Langmuir isotherm Eq. (1.85) are combined
with species information from a geochemical modeling code to predict the
likely fate of a contaminant in an aquifer. Both site-specific chemical information, possibly obtained by chemical analysis from samples collected at the site
and fundamental thermodynamic data on chemical species of a contaminant,
may become combined to support far-reaching decisions, for instance, closure
of a well, size of groundwater body with associated limitations in agricultural
production and obligation to compensate damages induced by the polluter to
others.

Those affected negatively by a decision have the right to question the decision. This possibility puts strain on the decision-makers to inquire for the
basis on which the data have been interpreted. In such situations of conflicting interests conventions will not reconcile the parties. On the contrary, the
weakness of the evidence will turn against itself (Walsh 2000).

Not only do the data from chemical analysis of individual samples serve
as a basis for important decisions. Chemical thermodynamics plays an important role, especially in environmental science and geochemical analysis.
Homogeneous solution equilibria of chemical entities are laws of nature by
their relationship with the fundamental laws of thermodynamics:

$$\Delta G_r^\circ = -RT \ln K \tag{1.86}$$

where ΔG_r° is Gibbs free energy of reaction r at standard condition, T is the
absolute temperature in K, R is the gas constants and K is the equilibrium
constant of reaction r under standard condition. A consequence of Eq. (1.86)
is the importance of thermodynamic data for chemical species. Such data are
investigated on a world-wide basis during the past at least 100 years. Due
to the difficulties with the complicated numerical and statistical problems,
the quality of the vast majority of data has not been assessed by objective
criteria (Meinrath et al. 2000b, 2004). In fact, depending on the method of data
evaluation, a variety of different "best fit" mean values may be obtained form

the same data set. In the majority of publications the numerical data and their statistical interpretation (if done at all) are not available for an independent assessment. For the view-point of metrology such data represent "rumours".

The sum of numerical operations necessary for an assessment of a large number of measurements and manipulations involved in the evaluation of thermodynamic data cannot be expressed by a closed mathematical formula. Thus, "classical" statistics on basis of linear progression-of-error (cf. Eqs. (1.25) and (1.26)) is not possible. If "metrology is becoming the language of the (inter)national marketplace" (de Bièvre 2001), means must be provided to adopt appropriate protocols to considered these situations (Meinrath 2001). Computer-intensive resampling statistics in combination with simulation procedures, as for instance applied in risk analysis (Burmaster and Anderson 1994; Helton 1994; Hoffman and Hammonds 1994; Ades and Lu 2003; Li and Hyman 2004), will do.

1.7.2
Non-Linearity, Correlation and Non-Normality: A Working Example

Interpreting non-linear models by linear equations results in bias (Box 1971). The magnitude of bias varies with the degree of non-linearity and magnitude of disturbance (Bates and Watts 1988). It might be in some situations possible to quantify the bias from model and experimental data provided the fundamental distribution would be known. As an example the following model function will be discussed:

$$F(x) = \frac{K}{x} + K\beta_1 + K\beta_2 x + K\beta_3 x^2 \tag{1.87}$$

where x is the carbonate concentration in a thermostated vessel holding an aqueous solution in a steady state with solid $UO_2CO_{3(s)}$. The atmosphere is pure CO_2. Parameters K, β_1, β_2 and β_3 represent the solubility product of the solid phase and formation constants β_n ($n = 1-3$) of solution species $UO_2(CO_3)_n^{(2-2n)}$. The regressors are free carbonate concentrations calculated from the thermodynamic formation constants of the CO_3^{2-} species and the measured pH. The measurand $F(x)$ is the total uranium(VI) concentration in solution at a given total amount concentration of CO_3^{2-}.

The chemistry of the system has been discussed elsewhere (Meinrath and Kimura 1993; Meinrath et al. 1996; Meinrath 1997). The solubility studies have been followed spectroscopically and the solution species are corroborated by following the spectral changes with changing pH, comparison with solubilities under different CO_2 partial pressures and comparison with fluorescence spectra (where available) (Meinrath et al. 1996a, 1998; Meinrath 1997a). There is no source of information in the data set giving an idea on the likely number of species legitimately to be considered as "relevant solution species". The experimental data displayed in Fig. 1.26 have been taken from the respective reference. An evaluation on basis of a least-squares optimisation criterion using weighted residuals (with the datum itself being the weighing factor) by a balanced bootstrap scheme (Gleason 1988) has been performed. Results are given below.

The model function in Eq. (1.87) is linear in its first term, while the other terms are non-linear. The magnitude of the parameters varies widely. While K has a magnitude of $10^{-13.5}$, β_1 is about 10^9, β_2 is $10^{15.5}$ and β_3 is 10^{22}. Such figures may pose numerical difficulties and, commonly, the logarithms are communicated. For the numerical evaluation, however, the optimisation criterion is important. Depending on the criterion (e.g. L_1 or L_2 with linear or logarithmic parameters), the results might be quite different. In particular, the cumulative distribution functions are strongly modified by an ordinate axis transformation. The median and the confidence limits, however, remain the same, irrespective of whether they are tabled in linear or logarithmic form. The mean values, however, may differ considerably.

It has been shown previously that a normal approximation, e.g. response surface approximations or likelihood regions, does not provide a reasonable description of uncertainty regions (Meinrath 2000). Progression-of-error concepts will also fail because the parameters are correlated. It is not possible to shift one parameter without affecting the position of another because corre-

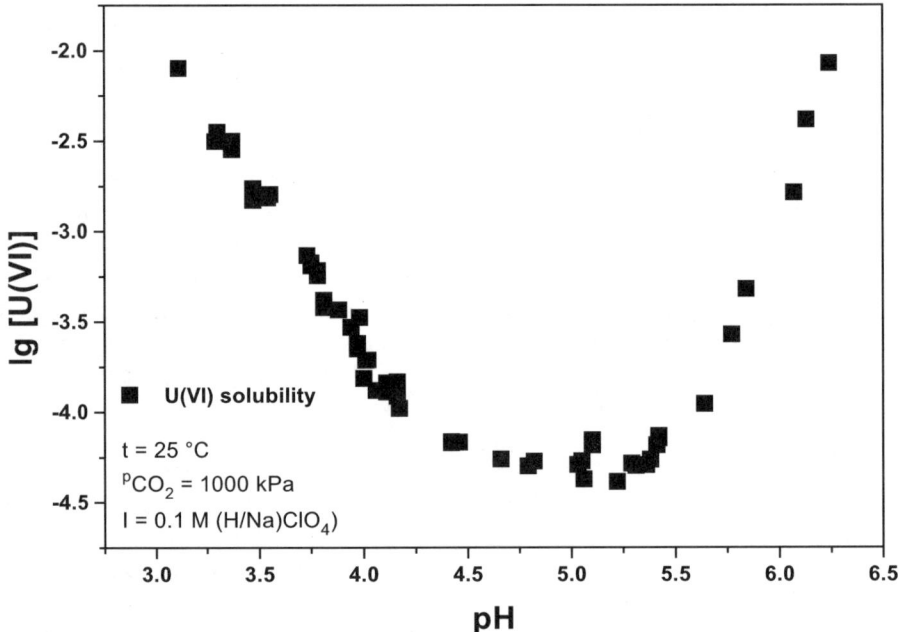

Figure 1.26. Solubility data of $UO_2CO_3(s)$ as a function of pH

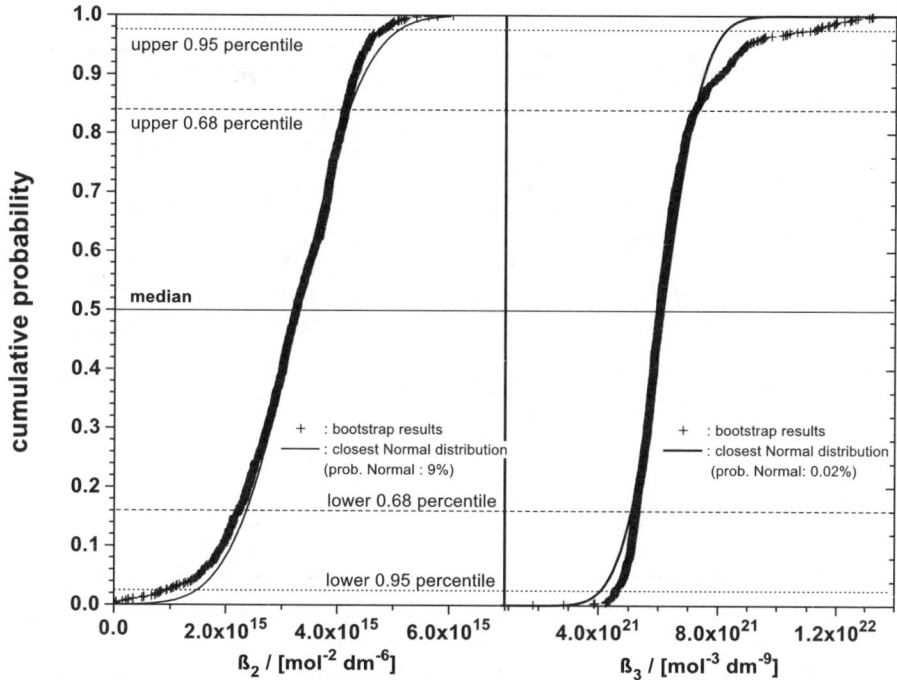

Figure 1.27. Cumulative probability distributions of parameters β_2 and β_3, compared to the closest fitting Normal distribution. *Horizontal lines* indicate the median and the upper and lower 0.68 and 0.95 percentiles, respectively

lation coefficients ρ (linearly approximated) are found to be in the range of $|\rho| \geq 0.6$.

The empirical probability distributions of the parameters (Fig. 1.27) reveal a seemingly normal distribution. A Kolmogorov–Smirnov test, however, suggests the following probabilities: K: 85%, β_1: 98%, β_2: 7% and β_3: 0.02%. The distributions of parameters β_2 and β_3 are characterised by distinct tails; β_2 has a distinct tail to lower values, β_3 has a distinct tail to higher values. Such tails look rather innocent, but may become a nuisance in a least-squares interpretation relying on normal distribution of the population under investigation. Table 1.12 summarises the median values, together with the upper and lower 0.68 and 0.95 confidence percentiles.

The asymmetry in the distribution of parameters β_2 and β_3 makes a rigid use of the symmetric coverage concept of the GUM unsatisfactory in these complex situations. It is quite possible to use normal statistics and express the uncertainty contributions from the solubility data by some figure for u, and an expansion factor $k = 2$ to obtain U. However, these symbols carry a meaning to represent 68% and 95%, respectively, coverage. In case of β_2, it remains unclear

Table 1.12. Median values and confidence percentiles of the parameters β_2 and β_3 (Eq. (1.87))

Statistics	Confidence percentile	Parameter	Value	Log (value)	Difference from median	Mean log (value)	SD
		β_2					
Lower	0.95		$9.348 \cdot 10^{14}$	14.9_7	0.54		
Lower	0.68		$2.234 \cdot 10^{15}$	15.3_5	0.16		
Median			$3.243 \cdot 10^{15}$	15.5_1		15.4_7	0.24
Upper	0.68		$4.110 \cdot 10^{15}$	15.6_1	0.10		
Upper	0.95		$4.765 \cdot 10^{15}$	15.6_8	0.17		
		β_3					
Lower	0.95		$4.570 \cdot 10^{21}$	21.6_6	0.12		
Lower	0.68		$5.245 \cdot 10^{21}$	21.7_2	0.06		
Median			$6.030 \cdot 10^{21}$	21.7_8		21.7_9	0.09
Upper	0.68		$7.176 \cdot 10^{21}$	21.8_6	0.08		
Upper	0.95		$1.081 \cdot 10^{22}$	22.0_3	0.25		

which side should express u. Using (on the log basis) $u = 0.10$ would result in $U = 0.20$ failing to cover a large part of the upper tail, while $u = 0.16$ gives $U = 0.32$, thereby oversizing towards the lower side.

On the other hand, the uncertainty contributions expressed in Table 1.12 by percentiles of empirical probability distributions provide only one influence factor to the complete measurement uncertainty budget. There have been numerous attempts to express meaningful uncertainty estimates for parameters from non-linear multi-parameter equations (e.g. Spendley 1969; Schwartz 1980; Brumby 1989; Caceci 1989; Alper and Gelb 1990, 1991, 1993; Kolassa 1991; Kragten 1994; Roy 1994). Nevertheless, the majority of publications in the field of chemical thermodynamics ignore the problem of communicating the doubt associated with numerical information. If available at all, the discussions were broadly restricted to account for reproducibility.

It has been, on the other hand, demonstrated that uncertainty contributions on the level of those given in Table 1.12 have an influence on the predictive capabilities of, e.g. geochemical speciation codes and reactive transport models (Nitzsche et al. 2000, Meinrath and Nitzsche 2000; Ödegaard-Jensen et al. 2003; Denison and Garnier-Laplace 2004). The second part of this treatise will offer a further discussion.

1.7.3
Cause-and-Effect Approach to Quantities from Complex Situations

The rules issued in the GUM do not preclude the assignment of a complete measurement uncertainty budget to a quantity resulting from a metrologically complex situation. The GUM states (GUM, Sec. 1.3.4) that a measurement uncertainty can be calculated by statistical means if all quantities influencing a measurement result, are varied. On the basis of this statement, a combination of Monte Carlo methods with computer resampling methods fulfils the request.

A cause-and-effect diagram for the data set in Fig. 1.26 is given in Fig. 1.28. The diagram has a different look to those familiar from standard situations (cf. Fig. 1.7). The fishbone part is present at the left-hand side, while a new feature appears on the right-hand-side. The left-hand side features comprise the uncertainties obtained by ISO type B evaluation (in short: ISO type B uncertainties). These uncertainties come from separate experimentation, e.g. from an analysis of measurement uncertainty associated with pH measurement and literature values for the formation of the CO_3^{2-} species from gaseous CO_2 in aqueous solutions. The right-hand side comprises the effects influencing values of parameters from the solubility data. These effects can be accounted for by computer-intensive resampling methods. These effects may be considered to be the ISO type A uncertainties. In case of complex situations it is not possible to separate the individual contributions. The model equation commonly describes a scientific relationship, not the definition of a measurand. Therefore, the central ray of Fig. 1.28 does not end in an arrow with the measurand at its end. In fact, the experimental data do not define a specific measurand. Instead, the quantities of interest (K, β_1, β_2 and β_3) are accessible only from a systematic study of an effect [variation of U(VI) amount concentration in steady state with a solid phase under well-defined conditions] as a function

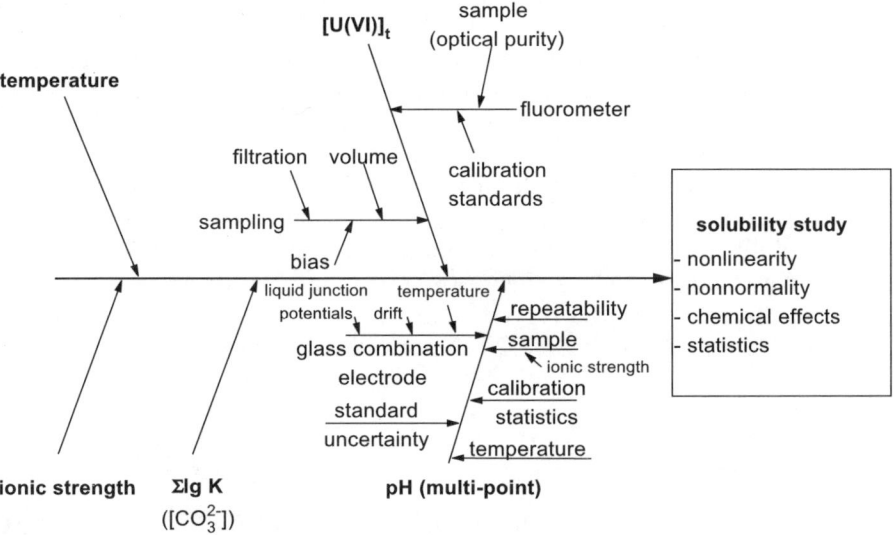

Figure 1.28. Cause-and-effect diagram illustrating the relevant influence quantities for the evaluation of thermodynamic formation constants for dissolution of $UO_2CO_{3(s)}$ and formation of solution species $UO_2CO_3^\circ$, $UO_2(CO_3)_2^{2-}$ and $UO_2(CO_3)_3^{4-}$

of one or more other parameters. The evaluation process also does not concentrate on the decomposition of the analytical process in individual groups, the identification of their magnitude of influence, and final combination of the respective variance contributions. In complex situations, the measurement process is simulated by resampling with the intention to derive the empirical cumulative probability distribution of the complete measurement uncertainty budget. This is the only reliable and comparable quantity.

This intention is not different from standard situations. However, in standard situations, the approximation has been made (mainly on basis of wishful thinking) that the empirical cumulative probability distribution is normal. It must be stated expressis verbis that the standard approach is not in contradiction to the approach outlined here, but is more appropriately understood as a drastic simplification of the process outlined here. For complex situations, the evaluation is necessarily computer-based. Thus, the cause-and-effect diagram serves as a basis for the coding of the measurement procedure in a computer model.

The measurement process consists of the simultaneous determination of two quantities: the total metal ion concentration and the value of pH. The number of influence quantities is quite limited. The main effects are temperature and the uncertainty associated with the thermodynamic constants for dissolution and dissociation of CO_2 in aqueous solution. There are plenty of values in the literature, the quality of which, however, has never been assessed in a trackable way. Derivation of traceable values for relevant thermodynamic quantities remain a task for the future. The prospects for traceable thermodynamic data will be discussed in the final section of this part of the book.

For the time being, reasonable estimates must replace the Type B influence quantities. The influences will be summarised as follows:

a) pH

pH has been determined by the multi-point calibration protocol using five different calibration standards. Partly as a consequence of the small sample size the variability of the uncertainty estimates may vary by 50–100% (Meinrath and Spitzer 2000). Therefore, the measurement uncertainty of $u_c(pH) = 0.037$ evaluated from a set of 50 representative pH measurements (Meinrath and Spitzer 2000) has been accepted. The uncertainty is normally distributed.

b) $\sum \lg K(CO_3^{2-})$

The concentration of CO_3^{2-} can be derived from the following overall equation within acceptable approximation:

$$\lg[CO_3^{2-}] = \lg K_H + \lg K_1 + \lg K_2 + \lg^p CO_2 + 2pH \qquad (1.88)$$

For data at ionic strength $I = 0$, the data from Harned et al. (Harned and Scholes 1941; Harned and Davies 1943; Harned and Bonner 1945) are still widely accepted with $\sum \lg K = -18.145 \pm 0.127$. Extrapolations to different conditions, notably salt concentrations, do carry considerable and unquantified uncertainties (Meinrath 2002b). Previously, a value of $\sum \lg K = -17.67 \pm 0.10$ has been used (Meinrath 2001). In the light of the varying values reported in literature, the uncertainty has been increased to $u_c \sum \lg K = 0.15$. The uncertainty is uniformly distributed.

c) $[U(VI)]_t$

The molar concentrations of U(VI) (denoted by square brackets) have been determined by fluorimetry, making use of the fluorescence of U(VI) in phosphate medium. The method has been cross-checked with ICP-MS measurements. At the time of experiment, traceable standards have not been available. The samples have been filtered prior to analysis by 220 nm cellulose acetate filters. By REM/EDX, residues on the filters could not been detected by various probes investigated at different pH values in the course of the experiment. The sample solutions have been clear except at the extremes of the solubility curve. For analysis, the samples were further diluted by water and uranium-free phosphoric acid. Thus, a relative measurement uncertainty $u_c([U(VI)]_t) = 8\%$ seems justified including the approximately 1% relative uncertainty associated with sampling by a mechanical pipette. The uncertainty is normally distributed.

d) Temperature

Ionic equilibria between actinides and carbonate are hard-hard interactions which are not sensitive towards temperature. The vessel has been thermostated to $\pm 0.1\,K$. Temperature is critical influence in pH measurement, where the electrode tip rests in a solution at, say, 25 °C while the head of the electrode including wiring is exposed to room temperature between 18 °C and 22 °C. Such temperature differences over a glass combination electrode may induce bias. In a solubility study of actinides and carbonate, temperature effects should not be detectable.

e) Ionic strength

Ionic strength may affect an ionic equilibrium. In a controlled environment where the loss of water from the vessel is minimised due to gas flows equilibrated with flasks of appropriate ionic strength, ionic strength may be expected to vary within a few percent, inducing only negligible effect. The situation, however, may be drastically different at the extremes of the solubility curves, especially in the alkaline region where high carbonate and $UO_2(CO_3)_3^{4-}$ concentrations are present. It may be supposed that the considerable increase in uncertainty of parameter β_3 may be at least partly due to such effects.

The effects summarised in the right-hand box will be cared for in the computer-intensive resampling cycle of the evaluation program. A random regressor scheme has been chosen avoiding accidental larger fluctuations in the relative use of each data point by a balanced bootstrap design (Gleason 1988).

1.7.4
Random Sampling by Uniform and Normal Deviates

The computer is designed to follow a set of given instructions without any variation. It seems a contradiction to ask this deterministic machine to engage into perfect random processes. Nevertheless several procedures exist to obtain random numbers from a computer (Marsaglia and Tsang 1984). The built-in random number generators of all higher language computer codes have a sufficient quality but are repetitive, either. Improvements have been proposed (Bay and Durham 1976). If randomness is crucial, an additional randomisation stage can be added by the following procedure:

1. Select a group of, say, 100 random numbers into a pool.
2. Select randomly one number out of this pool.
3. Replace the selected random number by a new one.
4. Repeat steps 2–4 each time a random number is needed.

Random number generators return a random number RND between 0 and 1. To obtain a integer number i between 1 and x, the pseudo code is

$$i = 1 + int(RND * x)$$

A square distribution is a distribution which is uniform between a lower value, lo, and an upper value, up. A value i that is uniformly distributed is obtained from RND by the pseudocode

$$i = lo + RND * (up-lo)$$

A normal distribution $N(1, 0)$ is obtained from the following sequence of operations (Box and Muller 1958), where u1 and u2 are independently normally distributed with mean zero and standard deviation 1.

repeat
$\quad u1 = 2 * RND - 1$
$\quad u2 = 2 * RND - 1$
$\quad s = u1 * u1 + u2 * u2$
until$(s > 0$ and $s < 1)$
$s = sqr(-2 \log(s)/s)$(Note: sqr = square root, log = natural logarithm)
$u1 = u1 * s$
$u2 = u2 * s$

Complete measurement uncertainty budget. The complete measurement uncertainty budget of the parameters is shown in Fig. 1.29 from which follows that the distribution is by no means normal. All probability densities are skewed. Table 1.13 summarises some relevant properties of the distributions for parameters β_1 and β_3. For K and β_1 the respective values are given in Table 1.13 for

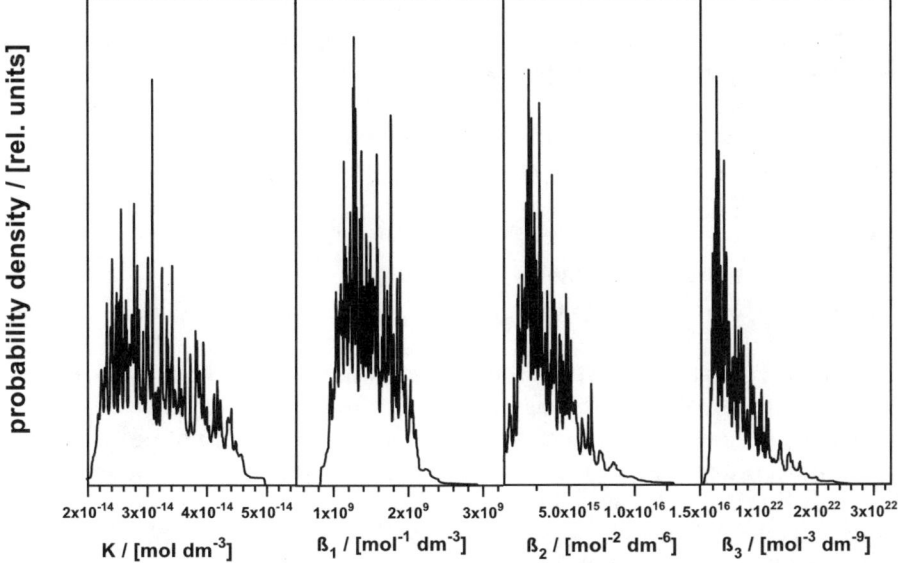

Figure 1.29. Probability densities (relative units) of the complete measurement uncertainty budget of solubility product K of $UO_2CO_{3(s)}$ and formation constants β_1, β_2 and β_3 of solution species $UO_2CO_3^\circ$, $UO_2(CO_3)_2^{2-}$ and $UO_2(CO_3)_3^{4-}$, respectively

Table 1.13. Median values and confidence percentiles of the parameters K and β_1 (Eq. (1.85))

Statistics	Confidence percentile	Parameter	Value	Log (value)	Difference to median	Mean log (value)	SD
		K					
Lower	0.95		$3.03 \cdot 10^{14}$	-13.5_2	0.03		
Lower	0.68		$3.14 \cdot 10^{14}$	-13.5_0	0.01		
Median			$3.27 \cdot 10^{14}$	-13.4_9		-13.4_9	0.02
Upper	0.68		$3.41 \cdot 10^{14}$	-13.4_7	0.02		
Upper	0.95		$3.55 \cdot 10^{14}$	13.4_5	0.04		
		β_1					
Lower	0.95		$1.15 \cdot 10^9$	9.0_6	0.08		
Lower	0.68		$1.26 \cdot 10^9$	9.1_0	0.04		
Median			$1.38 \cdot 10^9$	9.1_4		9.1_4	0.04
Upper	0.68		$1.49 \cdot 10^9$	9.1_7	0.03		
Upper	0.95		$1.60 \cdot 10^9$	9.2_1	0.07		

sake of completeness. Parameter K is linear, while the factor holding parameter parameter β_1 is a constant. Therefore both parameters have a quite narrow distribution. The narrow distribution is further caused by the low scatter in the experimental observations.

Mean and median are the same for both parameters within the resolution capabilities of this study. Despite this narrow distribution, the median values attributed to the four parameters have an only limited comparability because the metrological traceability between the uranium measurements performed during the present study and other investigations of the same chemical system remains unclear.

1.7.5
Correlation Within the Disturbances and Parameters – The Moving Block Bootstrap (MBB)

In standard situations, the disturbances are assumed to be identically and independently distributed. This assumption is often not fulfilled. Even in simple situations, it may become necessary to assume correlation not only with the parameters of a multi-parametric model but also within the disturbances (Hässelbarth and Bremser 2004). In complex situations accounting for correlation within the disturbances is required because the effect of correlation is difficult to be predicted and to be quantified a priori. Hence, preserving the correlation structure also within the disturbances keeps a data analysis on the safe side.

The linear relationship between two separate observations of the same random variable is determined by autoregression analysis. Autoregression (AR) assumes that a certain observation is not independent from another (commonly previous) observation over a distance of d neighbours. An autoregression analysing the influence of an observation at a distance d is abbreviated AR(d). The distance d is called the "lag". Assuming a lag of $d = 2$, the numerical expression is given by Eq. (1.89):

$$\varepsilon_t = \beta \varepsilon_{t-1} + \gamma \, \varepsilon_{t-2} + z_t \tag{1.89}$$

This equation assumes that there are two non-zero parameters, β and γ, which depend on the first and second neighbour of an observed disturbance ε at position t. The residual is z (here the term "residual" is used to distinguish the parameter z from the disturbances ε which form our observations in this analysis. Ambiguity is thus avoided).

To demonstrate autoregression, a set of 200 disturbances is created by random draws from a normal distribution $N(0.01, 0)$. To this data set, model (1.89) is fitted using a SIMPLEX algorithm. The resulting parameters are $\beta = -0.0554$ and $\gamma = 0.0810$. Both parameters are small but non-zero. Hence the question arises: are these figures significantly different from zero? An answer can be given on basis of a bootstrapping scheme using resampling from the residuals z according to the fixed regressor scheme (cf. List 1.6.2.2). For the present set of normally distributed residuals the bootstrap results are $\beta = -0.053 \pm 0.073$ and $\gamma = 0.096 \pm 0.072$. The β mean value is smaller than one standard deviation and thus not significantly different from Zero. In case of γ mean value the standard deviation is a bit smaller but a significance test would accept the null hypothesis on the 95% confidence level because the γ mean value is smaller than two standard deviations.

The autoregression scheme will be applied in the following to disturbances obtained from a UV-Vis spectroscopic analysis. By spectroscopy, a large number of individual observations are collected providing sufficient material for an autoregression analysis. Figure 1.30 shows that the parameters $\beta = 0.6163 \pm 0.069$ and $\gamma = 0.3405 \pm 0.0708$ are capable of reproducing the pattern of the observed disturbances. The mean values of both AR(2) parameters β and γ are significantly different from zero.

The presence of a correlation within the disturbances has a consequence for assessment of measurement uncertainty by computer-intensive resampling techniques because both the random resampling scheme as well as the fixed regressors resampling scheme would destroy this pattern. Classical linear regression will misplace the confidence regions and forward biased mean values. These effects of correlation in the disturbances will be demonstrated in a second case study.

The AR(2) regression scheme is the most simple autoregression model. A full-fledged autoregression analysis would increase the lag until the parameters get insignificant. Such an analysis is, however, outside the scope of the present discussion. The purpose of the AR analysis was to provide evidence for correlation in the observed disturbances. Figure 1.30 gives visual evidence for the correlation because the AR(2) scheme is evidently able to create the pattern in the observations with only small residuals.

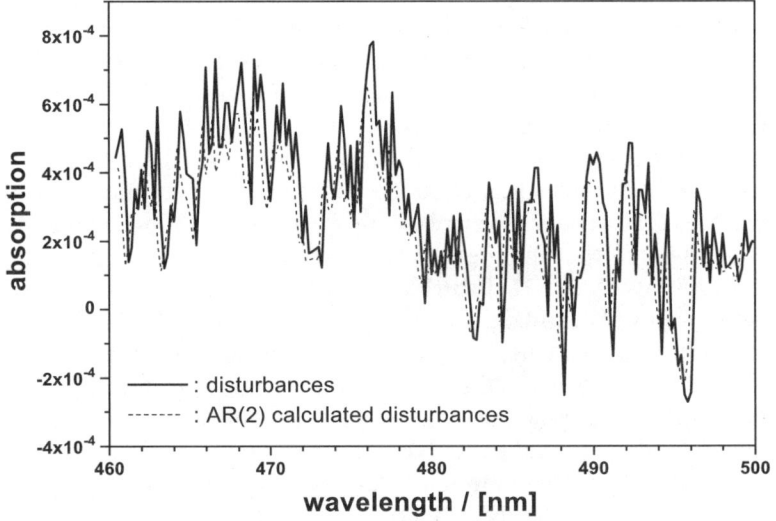

Figure 1.30. Observed disturbances and generated disturbances from parameters obtained by an AR(2) autoregression scheme

Some interesting observations can be made by using synthetic data sets. In Fig. 1.31a, two absorption bands are shown which do not overlap. The fundament of quantitative spectroscopy is the Bouguer–Lambert–Beer law:

$$a = \varepsilon c d \tag{1.90}$$

where a is the absorption at a wavelength λ, c is concentration of an absorber at wavelength λ with molar absorption coefficient ε and d is the pathlength of the light beam through the absorber. The linear relationship in Eq. (1.91) is commonly valid up to a limiting concentration. Above that concentration, deviations from linearity occur which commonly render the usefulness of the Bouguer–Lambert–Beer law void.

If the intensity of a light beam is weakened by m absorbers, Eq. (1.90) is extended:

$$a = \varepsilon_1 c_1 d + \varepsilon_2 c_2 + \ldots + \varepsilon_m c_m d + \xi \tag{1.91}$$

where ξ again represents a disturbance.

Figure 1.31a shows an example for $m = 2$. The both absorption bands are generated from Gaussian curves with known center positions and band widths. The both curves do not overlap and, hence, the correlation between the parameters c_1 and c_2 is negligible.

Synthetic data have the advantage that the true values of the parameters are known. In the case shown in Fig. 1.32, the concentrations of the hypothetical species are $1 \, \mathrm{mol \, dm^{-3}}$. These spectra may be analysed by least squares regression. From this analysis the mean value and joint confidence regions may be obtained. Results are shown in Fig. 1.32 for two different treatments of the disturbances. In one case, the disturbances have been randomised by random resampling. Random resampling destroys the residual correlation but does not alter the overall frequency in the set of disturbances. In the other case the disturbances have been left unchanged.

Both 95% confidence regions clearly differ. The randomised disturbances in the present case create a much larger confidence region. In fact, randomisation of the disturbances may produce quite varying sizes of the confidence region. It should be emphasised that the mean value of the disturbances is zero and the standard deviation of all 751 disturbances used for the synthetic data is equal for randomised and original disturbances. The reason for this variability in the confidence region calculated by least-squares regression is the fact that the disturbances are not normally distributed and not uncorrelated. Thus, by relying on least squares regression techniques a rather wide distribution of uncertainty values for calculation of complete measurement uncertainty budgets should be expected. Here, too, computer-intensive resampling methods may provide more stable results (Meinrath 2000).

The synthetic spectroscopic data also allows investigation of the effect of parameter correlation. The correlation in the concentration parameters can be

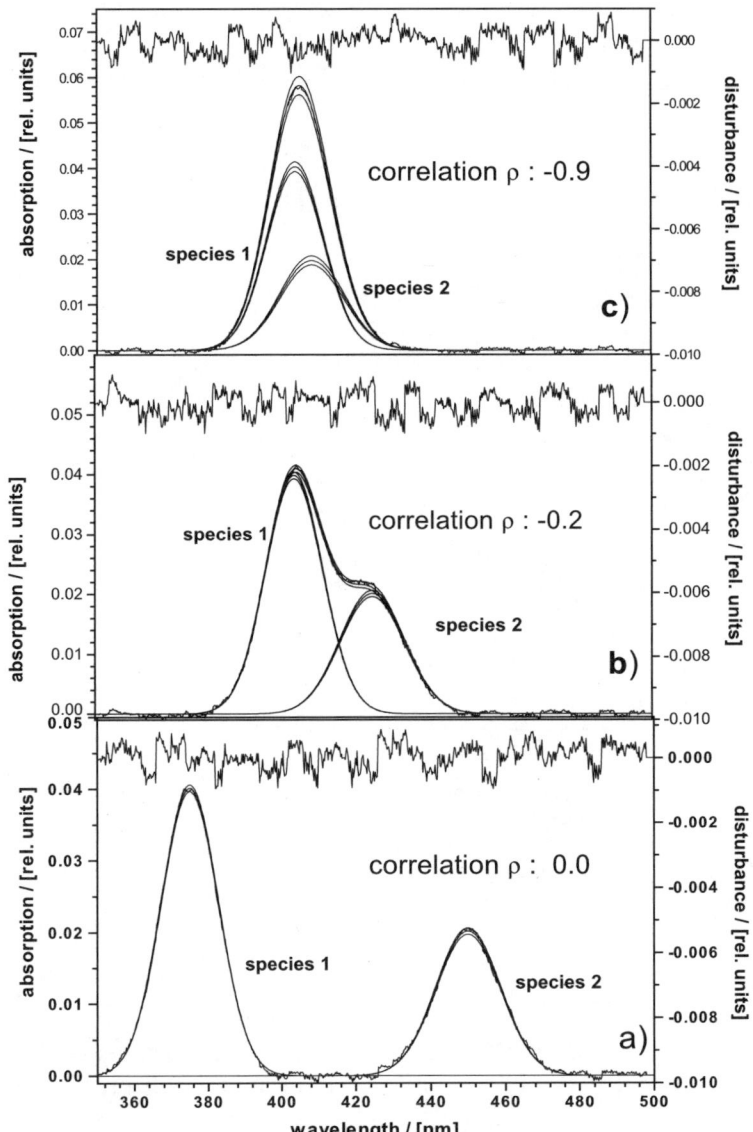

Figure 1.31. Synthetic spectral data of a two-species system with disturbances (*right side axis*) borrowed from an experiment. The Gaussian absorption curves are successively shifted together to obtain spectral overlap with increasing parameter correlation: **a** correlation $\rho = 0$; **b** correlation $\rho = -0.2$; **c** correlation $\rho = -0.9$

varied by increasing the overlap between both spectral bands. The result of this procedure is shown in Fig. 1.31a–c, where Fig. 1.31a shows a situation with a parameter correlation of zero, Fig. 1.31b with correlation $\rho = -0.2$ and Fig. 1.31c

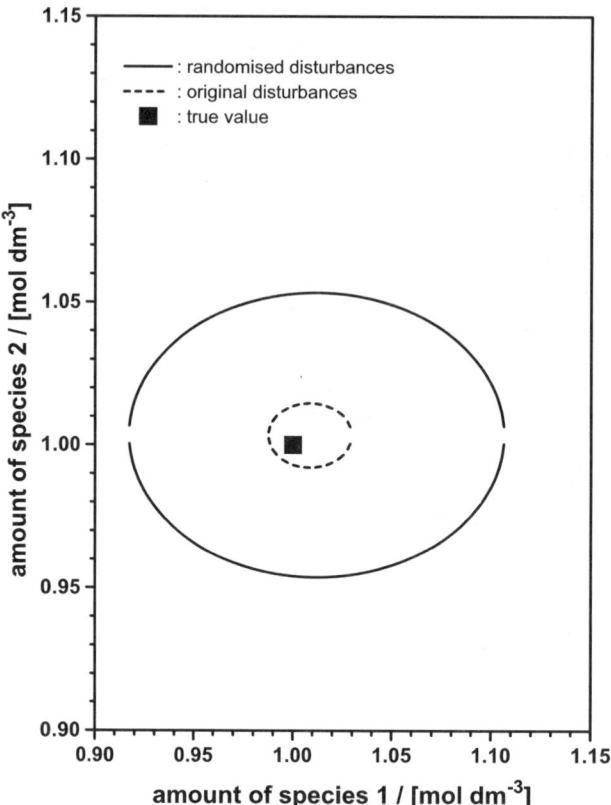

Figure 1.32. Effect of correlation-destructive randomisation of disturbances on the 95% confidence regions for the mean value of concentration parameters (cf. Fig. 1.31a) of synthetic UV-Vis absorption data

with correlation $\rho = -0.9$. Some 95% confidence regions are calculated in Fig. 1.33, together with the least-squares means and the bootstrap means.

The size of the least-squares 95% confidence regions varies considerably. There is no trend with the correlation. This observation underpins the previously given statement that the magnitude of a least-squares regression uncertainty may vary considerably, depending on the structure of the disturbances.

Another relevant observation is the rather wide scatter of the least-squares regression mean values, especially when compared to the moving block bootstrap (MBB) results. Evidently, bootstrapping provides results that are closer to the true value and less sensitive to parameter correlation than least-squares regression results. In studies on the basis of synthetic data, the MBB results were always found close to the true values, provided the disturbances were not too large compared with the measurement signal(s). An essential requirement to obtain stable bootstrap results is a resampling method which preserves the

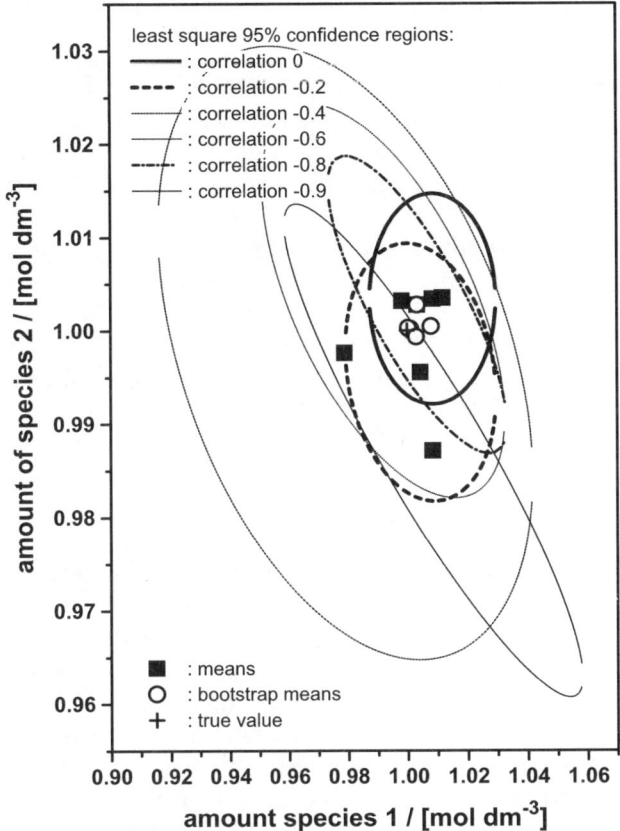

Figure 1.33. Elliptical 95% confidence regions for synthetic spectral data (cf. Fig. 1.31) with parameter correlation 0, −0.2, −0.4, −0.6, −0.8 and −0.9. The mean values from the least-squares regression analysis (original disturbances) are given together with bootstrap means obtained by a correlation preserving Moving Block Bootstrap analysis

correlation structure within the disturbances. MBB and threshold bootstrap are such approaches.

The classical bootstrap, implemented either as a fixed regressor scheme or a random regressor scheme, randomly resamples either from the disturbances or the experimental data set. In the case of correlated disturbances this correlation is destroyed. The consequences of this destruction are difficult to predict. A possible consequence is a bias, or an over-estimated measurement uncertainty contribution. However, it is acknowledged in statistical literature that assessing the effect of correlation in the disturbances may be quite difficult (Carlstein 1986; Künsch 1989). Consequently methods are required to preserve the correlation structure if possible.

The MBB is a statistically sound way to generate resampled data sets having the favourable properties of a bootstrap sample and at the same time preserving correlation. In practise, the block size is much larger than four if a sufficient number of data are available.

Figure 1.34 also illustrates one of the two major theoretical weaknesses associated with the MBB: often the number of disturbances in a data set is not a multiple of the MBB block size. Hence, at the end of the series one or more data points are not included in the bootstrap scheme. Furthermore, the last $n - 1$ data points (where n gives the block length) of a set have a much lower chance of being included into the resampling procedure. The second problem of the MBB is the arbitrary selection of a block size r. These deficiencies, which are usually not severe but theoretically unsatisfactory, have given rise to other resampling schemes, for instance, the threshold bootstrap.

A result of applying the MBB to the synthetic data set (cf. Fig. 1.31) is shown in Fig. 1.35 for the same correlations as in Fig. 1.33. The point clouds created by the MBB are plotted over each other using a different symbol for each data set. It is obvious that the MBB confidence regions are much more homogeneously scattered around the true value compared to the least-square 95% confidence ellipses. The elongation of the confidence region with increasing magnitude of the parameter correlation cannot be avoided, since this effect is a property of the investigated system.

While the MBB is relying on a user-defined lag (in the present situation lag is 20), the threshold bootstrap has a variable lag depending on the structure of the disturbances. The threshold bootstrap has been implemented into an analysis tool for spectroscopic data on basis of factor analysis, the computer-assisted threshold bootstrap computer-assisted target factor analysis (TB CAT) (Meinrath and Lis 2001). There, a detailed description of the threshold bootstrap is given. TB CAT evaluates a chemical system followed by spectroscopy and provides the empirical distribution function of the complete measurement uncertainty budget of relevant chemical information, for instance the formation constants of identified species, the single component spectra and the component concentrations. TB CAT also identifies situations where multiple

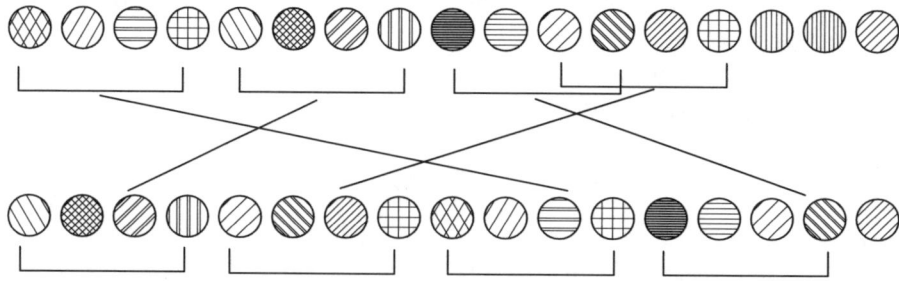

Figure 1.34. A schematic representation of the moving block bootstrap scheme. The *17 circles* represent specific values of a series of disturbances. *Blocks* (here with block size 4) are randomly chosen from the original series (*top*) and arranged to form a new sequence of disturbances (*bottom*)

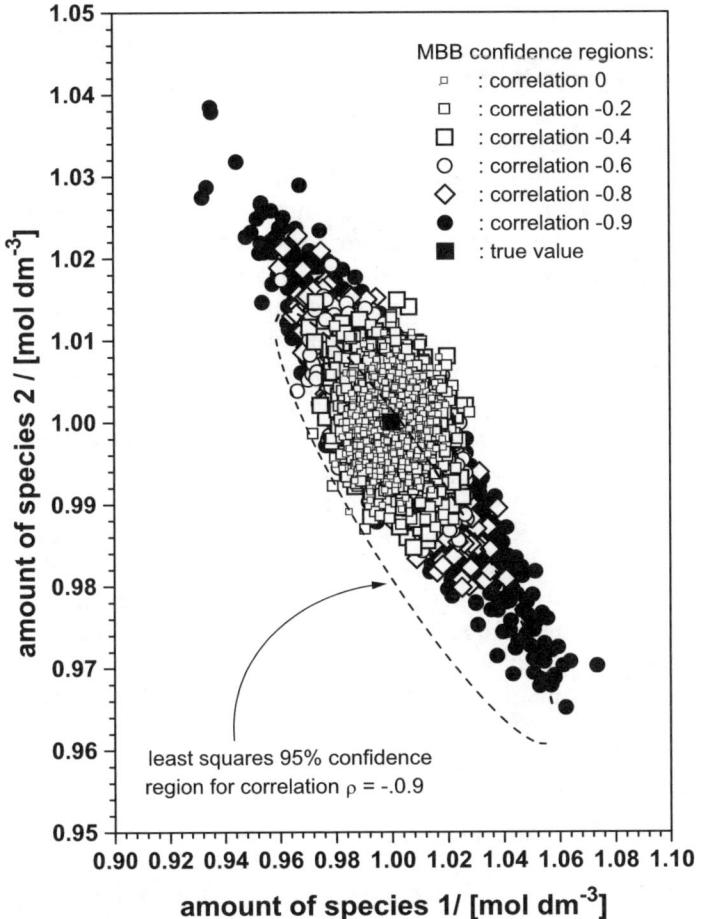

Figure 1.35. Moving block bootstrap generated confidence regions. For comparison with the least-squares 95% confidence regions (cf. Fig. 1.33) the least-squares confidence ellipse for $\rho = -0.9$ correlation is included

interpretations of the same data set are possible. In presence of measurement uncertainty straightforward statements on the composition of a chemical system are not possible on basis of a given data set. Ignoring measurement uncertainty in the assessment of a system, incorrect conclusions may be drawn.

1.7.6
Predictions in Presence of Measurement Uncertainty: Efficient Sampling from Multidimensional Distributions

The complete measurement uncertainty budget is an essential element to achieve comparability of measurement results on basis of traceable values.

Metrological traceability cannot be achieved without an appropriate assessment of the complete measurement uncertainty budget (Meinrath and Kalin 2005). The derivation of the uncertainty budget seems straightforward for standard situations where the measurement process can be expressed by a mathematical formula. As soon as the measurement process becomes more evolved, the quality assessment of measurement values becomes also more complex. The unified statement of the quality of numerical information obtained by experimental measurements is an important element in scientific communication. A numerical value of the complete measurement uncertainty budget may ease, on the one side, comparison of information obtained by different methods and validate claims on improvements in the measurement process. Thus, the complete measurement uncertainty budget will take its place in assessing, for instance, progress in analytical measurement.

Much more attention needs to be directed to the numerical and statistical procedures applied in the evaluation process. Complex measurement situations where the measurement process cannot be expressed by a mathematical equation or situations where different types of correlation are involved occur frequently in chemical measurement. Often standard recipes are not available. The insight into the numerical and statistical effects influencing the evaluation process of a measurement value will be helpful at least to ask the right questions and to find advice. A measurement value being not traceable to accepted units or standards is meaningless outside a very narrow local and temporal range of its determination.

1.7.6.1
Handling Data with Assigned Measurement Uncertainty Budget

In many situations, the interest in a quality assessment of measurement values goes beyond direct comparison of measurement values obtained at different places and times. Measurement values of chemical quantities often serve as a basis for subsequent decisions. These decisions may affect many people, e.g. in food chemistry or medical science. Drinking water quality is another important example. In other situations, measurement values will enter as ancillary information for further evaluations. An example is the use of measured pH values in E_H/pH diagrams or calculation of HCO_3^- concentrations in ground water or tap water.

Prediction is an important application of measurement data. From an understanding of the fundamental behaviour of physicochemical processes and their formal, usually mathematical, description an extrapolation into the future may be attempted. Quantum chemical modeling and geochemical transport simulation are typical examples of such applications. Especially with geochemical applications, measurement values, e.g. formation constants of relevant chemical species, play an important role. The processes may have time spans and spatial dimensions that cannot be studied experimentally. The geological settings where the processes evolve are often located inaccessibly in the sub-surface. Human interference, e.g. the construction of monitoring wells, may alter these geological settings. Thus, prediction by reactive transport modeling is of great relevance. Of course, the fact that measurement values cannot be obtained with arbitrary accuracy also needs to be taken into account in these applications (Nitzsche et al. 2000).

While it is quite possible to account for measurement uncertainty in a meas-ured value by methods outlined in the previous chapters of this book, e.g. by sampling from a normal distribution (using, for instance, Box and Muller's method) or the empirical distribution function if available, this classical Monte Carlo approach (Rubinstein 1981) requires a larger number of repetitions of the same simulation, depending on the accuracy the simulation should provide.

Resampling from a given distribution by Monte Carlo resampling carries an inherent problem due to the fact that most resampled values will come from the center of distribution, while the tails of the distribution will contribute only a small number of values. This simple fact is shown by Fig. 1.36, where two regions with the same probability are given.

The corresponding abscissa interval in the center of distribution is consid-erably smaller compared with the interval in the tail region of the distribution. This imbalance in the resampling probability increases further if a region even further to the tail of the distribution had been chosen. Hence, in order to appropriately include the tails into a Monte Carlo sample of values from the distribution, a larger number of Monte Carlo samples needs to be drawn.

It should be remembered that this inequality in the probability to obtain values from different parts of normal distribution is the rational for the Stu-dent t correction for small samples. Because the higher probability to obtain values predominantly from the center of a normal distribution, there is a risk to underestimate the spread of the distribution. This risk decreases with sam-ple size. Consequently, Student t depends on the degrees of freedom (df). It decreases with increasing sample size and approaches the normal distribution with df $\sim$ 30.

The number of Monte Carlo samples further increases with the number of parameters. If, for instance, a two parameter system is to be studied and a number of 60 Monte Carlo samples from one distribution are found satis-factory to achieve the desired accuracy, then 60 times 60 = 3600 samples are required to achieve the same level of precision for a two-parameter system. If a system with five and more parameters needs to be considered, Monte Carlo resampling is not feasible any longer even on high-performance main frame computers. A concise discussion focusing on aspects of uncertainty analysis in geochemical modeling is given, e.g. by Ekberg (1996).

There exist alternative and more efficient approaches to sampling from multidimensional distributions. Without going into much detail, the Gibbs sampler should be mentioned (Smith and Gelfand 1992; Smith and Roberts 1993). Implementing the Gibbs sampler, however, is a complicated task.

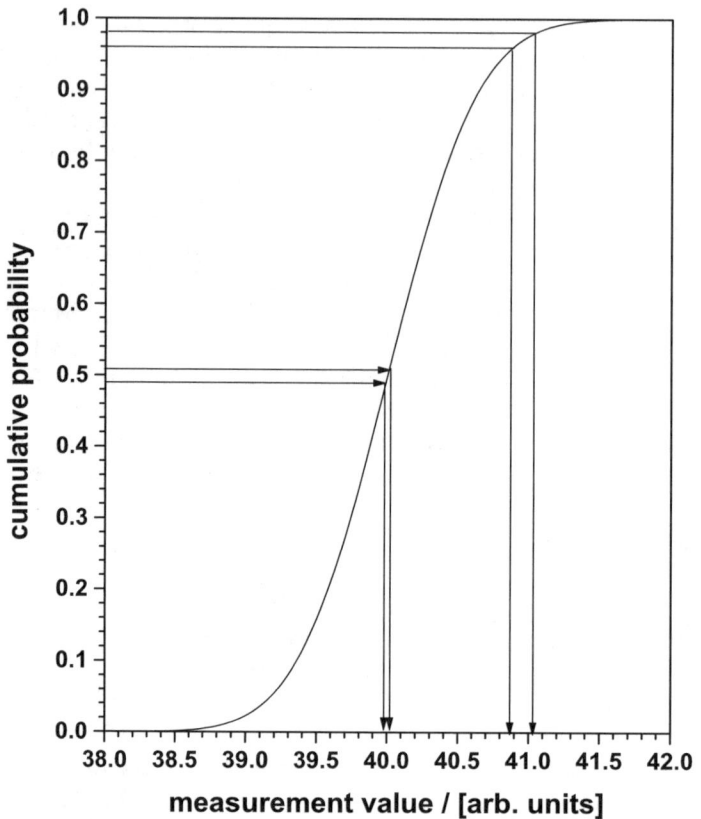

Figure 1.36. Cumulative probability curve $\Phi(0.5, 40)$ (cf. Fig. 1.9) showing two abscissa intervals with the same cumulative probability. In the tails, the interval is much larger compared to the distribution center. Hence, when sampling from a distribution by Monte Carlo methods, the majority of data will be sampled in the distribution center. To ensure inclusion of sufficient values from the tail regions of a distribution, a larger number of samples needs to be taken

Latin Hypercube Sampling (LHS) in an alternative which should be considered in most situations encountered in chemistry (McKay et al. 1979). An implementation of the LHS is the LJUNGSKILE code (Ödegaard-Jensen et al. 2004). LJUNGSKILE code calculates speciation diagrams from thermodynamic data of solution species affected by measurement uncertainty. An example is given in Fig. 1.37 for an iron system in natural waters at pH 7.95. The distribution is shown as a modified Box plot: the central line gives the mean values of the species concentrations while the square encloses the 68% confidence region. The whiskers enclose the 95% confidence regions. Table 1.14 gives the species, their respective formation constants, the associated uncertainty and the distribution of the uncertainty.

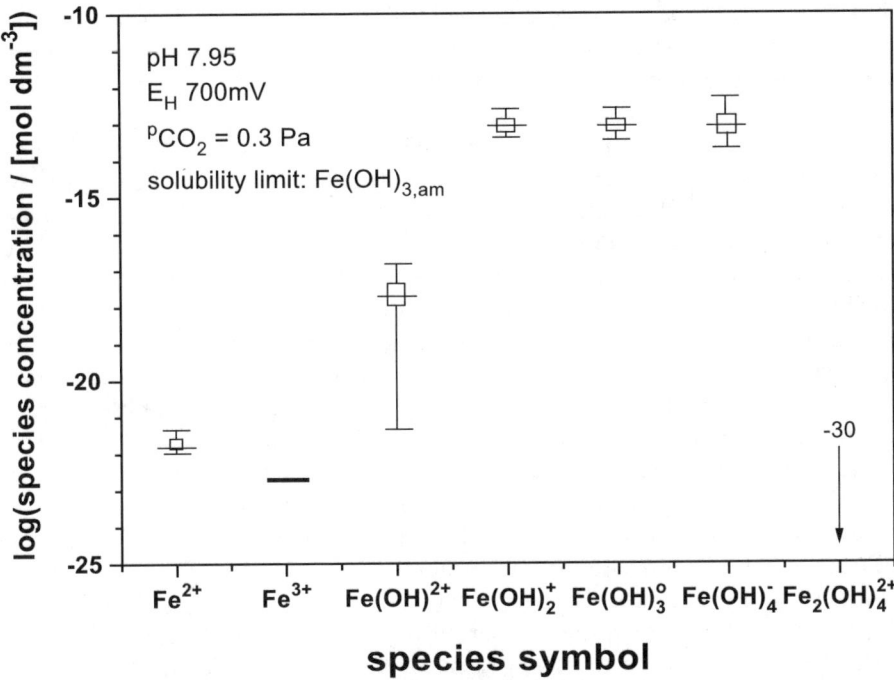

Figure 1.37. Distribution of iron species in a groundwater at pH 7.95 calculated by the LJUNGSKILE code on basis of formation constants from JESS database given in Table 1.14 (May and Murray 1991b, 2001; Meinrath and May 2002). The *central line* represents a mean values, the *boxes* enclose 68% confidence limits while the *whiskers* enclose 95% confidence intervals. (Species Fe^{3+} is a master species and, therefore, has no uncertainty)

Table 1.14. Species, formation constants and uncertainties for calculation of Fig. 1.37 by the LJUNGSKILE code (data from JESS data base (Meinrath and May 2002))

Species	Log (formation constant)	Uncertainty (1σ)	Distribution
Fe^{2+}	0	0	Master species
Fe^{3+}	−13.0	0.2	Normal
$FeOH^{2+}$	−15.2	0.2	Normal
$Fe(OH)_2^+$	18.7	0.1	Normal
$Fe(OH)_3^\circ$	26.6	0.1	Normal
$Fe(OH)_4^-$	34.6	0.2	Normal
$Fe2(OH)_4^{2-}$	29.0	0.2	Normal
Solid phase	Solubility product lg K_s	Uncertainty	Distribution
$Fe(OH)_{3,\,am}$	0.89	–	–

Despite the fact that the uncertainties in Table 1.14 are in the order of only 0.1–0.2 log units, the consequences for the uncertainties in the species concentrations are considerable. Species above 10^{-7} mol dm^{-3} are Fe(OH)$_3$

and $Fe(OH)_4^-$. On the 68% confidence level, the uncertainty interval for the concentration is about 1 order of magnitude. It is important to realise the different levels of information that are carried by Fig. 1.37. That information is giving rise to multiple questions.

First, the numerical speciation on basis of numerical data obtained commonly in laboratory environments at much higher concentrations is the only way to obtain at least some information on the species distribution in an aqueous medium with the composition of a natural ground water. For a direct speciation analysis, the concentrations are much too low. Hence, it is unlikely to obtain experimental verification of the simulation shown.

Second, the speciation calculation depends on information provided in literature which is not homogeneous. The existence of $Fe(OH)_4^-$ or dimmer species $Fe_2(OH)_4^{2+}$ is questionable. The reported evidence for these species often comes from experiments where a statistical data analysis has not been performed at all. In the vast majority of cases, the "best-fit" criterion on basis of mean value fitting using linear least-squares models has been applied. Such an analysis can be rather misleading, especially if the varying confidence ranges of least-squares approaches are not properly accounted for. The weakness of statistical analysis in most chemical analysis is well documented (for instance, Thompson 1994). In fact, there is very little discussion on the minimum statistical criteria required to accept a solution species (Filella and May 2005). Statistical misfit alone is a rather poor criterion – but which other criteria could be applied? it is it not much more important to attract attention by announcing the detection of a new species instead of starting a tedious discussion about acceptance/rejection criteria for solution species?

Third, the calculation depends crucially on the algorithms applied in the computational code. Thus, the LJUNGSKILE code is based on the PHREEQC code made available by the US Geological Survey (Parkhurst 1995). PHREEQC is based on the master species concept and uses linear least-squares procedures to minimise the optimisation criterion. In Fig. 1.37, Fe^{2+} is used as a master species. What solution composition would be calculated if other solution species would be used as master species? Ionic strength corrections are built into PHREEQC and also affect the outcome shown in Fig. 1.37. Due to the limited understanding of complex electrolyte solutions, alternative approaches to ionic strength correction exist. Furthermore, it is unclear to what degree a LJUNGSKILE-type calculation depends on the underlying geochemical code.

Fourth, the uncertainties given in Table 1.14 are mere estimates. Thermodynamic data are mostly reported without an assigned uncertainty estimate. Even though uncertainties of $0.01 - 0.2$ log units are not uncommonly reported, these 1σ uncertainties correspond to a span of a $0.5 - 1$ log unit on the 95% confidence level (4σ) for each input distribution. Four of the Fe(III) species reach the maximum concentration of $2.5 \cdot 10^{-6}\,mol\,dm^{-3}$. Anything goes? If so, what is the raison d'être of a determination of thermodynamic constants. Pondering on the magnitude of all influence contributions to the measurement uncertainty budget, is it possible at all to determine thermodynamic formation constants with higher accuracy? Or, said otherwise, what are the target uncertainties a measurement value needs to be associated with in order to be a meaningful contribution to chemical simulations?

Fifth, if already a simulation at a given pH carries such widely varying concentration estimates, what might be the result for a speciation diagram over a wider range of pH? Such diagrams are commonly given in literature but almost exclusively show mean concentration values only (e.g. Hartley et al. 1980; Sigg and Stumm 1996).

Sixth, if the simulated species concentrations may vary that widely why aren't there more codes with at least the capabilities of the LJUNGSKILE code?

The fifth question can be answered by another LJUNGSKILE calculation doing multiple runs in the range pH 4–9.4 in intervals of 0.2 pH. The result is given as a species distribution plot in Fig. 1.38. Three different informations are given: the mean values of the species distributions, the upper and lower 68% confidence limits (dashed lines) and the corresponding 95% confidence limits (dotted lines). Thus, the diagram holds multiple curves. The clear picture communicated by the common mean value graphs is replaced by a diagram requiring more careful inspection.

On a mean value level, the LJUNGSKILE simulation indicates $Fe(OH)^{2+}$, $Fe(OH)_2^+$ and $Fe(OH)_4^-$ as the species of major importance, provided an amphoteric Fe(III) species exists. $Fe(OH)_3^\circ$ plays a minor role in the range pH 7–9. The question on the dominating species can be reasonably well answered on the 68% level. However, on the 95% level, the species $Fe(OH)_2^+$ and $Fe(OH)_3$ compete with each other in the range pH 4–8!

A critical comment on Figs. 1.37 and 1.38 immediately follows: why is it necessary to calculate such diagrams if they cannot provide a clear answer? And how meaningful can such diagrams be if metrologically sound thermodynamic data are not available?

Such questions are highly justified. On the current state of the discussion, it is important to realise that a mere mean-value based discussion can be extremely misleading (Finkel 1994). If, for instance, speciation diagrams are interpreted without a clear understanding of its limitations, completely wrong conclusions can be derived. Speciation calculations are, for example, an important element in the derivation of sorption data (Jakobsson 1999; Meinrath et al. 2004a). Taking the surface complexation approach of metal ion-surface interaction as an illustrative example, it is commonly assumed that a specific form of a metal ion binds to a reactive site on a surface. This interaction is often assumed to be comparable to a complex in homogeneous, aqueous phase and described by surface complexation constants. Such calculations can be meaningful if a species' concentration in solution can be assessed with appropriate accuracy. However, despite the fact that the determination of surface complexation constants is a rather active field of research, an assessment or, at least, a discussion of the limitations of the approach due to metrological re-

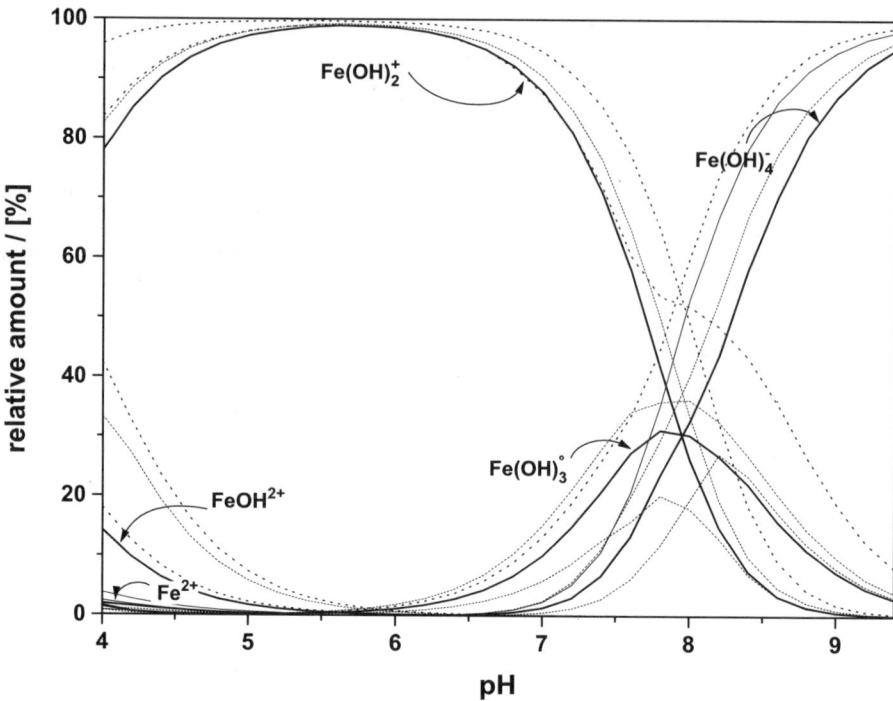

Figure 1.38. Simulated species distribution of in the range pH 4–9.4 for a natural ground water in equilibrium with atmospheric CO_2 partial pressure and a solid phase $Fe(OH)_{3,\,am}$. Data from Table 1.14 are used with the LJUNGSKILE code

strictions (cf. Figs. 1.37 and 1.38) is virtually absent. How precise and accurate can the surface area be determined? How precise and accurate can the solution composition be assessed? How precise and accurate can the total amount of the metal ion be determined? What are the inherent uncertainties in the numerical approach fitting model functions to measurement data? The list of such questions can be extended. If an agreed and proven concept for the assessment of measurement uncertainty in solution does not exist, there is no use to ask for such a concept in case of heterogeneous equilibrium models in aqueous chemistry. Some attention to this point will be devoted in Part II of this treatise. The answers to these questions are of interest to those having to make and to defend decisions based on chemical measurement data.

1.7.6.2
Stratified Sampling

The further discussion here will focus on the machinery providing the information displayed in Figs. 1.37 and 1.38. Table 1.14 shows six species with their formation constants and an estimated uncertainty (next to Fe^{2+} which is free of uncertainty). To achieve a simulation precision of approximately 95% by Monte Carlo resampling, about 10^{12} runs at each pH value would be necessary. Because solution composition has been simulated in 0.2 pH steps, the number of total number of repetitions by a Monte Carlo approach would be in the order of magnitude $10^1 \times (10^2)^6 = 10^{13}$ runs. Obviously, a Monte Carlo approach is not feasible.

In the case of the normal distribution, a Monte Carlo approach requires a large number of samples because the samples are rather inhomogeneously distributed over the axis of values. Most samples are coming from the center of the distribution, only few from the tails. This imbalance in the sampling probability exists not only for the normal distribution but for all distributions having one or more modes and tails. A possible remedy is to balance the sampling probability (Satterwhite 1959). Whatever is done to reduce the imbalance in the sampling probability from given distributions must allow calculation of a mean and some measure for the spread of the outcome (McKay et al. 1979; Garthwaite et al. 2002) in a statistical sound way.

A strategy is stratification. The distribution is divided in various sections. Subsequent sampling is made from the different sections. If a distribution is stratified into strata of equal probability the problem of imbalance in sampling probabilities can be drastically reduced. Figure 1.39 shows a cumulative distribution function stratified in ten strata of equal sampling probability of 0.1.

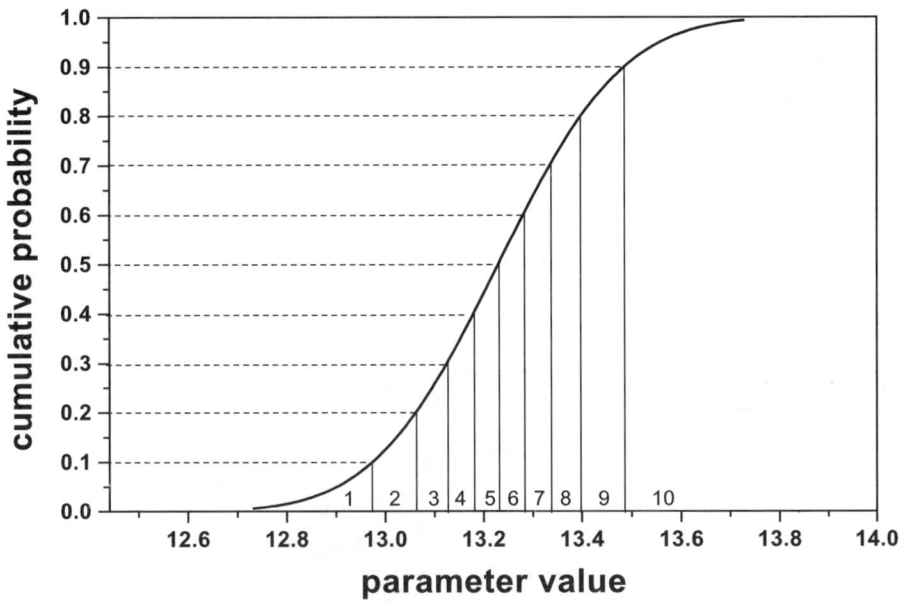

Figure 1.39. A stratified distribution divided into ten strata with equal sampling probability. Note that the range of parameter values covered by each stratum narrows in the center of the distribution and widens toward the tails

It is now possible to sample from each sections of the distribution with a largely reduced risk to introduce bias by neglect of one section of the distribution. Within each stratum, a specific value is randomly chosen, representing its stratum in the further procedure.

If sampling has to be made from n distributions, the outcome of the stratification procedure are n stratified distributions. It is now possible to select randomly values from each distribution combining them to a vector of input data into a simulation run. In the Fe hydrolysis examples (cf. Figs. 1.37 and 1.38), $m = 6$ parameters have been used, each parameter having assigned a normal distribution. Staying for the time being with $n = 10$ strata per distribution, a total of $m \times n = 60$ values $R_{ij}(i = 1 - m, j = 1 - n)$. With these 60 values, a total of ten simulation runs can be performed. One of a large number of possible combinations of parameter values is shown in Fig. 1.40, whereby the Rs are selected randomly.

The output of $i = (1, q)$ Monte Carlo simulations, Y_i, has a mean value

$$\overline{Y} = \frac{1}{q} \sum_{i=1}^{q} Y_i \tag{1.92}$$

and variance

$$s_y^2 = \frac{1}{q-1} \sum_{i=1}^{q} (Y_i - \overline{Y})^2 . \tag{1.93}$$

run j

parameter i	1	2	3	4	5	6	7	8	9	10
1	R_{17}	R_{12}	$R_{1,10}$	R_{15}	R_{11}	R_{13}	R_{19}	R_{14}	R_{16}	R_{18}
2	R_{23}	R_{29}	R_{21}	R_{24}	R_{27}	R_{22}	R_{26}	$R_{2,10}$	R_{28}	R_{25}
3	$R_{3,10}$	R_{35}	R_{39}	R_{38}	R_{33}	R_{31}	R_{37}	R_{32}	R_{34}	R_{36}
4	R_{48}	$R_{4,10}$	R_{43}	R_{41}	R_{46}	R_{49}	R_{42}	R_{47}	R_{45}	R_{44}
5	R_{56}	R_{58}	R_{52}	R_{57}	R_{55}	R_{54}	R_{51}	R_{59}	R_{53}	$R_{5,10}$
6	R_{69}	R_{66}	R_{67}	R_{62}	R_{68}	$R_{6,10}$	R_{64}	R_{65}	R_{61}	R_{63}

Figure 1.40. A stratified random lattice. Note that some column indices appear multiply in a column, for instance in case of R_{22}, R_{52} and R_{62} in run 6. This situation represents stratified random sampling where no restriction exists for distributing the column indices within the lattice

These statistics will be denoted Y_R and s_R, respectively where the subscript R stands for random. The following statistics are given for stratified random sampling (McKay et al. 1979).

The mean is given by Eq. (1.92) for n strata with equal probability $p_i = n^{-1}$ and one representative value randomly chosen from each stratum. The corresponding variance s_S^2 is given by Eq. (1.92):

$$s_S^2 = s_R^2 - \frac{1}{q^2} \sum_{i=1}^{q} (\bar{Y} - Y_i)^2 \, , \tag{1.94}$$

where s_S^2 is the variance of the stratified random sampling scheme. Stratified random sampling is more efficient compared to Monte Carlo sampling because the second term in Eq. (1.94) is always ≤ 0.

Upon inspection of Fig. 1.40, columns can be found where certain column indices appear twice or even triply, for instance the column index 2 in column 6: R_{22}, R_{52} and R_{62}. The lattice is similar to the one shown in Fig. 1.40, but with the additional restriction that no column index may occur twice in a column and a row is a j-dimensional extension of a Latin lattice, a Latin Hypercube lattice. Figure 1.41 shows a square Latin Hypercube lattice. Note that in each of the six runs all strata of the sample distribution are equally present. By evaluating the simulation output based on all six sets of input parameters, the distributions of the input parameters are exhaustively sampled with a minimum of total runs.

Thus, the number of uncertainty affected input parameters of a simulation defines the minimum number of runs required to sample the complete parameter space. However, it is rarely sufficient to perform just this minimum number of simulation runs because the variance of the output decreases with the number of runs. Figures 1.37 and 1.38 are calculated on the basis of 100 runs. Therefore, the Latin Hypercube lattice was a 6×100 lattice. The variance estimator s_L^2 for Latin Hypercube Sampling is:

$$s_L^2 = s_R^2 + \left(\frac{q-1}{q} \right) \left(\frac{1}{q^m (q-1)^m} \right) \sum_{R} (Y_h - \bar{Y})(Y_g - \bar{Y}) \tag{1.95}$$

where R denotes the restricted subspace of all pairs Y_h, Y_g having no column coordinates in common. For small lattices where $n < 2m$, such a situation cannot occur. With larger n, however, such pairs exist and R increases with n. The variance of a Latin Hypercube sampling scheme is smaller than a Monte Carlo sampling scheme only if the sum over subspace R is negative. This is true in almost all practical situations with a mathematical proof given by McKay et al. (1979).

The LJUNGSKILE code is not a ready-to-use program, but is intended to show the relevance of progressing uncertainties in thermodynamic data to the computed simulation output. There has been no other code available with abilities comparable to the LJUGSKILE code, despite the fact that speciation diagrams such as Fig. 1.38 can be found abundantly in the respective literature (on a mean value basis). A preliminary assessment of the impact of uncertainty on the results of numerical speciation in the field of uranium mining remediation in the eastern part of Germany indicated the limited precision of speciation calculations being not fit-for-purpose of improving communication between operators and public authorities (Meinrath et al. 2000b).

As shown above (cf. Fig. 1.29 and related discussion) the empirical distributions of the complete measurement uncertainty budget may be non-normal. Nevertheless, LJUNGSKILE currently does not allow to specify distributions other than normal or rectangular. The number of available empirical distributions is yet too small to justify such an extension. Empirical distributions can be approximated for instance by Chebyshev polynomials and reported by a small number of polynomial coefficients. It is important to note that the statistics given in Eqs. (1.92)–(1.95) are independent of the cumulative distribution functions, provided these are monotonous (McKay et al. 1979).

Currently, the LJUNGSKILE code allows the definition of water composition, solution species, a solid phase, temperature and a CO_2 partial pressure. If multiple runs are performed to create curves as a function of a running variable, almost all component and physical parameters can be used as abscissa values. The ionic strength correction is done on the PHREEQC level. The Davies equation approach is used limiting the applicability of the code to moderate levels. An uncertainty for the solubility product of the solid phase cannot be

run j

parameter i	1	2	3	4	5	6
1	R_{14}	R_{11}	R_{16}	R_{12}	R_{15}	R_{13}
2	R_{23}	R_{25}	R_{21}	R_{24}	R_{26}	R_{22}
3	R_{32}	R_{33}	R_{35}	R_{36}	R_{31}	R_{34}
4	R_{41}	R_{42}	R_{44}	R_{45}	R_{43}	R_{46}
5	R_{56}	R_{54}	R_{53}	R_{51}	R_{52}	R_{55}
6	R_{65}	R_{66}	R_{62}	R_{63}	R_{64}	R_{61}

Figure 1.41. A square Latin Hypercube lattice. Note that no column index appears twice in a column and a row

specified. The evaluation of the probabilistic output however is performed in a rather short time (depending of course also on the clock rate of the computer) in the range of a few minutes to a few hours for larger jobs.

1.7.7
Deciding when a Limit Value has been Exceeded: Measurement Uncertainty and Law

1.7.7.1
Compliance with Legal Limits

National and international regulations require give limits for concentrations of chemical substances in various materials and media. Exceeding such values may be heavily sanctioned. In an environment where mean values only were considered, the keeping of a limit could be unambiguously surveyed. A measurement value below the limit was accepted, a measurement value above the limit was sanctioned. Non-compliance with legal limits can have drastic consequences, e.g. in international trade (Källgren et al. 2003; Marschal 2004). The risks and the difficulties in assessing a given situation may be severed by differences in legal and scientific concepts, e.g. of truth (Rechberger 2000; Walsh 2000). In this way, it is often unclear how non-compliance with a limit value is assessed if it is associated with a measurement uncertainty.

As long as the limit value and the measured value differ widely, this decision in most situation will be made in consent. However, if the both values are close, the probability increases that a conflict of interest is present (cf. Chap. 1.2.1). In such situations, a conventional concept will not provide a solution. Hence, protocols and concepts are required which can be accepted by both parties despite the conflict of interest.

A limit value may be understood as the maximum permissible concentration of a substance in a particular compartment (e.g. sample) which may not be exceeded. Justification of the limit value concept on the legal level is based on the assumption that its application generates legal certainty, allows more rapid judgement, and eases the civil acceptance of authority decisions (Neidhardt et al. 1998).

Limit values are mostly set by public authorities. However, also in the scientific environment limit values exist. Examples are limits of detection, limits of identification and limits of determination. The relevant issue with these limit values is not to comply with them but to evaluate them appropriately. Significant differences in these values may result by different evaluation procedures (Vogelsang and Hädrich 1998) with consequences for the economic situation of the issuing bodies in a competitive environment (Kaus 1998).

A limit value makes only sense if compliance can be assessed with a minimum of unambiguity. Hence, measurement values have to correspond to the true (but almost always unknown) concentration. The concept of "true values" is not without critical inquiry (de Bièvre 2000; Meinrath 2002); even so, it forms an essential basis of all metrological sciences. The legal system requires "fact finding beyond all reasonable doubt" (Rechberger 2000) while for a metrologically trained chemist, doubt in a measurement value is an essential element. Mathematically speaking, a normal distribution never becomes truly zero even at wide distances from its mean. Hence, the discussion about limit values, compliance and sanctions rests with the word "reasonable".

Since limit values form an interface between science and law, the different legal systems in different parts of the world interfere. One reason for this mutual interference is globalisation. Liability for non-compliance with strict limit may be costly as is the surveillance. Insurance costs may become a non-negligible part of the calculation. On the other hand, production, trade and commerce may profit from generous or even non-existing limit values. Therefore, on a competitive global market national policy with regards to environmental protection and labour security may become an important factor. In a country requiring strict compliance with ISO 17025, greater care need to be given to limit values compared to countries where reporting mean values is sufficient. Another reason is the different legal systems. Sanctions are easier to avoid if "fact finding beyond all reasonable doubt" is required instead of convincing a jury that decides by majority vote.

Limit values are therefore an important issue in metrology in chemistry. Due to the strong relationship of the current measurement uncertainty concept to the normal distribution, the following discussion will be based on the concept of complete measurement uncertainty budget expressed by an expanded standard uncertainty U ($k = 2$).

A limit value is commonly stated without uncertainty, while the measurement value carries a measurement uncertainty. A variety of situations may exist, shown schematically in Fig. 1.42.

Situation A is safely below the limit value. No further analysis of the situation is required. Two cases are shown for situation B. The mean values of both measurement results are below the limit value, but the expanded standard uncertainty states that the right hand side value may exceed the limit value. The left-hand-side measurement value has a higher mean value, but its measurement uncertainty is lower. Situation B shows a first complication in assessing the compliance with the limit value in presence of measurement uncertainty. The complication increases for situation C. The measurement value is above the limit value. However, there is doubt whether the limit value is exceeded. If sanctions are imposed by the authorities, the polluter would probably disagree. Situation D is likely to cause more intense disputes. If authority would accept situation D as compliance, the polluter would be able to use a method with much higher standard uncertainty and by this way extending the acceptability range of a measurement value far above the limit value, as is shown on the right hand side of situation E. The left hand side of situation E shows a clearly non-complying value. By expanding the measurement uncertainty (e.g. by using a less precise method), the polluter would be able to

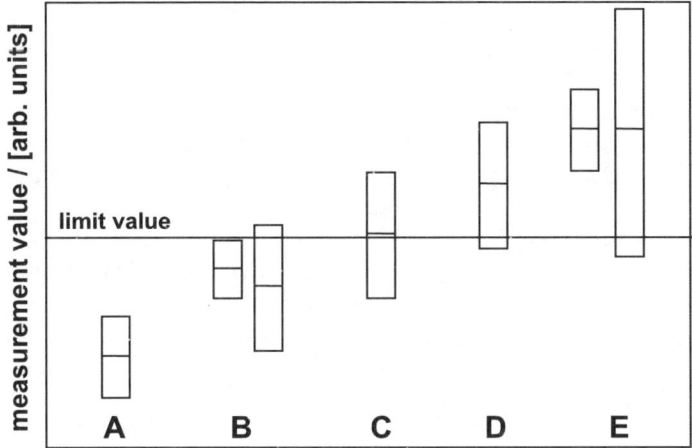

Figure 1.42. Five possible situations (A–E) to be encountered in comparing a limit value with uncertainty affected measurement values. The measured value is given as a *horizontal bar*. The size of the enclosing *boxes gives* the expanded combined standard uncertainty U ($k = 2$) of the respective measurement value

claim compliance because the non-compliance is not beyond all reasonable doubt.

Figure 1.42 illustrates existence of a critical region about a limit value where non-compliance is difficult to assess and, consequently, to enforce. It is therefore necessary to find logical criteria to assess compliance and non-compliance. This issue is the more urgent if a third party is affected by the decision made by authority and polluter. Such a third-party may be the population around the locality of concern, agricultural production, animal habitat etc. Limit values are usually issued to protect important values.

A possible approach might be to assign a target uncertainty to a limit value. A limit value with associated target uncertainty will be able to avoid an unpleasant situation E where non-compliance is covered by an inadequate analytical method with large measurement uncertainty. Taking into account the often time-consuming efforts to establish a limit value, further complications are to be expected if additional target uncertainties would be required. In the end, a situation might be reached where "one has to be careful not to design a diabolic machine of complexity which will lead to the practical impossibility to make any decision and take any action" (Shakespeare's razor) (Shakespeare 1596; Marschal 2004).

1.7.7.2
Assessing Compliance on Basis of Normal Approximation

The complexity of compliance testing may be appreciated from the discussion of Christensen et al. (2002). These authors suggest two complementary approaches. The first approach judges the measurement value for compliance or non-compliance, depending whether the risk resulting from a wrong statement is higher in case of compliance (stating compliance incorrectly) or non-compliance (stating non-compliance incorrectly). If the situation occurs that neither compliance nor non-compliance can be stated beyond all reasonable doubt, statistical tests are suggested on basis of a normal distribution of the measurement uncertainty. The second approach combines two different procedures. The first procedure is the one-stage procedure which consists of the first approach given above. The two-stage approach requires the determination of additional measurement values if neither compliance nor non-compliance can be stated. As an example for the sophistication of these rules, the number n of measurements necessary to declare compliance (one-stage procedure) can be calculated from the following four factors:

1. The standard deviation of the measurement result (here: combined standard uncertainty u_c).
2. The pre-set degree of confidence for the uncertainty interval $(1 - \alpha)$.
3. The closeness of the measurement value, μ, to the limit value, LV at a pre-set probability $(1 - \beta)$.
4. the probability $1 - \beta$ of declaring conformity for a value of $(\mu - \text{LV})$.

Then,

$$n = \left(\frac{(u_{1-\beta} + u_{1-\alpha/2})u_c}{\text{LV} - \mu} \right)^2 . \tag{1.96}$$

The equations for assessing compliance or non-compliance can be found in the cited paper. It might be difficult to communicate a cyclist's exclusion from a competition because his blood hematocrit is higher than 50% (set by the Union of International Cycle Racing) established by a reasoning including equations such as Eq. (1.96). This approach is another example of "Shakespeare's razor". Only statistically trained metrologists will be able to follow such argumentation. It is important to note that approaches and protocols are urgently needed. The logical, mathematical and statistical arguments given by Christensen et al. are correct and trackable.

1.7.7.3
Loophole

While legal limits are set with the expectation to comply with them, ensuring compliance may become a headache. A reason for the headache may be found in the definition of convention presented in Chap. 1.2.1. If measurement values are accepted without measurement uncertainty the following convention applies: Compliance with a limit is demonstrated by forwarding a measurement (mean) value below the legal limit. When, however, measurement values have to carry a measurement uncertainty, demonstration of compliance will be less straightforward. The measurement uncertainty is set by those forwarding the measurement value. A situation shown for case E in Fig. 1.42 states that the measurement mean value is above the legal limit. A possible consequence might be to carry out costly remediation activities for the party being responsible for exceeding the limit value (polluter). For the polluter, this unfortunate situation can be "improved" by extending the measurement uncertainty (right hand side of case E). Thereby, it should be noted that the error bars not necessarily need to be symmetrical. Because it is the interest of the (accused) polluter to avoid remediation operations it may be simpler to find arguments increasing the error bar of the measurement value. Figure 1.42 and related discussion should prompt a considerably higher interest of the law-makers in metrological issues than shown up to now. At present, the metrological discussion offers some strong points for (accused) polluters to circumvent the consequences of their action.

The situation may require, despite all scepticism, to introduce target uncertainties for analytical methods and measurands (de Bièvre 2004b; Marschall 2004). Proficiency tests and a strict application of the requirements of ISO 17025 will drastically reduce the possibilities to arbitrarily increase the measurement uncertainty. National standards and norms will form a framework within conformity, and non-conformity with limit values has to be demonstrated. The concepts of Good Laboratory Practice (GLP) can form a further defence line against weakening the role of stated limit values if properly applied and enforced.

1.7.8
Further Considerations

Most discussions of compliance assessment with limit values are based on a normal distribution of the measurement uncertainty. What happens in situations where a polluter claims non-normality of the measurement result? Can a regulator judge a suspected pollution on basis of the EURACHEM/CITAC guide (EURACHEM/CITAC 2002) if the presumed polluter shows that the pollution is inferred (quasi as an artefact of the assessment procedure) only due to the EURACHEM bottom-up approach and would not be judged as non-conformity if a different evaluation procedure, e.g. on the basis of a complex method as outlined in the respective sections of this chapter, is used? If skewed distributions or even multimodal distribution come into consideration, the topic of compliance with limit values may become highly confusing.

The discussion of compliance with limit values has many facets. The probabilistic approach briefly outlined above is familiar because it is built upon classical t-test acceptance/rejection argumentation. Such studies are at least occasionally done by most chemists during their professional life. In chemistry such argumentation, however, faces criticism (Hibbert 2001). A common basis for these concepts is the commitment of the (suspected) polluter or a scientific attitude towards the evaluation of the measurement value and its associated combined standard uncertainty u_c. Given such boundary conditions, the numerical problem can apparently be solved.

In the real world, however, the conflict of interest between authority/regulator and (suspected) polluter puts a different taste to the situation. Because metrology is becoming the language on the international market place, the partners on this market place need to explore and to use the margins of their field. In contrast to the measurand, which is usually supposed to have a "true" value, the measurement uncertainty is an element of psychology. It is assessed and evaluated according to principles in many situations also involving personal judgement. The margins for "professional judgement" are different to be set. On the other hand, it should be kept in mind that metrology and market have a long and intense relationship. This relationship has worked out to the better of all partners including customers, in manufacturing industries and commodities, if not (yet) in chemistry.

At the practical level of the discussion, mutual agreement and logical argumentation seems to be a practical way to fulfil the intentions associated with the issue of limit values: to ensure protection of values, habitats and quality of life. The main driving force to comply with limit values, to establish quality assurance systems, to invest into accreditation and participate into proficiency testing, is the desire to create trust; trust of clients in the abilities of the producer, trust of the authorities into the practices of the manufacturers and, last but not least the manufacturers into their own abilities.

On the theoretical level, however, compliance with limit values remains a challenge and a touchstone capable of guiding the discussion of metrology in chemistry in the years to come.

1.8
Deficit Analysis of Existing Thermodynamic Data for Environmental Simulation

"The greatest friend of truth is time, her greatest enemy is prejudice, and her constant companion humility."

(C.C. Colton)

Environmental modeling is an important application field of values obtained from complex measurement processes. These data are mostly formation constants of chemical species, solubility products of solid phases and solubility data of gases in aqueous solutions. The reasons for the interest in thermodynamic data is associated with Eq. (1.86): thermodynamic data are constants of nature. Even though most of these data are determined in laboratories under controlled conditions in concentration ranges and in media which do not occur abundantly in nature (e.g. perchlorate media), Eq. (1.86) justifies their application under widely differing conditions. It is understood that the data must be adequately corrected, e.g. for temperature and ionic strength effects but their almost universal applicability is not questioned.

1.8.1
Uncertainty Contributions from Methodology and Practice

Most topics of the first section of this book have, up to now, never been al-lowed to play a role in the discussion of quality assurance of thermodynamic data. The versatility of, for instance, computer-intensive statistical approaches to compare different data sets at least on the basis of repeatabilities from lack-of-fit (Meinrath et al. 1999) is shown, and the concept of complete meas-urement uncertainty budgets has been discussed also for thermodynamic data (Meinrath 2001). The vast majority of thermodynamic data, however, were determined decades before ISO's GUM has been published. There have been various attempts to "smooth" the available data according to some criteria. As the result of such an approach, the Joint Expert Speciation System (JESS) data base is available (May and Murray 2001; JESS 2006). JESS is a rather unique system, as the data available in literature enter the procedure to simu-late the desired information without pre-selection, expert reviewing etc. The commercial JESS program package is designed to reach internal consistency automatically, in part depending on the boundary conditions of the given simulation task. "It attempts to deal with the serious discrepancies that often occur between published equilibrium constants, allowing a consistent set of mass balance equations for chemical speciation modeling to be produced by systematic means" (Filella and May 2003). The problem of discrepancies is abundant within thermodynamic data. Hefter (Bond and Hefter 1984) states: "It is an unfortunate, but only too well known, fact of solution chemistry that different techniques may yield different values of stability constants. In-tending investigators therefore need to recognise and understand the known limitations of any method they are considering to use and to bear in mind the possibility of unknown limitations". Grauer's (1997) analysis is almost a dia-tribe against the varied activities in chemical thermodynamics in the view of extremely few reliable information. It is interesting to balance his statements and examples with the claims and interpretations by other authors in the re-spective publication. In some cases, the search for obvious inconsistencies by scrutinising the existing universe of thermodynamic data yields clarification (Hummel 2000), but these cases are rare. Some reasons for the existence of a wide variety of data, augmented by numerous recalculations, medium and temperature corrections, reinterpretations of the species compositions, criti-cal evaluations, inter- and extrapolations, are given by Filella and May (2005): "The specific problem addressed in this paper is that many formation con-stants now being published in the literature are calculated and reported in disparate ways so that it is difficult to incorporate them into thermodynamic databases and to make valid comparisons between the results from different investigators". Further: "There are at least three reasons why this difficulty has become acute in recent years. First, there has been a proliferation of computer programs to determine formation constants (even though some of these codes contribute little, or nothing, new). Secondly, many possible methods of calculation, each capable of producing significantly different answers, have become readily available. Thirdly, and perhaps most importantly, formation-constant determination remains necessary, but is less fashionable than it used to be. So, results are now often performed by researchers with less background and experience in this field." Available reference manuals are either old and do not encompass the use of computer calculation-based approaches (Rosotti 1978) or are restricted to a particular calculation program (Martell and Motekaitis 1992).

These few selected voices from experienced representatives of the scientific community (alternative choices are available) highlight the key issue: existing thermodynamic data are not reliable, are inconsistent and/or produced by inexperienced experimenters. The statements of Filella and May in their frankness focus on the effect of overoptimistic (uncritical) use of computation tools, increasing the variance within thermodynamic formation constants. But even in their analysis (highly recommended for reading), the terms "metrology", "comparability", "traceability of measurement values" do not occur. Alternative concepts are not presented.

The advent of computers in the 1980s increased interest in thermodynamic data of environmentally relevant metal ions and their species. This interest was further motivated by the increasing ease of determining metal ion concentrations in aqueous solution due to more and more sensitive and rapid automatic analytical equipment. Over the past century, analytical chemistry has taken a strong development from a more or less qualitative subject to a quantitative science encompassing single atom detection capabilities, destructive and non-destructive methods up to speciation options in inorganic and organic matter, including biosubstances. To a large extent, this development was made possible by instrumental methods in chemical analysis. These methods have originated largely between 1970 and 1985. The period 1985–2000 was largely devoted to refining the instrumental tools and to the development of methods that could cope with the increasing amount of data generated by these instruments. Chemometrics provides powerful data treatment tools to get "the best out of the complex data structures" generated by these methods. The question of the reliability of these data was overpowered by belief in the computer's abilities and the increasingly sophisticated graphical data displays (Chalmers 1993; de Bièvre 1999). It is teaching for chemists to follow the discussion of some chemometric regression tools (Frank and Friedman 1993a) together with invited comments by fellow statisticians and chemometricians. While statisticians worry about the reliability and variability of parameters derived from ridge regression, factor analysis (PCR) and partial least squares

(PLS) (Hastie and Mallows 1993), the chemometrician emphasises interpretation and insight gained from measurement data otherwise not accessible (Wold 1993). The chemometrician argues on basis of "independent variables that are correlated". The problem of reliability is of less importance to the chemometrician. He defends both PCR and PLS despite the fact that both techniques give different solutions for the same data set (Frank and Friedman 1993). The statisticians point out the importance of assessing the variability of all estimated quantities before they are interpreted (1993b), suggesting bootstrap methods (Efron and Tibshirani 1986).

The computer offered an enticing perspective to aqueous chemistry: having the thermodynamic data for all possible solution species at hand makes it possible to calculate the species composition of a solution from measured component concentrations. The speciation diagrams shown in Figs. 1.37 and 1.38 even though without the uncertainty limits, give examples for such calculations. By hand it is almost impossible to perform the necessary calculations even for the simpler situation of Fig. 1.37.

At the same time, groundwater models were developed capable to predict the transport of water in porous media. Reactive transport models combining the ground water flow with numerical speciation were created with the hope to simulate and predict the transport of matter in the subsurface formations. The situations was described by CI Voss (1998) for the hydrological models: "The fact that a ground water model is really a simple device is often obscured by the manner in which it is applied. Uninitiated model users and even many experienced model users believe that the numerical model can be made to give an accurate representation of reality if only enough detail and data are included." In chemistry, a large number of thermodynamic data points were determined with the intention to predict future developments. The disposal of nuclear wastes in deep geological formations is a typical, an important application where the time-scale is by far too great to be explored by experiment. A large number of thermodynamic data aim at the safety assessment of nuclear waste repositories.

Almost all of these data have been determined without even the least analysis of their reliability. They have also determined without an idea on the required quality. It is, even today, completely unclear what target uncertainty a thermodynamic datum is required to have to be fit-for-purpose (Thompson and Fearn 1986) in performance assessment of nuclear wastes.

The main goal of metrology is to gain trust in a measurement value. The users of chemical information should trust the data produced by a measurement procedure. Of importance also is the trust the producer gains into his own abilities and the experience of his limitations. In those situations where the decisions based on certain data may affect others, academic freedom is probably not an acceptable attitude in the determination of these data.

It happens often that chemists complain about their low reputation in the general public. In Germany, which merits a lot to chemistry and chemical industry, the reputation of chemistry and chemical education may be appreciated from the German Law on Chemicals. This law states that the required knowledge to handle hazardous chemicals should be held by: pharmacists, pharmaceutical engineers and pharmacy assistants, pharmaceutical-technical assistants, and qualified pharmacists and pest controllers. Others, among them academically trained chemists, have to prove their knowledge by passing a separate examination. Does this law underpin a high esteem for chemical education?

The decision not to accept an academic degree in chemistry as a sufficient proof for the ability to handle hazardous chemicals may be considered as an indication of poor lobbying. However, a recent textbook on Analytical Chemistry (Kellner et al. 1998) co-authored by a renowned German analytical chemist and based on FECS's Eurocurriculum of analytical chemistry does not give even the most basic informations on concepts of metrology in chemistry (Plzák 2000). The EURACHEM/CITAC guide "Quality Assurance for Research and Development and Non-routine Analysis" (EURACHEM/CITAC 1998) does not play a sufficient role in academic analytical chemistry. The lack of quality management in research and education is well known (Cammann and Kleiböhmer 1998; Prichard 1999). An important factor is, obviously, time (Libes 1999). Education towards a critical attitude and living with uncertainty and doubt is certainly a difficult task. Society life is based on convention. The nail that sticks out is the one that is hammered, in almost all societies. Therefore, it is of outmost importance to scrutinise all information on which important decisions are built whether it is universal (reproducible within given limits independent of space and time), site-specific (reproducible only within a certain local and/or temporal interval) or conventional (an agreement within a certain group having an advantage from adhering to the convention) (Price 2001).

It is important to recall the definition of convention given in Chap. 1.2. The fact that the group adhering to a convention is large cannot replace trust generated by comparability of measurement values. Here, the focus is on "measurement value". It is not acceptable to modify measurement values independent from the detailed information of their measurement. Measurement results are information. Without knowledge concerning their determination and their uncertainty, however, they are just rumours. Subsequent corrections and modifications, in case of thermodynamic data, e.g. by ionic strength corrections, require detailed information about the measurement process and the evaluation of the value.

1.8.2
Uncertainty Contributions from Ionic Strength Corrections

Rumours cannot be "corrected" into reliable measurement values. If measurement values are transformed, corrected or modified, the uncertainty contribution resulting from the procedure must be adequately assessed and included into the combined standard uncertainty of the measurement value. A typical modification that thermodynamic data are subjected to is ionic strength correction. Unfortunately, there is little knowledge about the uncertainty introduced by ionic strength corrections.

There are a small number of ionic strength correction equations in use. The basis for ionic strength correction is the Debye–Hückel equation (1.97):

$$\lg \gamma_i = -Az_i^2 \sqrt{I} \,. \tag{1.97}$$

The mean activity coefficient γ of substance i results from a constant A, charge z_i of substance i and the ionic strength I of the medium

$$I = \frac{1}{2} \sum_i m_i z_i^2 \,, \tag{1.98}$$

whereby m_i is the molal concentration of substance i (mol kg^{-1} solvent). With Eq. (1.97), electrolyte solutions at very low concentrations can be described well. Above $\sim I = 10^{-3}$ mol kg^{-1}, the Debye–Hückel (DH) equation is not satisfactory any longer. There are several modifications of the DH equation that try either to overcome certain approximations in the derivation of the DH equation, or to create a broader range of applicability by empirical extensions. The so-called extended DH equation

$$\lg \gamma_i = \frac{-Az_i^2 \sqrt{I}}{1 + Ba_i \sqrt{I}} \tag{1.99}$$

with $A = 1\,824\,928 \cdot 10^{106} \, \rho_0^{1/2}(\in T)^{-3/2}$. $B = 0.3(\in T)^{-1/2}$. ρ_o gives the density of water, $\in$ the dielectric constant of water and T is the absolute temperature in Kelvin. At 25 °C, $A = 0.509$ and $B = 0.3283$. Term a_i is the effective radius of the solvated substance i (in units of 10^{-8} m). The effective size of a metal ion is $2.5-4.5$ for most univalent ions, $4-8$ for divalent, $4-9$ for trivalent and $5-11$ for tetravalent ions. Typical magnitudes for effective sizes are: $a_{(H_2O)} = 3.5$, $a_{(ClO_4^-)} = 3.5$, $a_{(CO_3^{2-})} = 4.5$ and $a_{(Cl^-)} = 3$.

While the extended DH equation is claimed to be applicable up to $I = 0.1$ mol kg^{-1}, the WATEQ-DH equation (Eq. (1.100)) is applied up to ionic strength $I = 1$ mol kg^{-1}.

$$\lg \gamma_i = \frac{-Az_i^2 \sqrt{I}}{1 + Ba_i \sqrt{I}} + b_i \cdot I \,. \tag{1.100}$$

This equation has further ion-specific factors b_i.

The Davies Eq. (1.101) is closely related to the extended DH equation

$$\lg \gamma_i = -Az_i^2 \frac{\sqrt{I}}{1 + \sqrt{I}} + 0.3 \cdot I \,. \tag{1.101}$$

Comparison of Eqs. (1.99)–(1.101) shows that Davies equation is closer to the extended DH, missing the constants B and a_i. Statistically speaking, the degree of freedom is reduced, while allowing a higher flexibility of the equation in fitting its curve to data. The SIT equation (specific interaction theory) has additional parameters added to the DH equation

$$\lg \gamma_i = -z_i^2 D + \sum_k \in (i, j, I) \, m_j \tag{1.102}$$

with Debye–Hückel term D : $\frac{0.5901\sqrt{I}}{1+1.5\sqrt{I}}$ at 25 °C. Terms $\in (i, j, I)$ represent so-called interaction coefficients of ion i with ion j. Summation is done over all counter ions j of a given ion i. The SIT equation (Eq. (1.102)) is almost exclusively used in the context of the thermodynamic database of OECD's Nuclear Energy Agency. The interaction coefficients are derived from given thermodynamic data. The statistical basis as well as details of its derivation remained unclear even after several contacts with OECD/NEA.

All these extensions of the DH equation fail at ionic strengths $I > 1 \, \text{mol} \, \text{kg}^{-1}$. Pitzer (Pitzer 1986; Pitzer and Simonsen 1986) has proposed virial extensions of the DH equation that are able to describe even highly concentrated electrolyte systems $I > 6 \, \text{mol} \, \text{kg}^{-1}$. Of course, all the parameters of Pitzer's equations (binary coefficients and ternary coefficients) have to be obtained from experimental measurements. Marshall et al. (1985) have given a statistical analysis of binary Pitzer coefficients on basis of linear statistics. Their analysis has been supported by a non-parametric study of statistical variability using bootstrap methods (Meinrath 2002b).

Due to the abundance of mean value-based analysis of the interaction and mixing parameters in Pitzer's equations, there is very little insight into the limitation of the Pitzer approach. Pitzer's equations are apparently complex (cf., e.g. Kim and Frederick 1988). Seen mathematically, these equations are linear and can be interpreted by classical linear statistics (Marshall et al. 1985). Such an analysis does not provide a complete measurement uncertainty budget of the experimental data from which the parameters of Pitzer's equations are derived. But an interesting insight into the variability of the parameters due to different assumptions regarding their distribution and different analysis methods can be obtained.

In this analysis, the single salt parameters $\beta^{(0)}$, $\beta^{(1)}$ and C_Φ for the single salts NaCl and KCl need to by used as auxiliary data from a previous analysis of experimental data obtained by different authors. The compilation by Hamer and Wu (1972) is a standard source for these data. A result is shown in Fig. 1.43, where the distribution of the binary parameter C^Φ of NaCl is shown as resulting from a bootstrap resampling analysis using two different optimization criteria: the classical least-squared residuals result (L_2) and the more robust least absolute residual (L_1) criterion. The both distributions are compared to the mean value given by Kim and Frederick Jr. (1988) (KF) and the normal distribution derived by Marshall et al. (1995) (MMH). The L_2 bootstrap distribution is

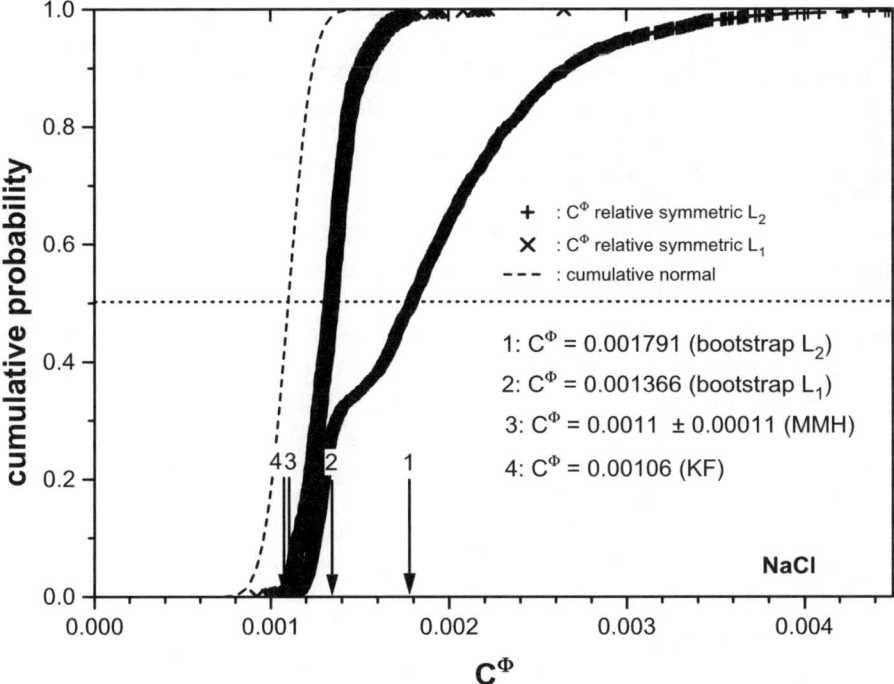

Figure 1.43. A comparison of values and distributions derived for the binary Pitzer parameter C^{Φ} of NaCl from the same data set under different assumptions

obviously non-normal while the L_1 bootstrap distribution nicely corresponds (with a slight shift) to the normal distribution derived by MMH. MMH's mean value is close to the value given by KF.

Again it is shown that even with the limited scope to study only repeatability, different values/distribution of values will be obtained from a given data set. This observation is not unexpected given the discussion in the preceding chapters of this book. Here, however, a different aspect is emphasised because the value(s) obtained for the binary Pitzer coefficients are required as auxiliary data in the evaluation of the ternary mixing parameters. Hence, these auxiliary information has to be derived from a six-dimensional empirical probability distribution.

If binary and ternary mixing parameters of all binary and ternary salt combinations in given complex salt mixture are available, the Pitzer approach will be able to predict the activity coefficients in these salt mixtures. The point of interest here focuses on the reliability of such an application. The evaluation of a complete measurement uncertainty budget for the Pitzer parameters is not possible, because the necessary information on influence quantities and their magnitude, which are a requisite to evaluate the cause-and-effect diagram, have never been reported. An interesting aspect, however, is the evaluation of ternary parameters in presence of statistical uncertainty coming from the auxiliary, binary parameters. To do so, a Monte Carlo approach is insufficient and the LHS approach has to be used. A total of 25 LHS runs were performed on basis of the EDFs obtained from analysing the binary salt mixtures. For each of the 25 runs, 2000 bootstrap samples were obtained amounting to 50 000 simulated data for the ternary parameters.

Experimental data on the activity coefficients of Na/K/Cl mixtures have been obtained from Huston and Butler (1969). This paper is the standard source from which a the respective ternary mixing parameters, $\theta_{Na, K}$ and $\psi_{Na, K, Cl}$, are derived. To allow a clear display of the results, only the 20 activities obtained by electrode C are included into this analysis. These data are given in Table 1.15. Kim and Frederick Jr. (1988) report $\theta_{Na, K} = 0.007$ and $\psi_{Na, K, Cl} = -0.0098$ from linear least-square fitting of all data in Huston and Butler (1969).

The result is shown in Fig. 1.44a and b. In Fig. 1.44a, the ternary parameters have been set to zero. In Fig. 1.44b, these parameters take the values given by Kim and Frederick Jr (1988). As can be seen immediately the effect of the ternary parameters is marginally to non-existent and supports the conclusion that the evaluation of ternary parameters for the given salt mixture interprets at large random effects from unaccounted measurement uncertainty.

Table 1.15. Activity coefficients of NaCl in ternary Na/K/Cl solutions (Huston and Butler 1969). Only data reported from measurements with "electrode C" are used

Total ionic strength	γ(NaCl)	x(NaCl)	Sample number
0.4805	0.68171	1	1
0.4834	0.70291	0.8564	2
0.4876	0.68234	0.6585	3
0.4911	0.68486	0.4722	4
0.4952	0.69599	0.2706	5
0.4983	0.69775	0.1145	6
0.9323	0.65826	1	7
0.9466	0.67655	0.703	8
0.9548	0.67999	0.5329	9
0.9703	0.66252	0.2111	10
3.1134	0.72044	1	11
3.0675	0.68992	0.7492	12
3.021	0.67671	0.4957	13
2.9522	0.65283	0.2667	14
2.9791	0.63753	0.1202	15
4.3393	0.81096	1	16
4.3006	0.78524	0.7884	17
4.2547	0.72344	0.5329	18
4.2129	0.68533	0.3099	19
4.1743	0.65615	0.0986	20

x(NaCl) mol fraction of NaCl in the NaCl/KCl mixture

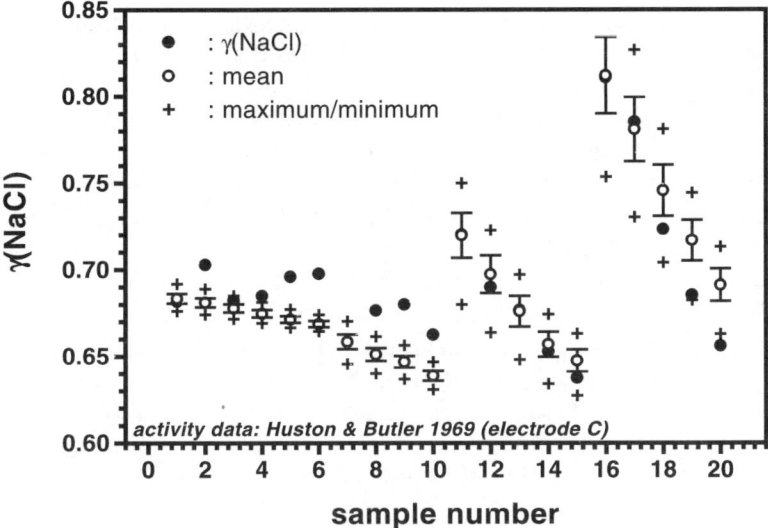

Figure 1.44a. An interpretation of the activity coefficients of NaCl-KCl mixtures by Pitzer's equations ignoring ternary mixing parameters. The distribution of the calculated activity coefficients is shown by *error bars* (1σ) and *crosses* (+) giving the observed extreme values.

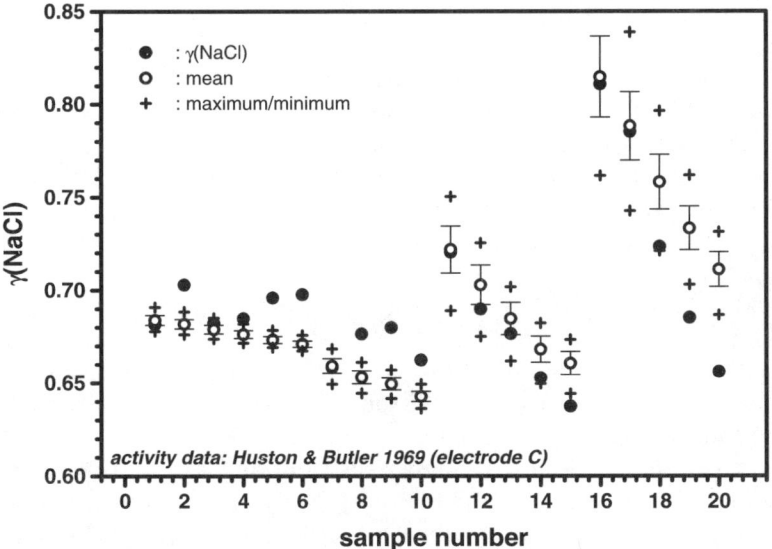

Figure 1.44b. An interpretation of the activity coefficients of NaCl-KCl mixtures by Pitzer's equations using ternary mixing parameters as determined by Kim and Frederick Jr. (1988). The distribution of the calculated activity coefficients is shown by *error bars* (1σ) and *crosses* (+) giving the observed extreme values

This example may unintentionally have been applied here to a salt combination where the ternary coefficients in fact were negligible. Further unpublished analysis, however, is available corroborating the results shown here for the Na/K/Cl system. The purpose of the discussion, however, has been to illustrate the inadequacy of merely mean value-oriented analysis of multi-parameter models. Pitzer's equations are complex, and a full analysis of the complete measurement uncertainty is a challenge. If these data are, however, applied for important decisions outside the academic world, a detailed analysis has to be performed. Pitzer's equations are applied for ionic strength correction in solutions at higher ionic strength were the other ionic strength correction methods (Eqs. (1.97)–(1.102)) cannot be applied. Ionic strength correction may have a non-negligible impact on the magnitude of the complete measurement uncertainty budget of thermodynamic data. The higher this budget, the more restricted is the reliability of simulations and predictions performed on a basis of these data. The discussion of simple iron hydroxide equilibria (cf. Figs. 1.37 and 1.38 and related discussion) is an example illustrating this aspect.

1.8.3
Uncertainty Contributions from pH

The traceability chain in case of pH measurement includes all stages shown in Fig. 1.6. The equivalence to measurement capabilities in the national metrological institutes in the EUROMET region has been tested and documented (Spitzer 1996, 1997, 2001; CCQM-9 1999). Key comparisons under the auspices of BIPM have established the measurement uncertainty of the highest metrological quality. The problems of establishing a traceability chain for the non-measurable quantity pH are well documented in the 1985 IUPAC compromise recommendation (IUPAC 1985) and the 2002 IUPAC recommendation on pH measurement (IUPAC). The primary standards for pH with stated uncertainty u_{prim} therefore are available. Calibration solutions traceable to primary reference materials are now commercially available.

Nevertheless, measured values of the quantity pH are reported without associated uncertainty. It is mostly unclear in which way these values have been derived. Some information on the likely measurement uncertainty of values of quantity pH comes from round-robin studies. An early example is provided by Metcalf (1987).

From 485 measurement values, the binned distribution Fig. 1.45 is obtained. The spread of the distribution is 0.05 pH units. Note that no information is given on the meaning of the "±" symbol: a standard deviation, perhaps, or a confidence interval?

In Fig. 1.46, results from a recent proficiency test are given (BAM 2003). The abscissa shows the laboratory codes while the data are represented as boxes with a central bar representing the mean value and the box giving the estimated standard uncertainty.

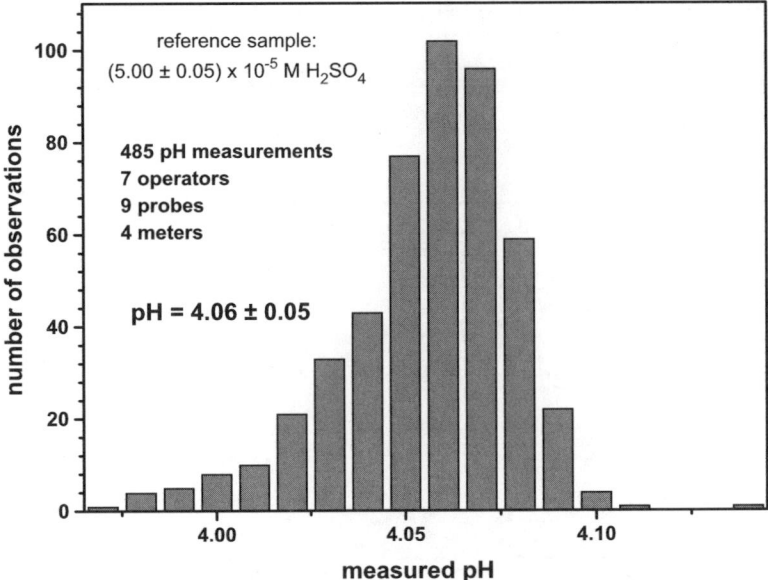

Figure 1.45. Distribution of (binned) values of quantity pH in solutions $(5.00 \pm 0.0) \times 10^{-5}$ M H_2SO_4 (Metcalf 1987)

The reference value (nominal true value) is given as pH 4.450. The tolerable limits ($|z|$ score < 3.000) is pH 4.152 and pH 4.752, resulting in a standard deviation of about 0.1 pH units.

A similar result is obtained by a French proficiency test, where values of pH in drinking water where determined (IUPAC WP 2007). The results of that study are given as a cumulative empirical distribution function, which is compared to its closed fitting normal distribution in Fig. 1.47. The respective normal distribution is N(0.186, 7.074). The spread is even wider compared with the BAM study (Fig. 1.46).

Numerical results from proficiency test can be used, e.g. within the Nordtest approach (Magnusson et al. 2004), to estimate the combined standard uncertainty of pH measurement. Thus, in combination with traceable calibration standards for pH measurement these data allow to link laboratory pH measurements to the primary standards and, ultimately, to the definition of the mole in SI.

Returning to thermodynamic data of hydrolysis reactions of metal ions in aqueous solution, it must be kept in mind that reference laboratories, for example for pH in the NMI, were unavailable before 1993. The majority of hydrolysis data, however, has been determined at times long before the concept of metrology reached the chemical sciences (Baes and Mesmer 1976). The importance of including uncertainty in pH measurement into thermodynamic data was shown (Meinrath 1997, 1998; Meinrath and Nitzsche 2000). Thus, the

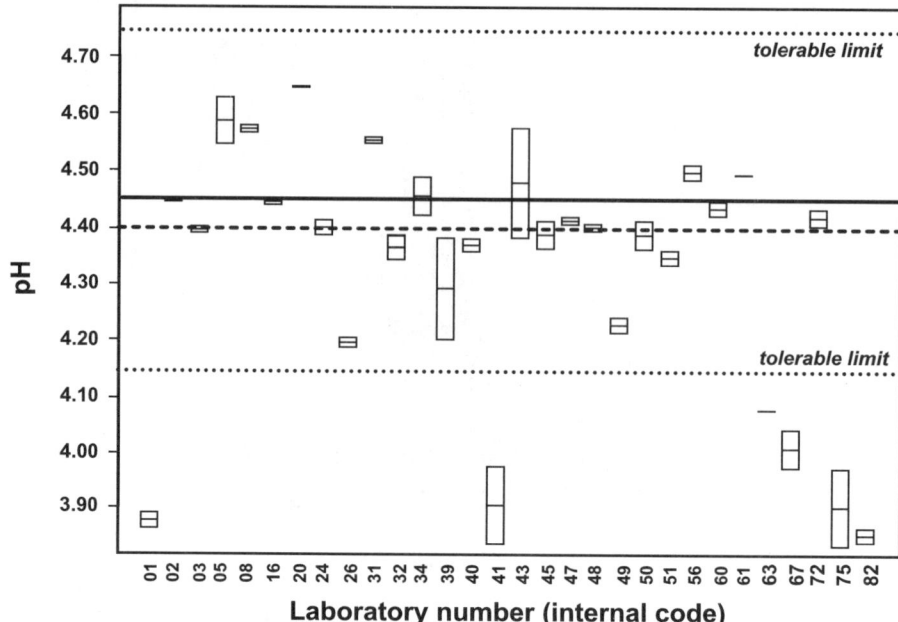

Figure 1.46. Results of a proficiency test on pH measurement with 30 participating laboratories (BAM 2003). The *boxes* given the estimated standard uncertainty as estimated by the participating laboratories

importance of including measurement uncertainty in pH into the evaluation of thermodynamic data is demonstrated in literature. Methods for achieving this goal are available (Meinrath 2000a, 2001; Meinrath and Spitzer 2000).

From 1983 (Covington et al. 1983) to 2002, two pH scales were in use. For the time before 1983, little information exists on the metrologically relevant details of pH measurement. The calibration materials often have been in-house produced solutions with nowadays unknown measurement uncertainty. The electrode design has improved during the past years considerably and the glass material of the electrode also has undergone considerable modification to improve stability and repeatability of potential measurements (Baucke 1994). A common calibration procedure included the use of the theoretical Nernst slope (59 mV at 25 °C) which is meanwhile understood to introduce considerable bias (Baucke et al. 1993; Baucke 1994; Naumann et al. 1994). Currently, the focus is on liquid junction potentials and design of the diaphragms (Baucke et al. 1993; IUPAC 2002). With the current IUPAC recommendation (IUPAC 2002), a unified approach on basis of metrological principles is available (Baucke 2002), and electrode characteristics as well as protocols for an evaluation of measurement uncertainty can be based upon this document (Naumann et al. 2002). Norms and standards including these principles into normative documents are currently under preparation.

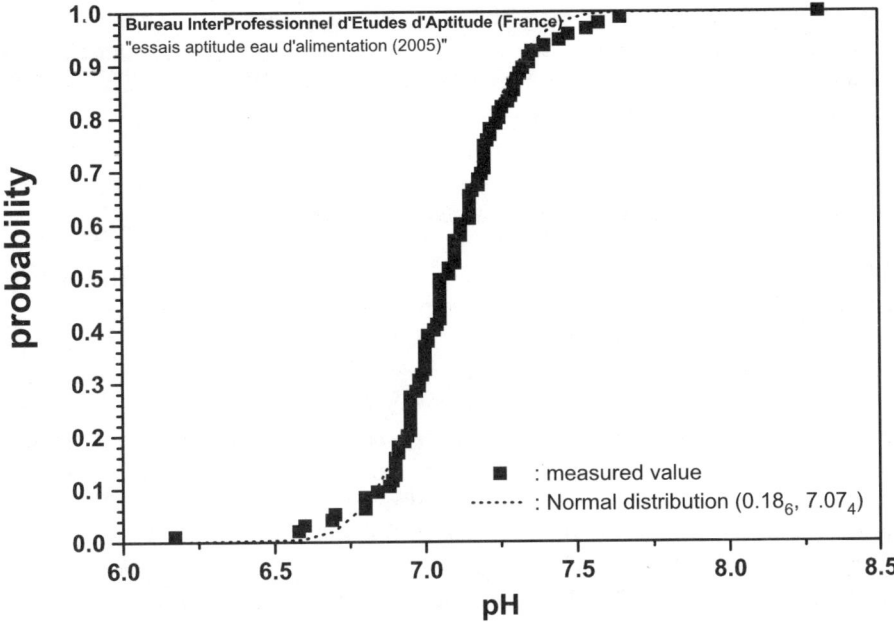

Figure 1.47. Results of a proficiency test "pH in drinking water" under the auspices of French national metrological institute, LNE (2005) with 95 participating laboratories. The results are given as empirical cumulative distribution curve together with its closest fitting normal distribution (using Kolmogorov–Smirnov test). The probability of normality is > 98%

For the existing data, however, these metrological traceability features have not been available. The round-robin results (cf. Figs. 1.45–1.47) give evidence that a value from a measurement does not need to correspond to the true (or conventional true) value. Small variations in an influence quantity may give rise to considerable variation in the measurement value derived for a measurand.

The quantity pH is an excellent showcase example on what is necessary to get an understanding of the accuracy and precision of the SI unit "amount of substance". The quantity pH is the most frequently measured chemical quantity, playing an important role in the medical field, in food and pharmacentical industries, in the management of the environments and in the control and preservation of resources.

It is of crucial importance to understand that measurement of pH values is not an issue for academic sand-box games, but strongly influences our daily life and important decisions from tap water surveillance to climatic change. Often, resources are either limited or the demand exceeds the supply. Various powerful social groups reflect on short-term exploitation of essential resources indispensable for the well-being of others (Price 2002). Within an over-populated world, conflicts can only be avoided or at least mediated by accepted criteria; proven and controlled measurement capabilities are indispensable. Therefore, metrological institutes are essential elements of sovereignty. Their relationships are fixed by international treaties (MRA 1999, IRMM 2000). Deficiencies in such capabilities must rise concern (Jenks 2003). Trust is easily lost but difficult to (re-)install (Zschunke 1998).

1.8.4
Uncertainty Contributions from Metal Ion Concentration Determination

While routine pH measurements are made in a broad range of fields by personnel with a widely varying expertise, measurement of amount of substance of metal ions in aqueous solutions is mostly left to trained analytical personnel. Instrumental methods of analysis, for instance ICP-OES (inductively coupled plasma optical emission spectroscopy) and ICP-MS (inductively coupled plasma mass spectrometry) are known for their often low detection limits at routine measurements. The little information on the measurement uncertainty of ICP-OES determination of metal ions indicates a relative expanded standard uncertainty of $1-3\%$ in the $100\,\mathrm{mg\,dm^{-3}}$ metal concentration region (Drglin 2003). Compared with pH measurements, measurement values from proficiency test of metal ion determination in moderately concentrated aqueous matrices may be expected give a rather homogeneous result.

The International Measurement Evaluation programme (IMEP) is a service offered by the Institute of Reference Materials and Measurement (IRMM), the metrological authority of the European Commission. The programme offers SI-traceable reference values, established by the primary methods of measurement, against which participating laboratories can evaluate their performance. The degree of performance is thus established against the most objective reference possible and available at present. Participating laboratories may work under normal conditions with their own choice of techniques, procedures, and instrumentation. IMEP is open to all laboratories on a voluntary basis, and full confidentiality with respect to the link between results and identity of each participant in guaranteed.

A result from IMEP-6 is shown in Fig. 1.48. In IMEP-6 a total of 180 laboratories from 29 countries participated in the analysis of 14 elements in a natural water. Figure 1.48 shows results for Pb from 75 laboratories declaring to follow guidelines from the quality management system EN 45000/EN 45001 (van Nevel et al. 1998). The central grey bars represent the certified Pb content displayed with the combined relative standard uncertainty U ($k = 2$). Without going too much into detail, the general message from these graphs is a considerable variation in the reported measurement results by participating laboratories, despite the fact that the measurement tasks should be a routine. In everyday life we would not accept, say, a kitchen balance giving values varying on a regular basis by 20%. Most people would sharply protest receiving a wooden stick of, say, 1.20 m when 1.35 m was ordered. Modern industry would not exist if

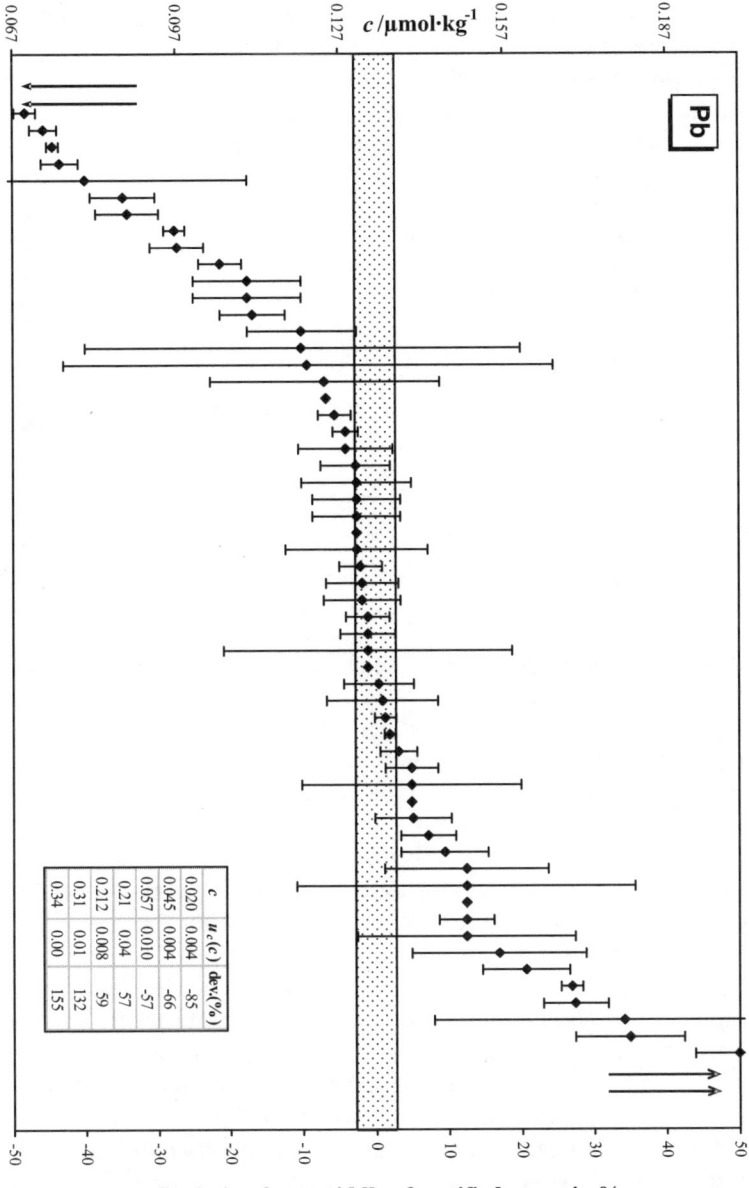

Figure 1.48. Results for Pb in the natural water material as obtained by "accredited/certified/authorised" (self-declaration) laboratories. Results which fall outside ±50% of the reference value are shown in the *text box*. *Grey bar* indicates certified reference value with uncertainty (van Nevel et al. 1998)

assessment of the quantities length, mass and time would carry uncertainties comparable with present-day chemical measurements.

The IMEP example shown in Fig. 1.48 is a typical result of such a study. These studies ask for cautions with routinely determined concentration values. It is certainly necessary to contrast the message of Fig. 1.48 with the evaluation reported by, for instance, Drglin (2003). The determination of metal ion contents in aqueous samples is an essential step in the analysis of solution equilibria. In case of the general equation

$$M^{z+} + L^{y-} \Longleftrightarrow ML^{(z-y)+} \tag{1.103}$$

with formation constant K

$$K = \frac{[ML^{(z-y)+}]}{[M^{z+}][L^{y-}]} \tag{1.104}$$

the value for the standard uncertainty of the metal ion concentration enters twice into the value of the measurand, K. The same is true for the standard uncertainty of the ligand, L^{y-}.

1.8.5
Uncertainties from Subsequent Manipulation, Selection and Recalculation

There is no doubt associated with obtaining a value for a formation constant K for a mean-value based evaluation. Such data are abundant in the literature, and collections are published, e.g. Bjerrum et al. (1957) or Smith and Martell (1989). The OECD/NEA is running a project to collate available data with potential interest in nuclear waste disposal. The OECD/NEA Thermodynamic Database project terms its data elicitation "critical" and guidelines for the evaluation process are issued (OECD/NEA 2000a,b,c). These guides do not include any reference to the normative and internationally agreed metrological documents. The word "metrology" is not found in the three guidelines. Uncertainty evaluation (OECD/NEA 2000b) is restricted to repeatability estimation (mean values and standard deviations). The guide holding the guidelines for data judgement (OECD/NEA 2000c) holds the unfortunate term "independent peer review", implying that there were actually "independent" reviewers available in the highly political field of nuclear waste disposal with a somehow superior ability to recognise "good" data if they see it. This ability must necessarily go with all reviewers of this project with its truly challenging dimensions.

The OECD/NEA thermodynamic data base project is a good example for the difficulties to judge measurement values without a stable framework of traceable references and carefully evaluated measurement uncertainty budgets. Again, the definition of convention is invoked (Lewis 1969) here. A conventional agreement is of course possible and may be a valuable tool in creating a unified data collection to perform comparable studies in the field of modeling and simulation. However, the rules and guidelines imposed by OECD/NEA are not able to create trust into the review process as shown by the heavy dispute about the judgement handed down to some publications on Np (Vitorge 2005). These arguments document the current state in discussing chemical data and are available also on the CD accompanying this book. The conflict of interest within the OECD/NEA reviewer group has not been solved by applying adequate protocols, but by exchange of the reviewer by hierarchical structures within the OECD/NEA data base project. Of course, people outside the population P (e.g. due to death, change of research subject, ignorance etc.) cannot interfere with the evaluation process were persons not involved into the measurement process of the respective measurand value judge mostly peer-reviewed publications upon merely subjective impressions on the basis of a supposed neutrality.

To be meaningful, elicited reviews require access to a comparable amount of information to that used by the author(s) in preparing the manuscript. Up to now, the original data of an experiment are not usually available for subsequent re-analysis. Even worse, as outlined by Peter Medawar (Medawar 1964) in an essay, a scientific paper is always an idealisation which from its very intention does not attempt to give an accurate account of the actual measurement process. It should be seen rather as an idealised account a researcher gives about the object of his studies.

This situation is not a new one. It is well known for measurement values from the era before the introduction of the basic quantities meter, gram and second. The discussion of measurement data did not address the question "which data are right" but "who is right". Data quality was bound to authority, not to criteria. The introduction of the metrological network by the Meter Convention in 1875 has given emphasis to the measurement process. Proper training in handling of measurement devices, demonstration of proficiency and independent calibration services form the current-day basis for trust in weight and measures.

Can we expect that any person accepts an unpleasant decision on basis of conventional data, for instance construction of a waste incinerator plant or even a nuclear waste disposal site in direct neighborhood? It is certainly interesting to ponder the question whether a car-driver would accept loss of his driver license for driving under alcohol if the measurement signal would be judged upon criteria similar to those of thermodynamic data base compilers, especially if it is known that the measurement instrument gives quite variable signals if exposed to the same amount of alcohol. Can trust in a measurement result can be built on authority? Metrology is the only system having proven its ability to create trust in measurement values. The MRA (MRA 1999) is the most obvious sign of this ability.

The SI is the result of the cultural and philosophical development in Europe over a period of several hundreds of years. It incorporates over 2000 years of inquiry into the fundaments of our world (von Weizsäcker 1990). The probabilistic nature of our world is a part of the results of this endeavour. The ISO "Guide to the Expression of Uncertainty in Measurement" (ISO 1993) is based on this experience. A consequence is shown schematically in Fig. 1.49. Two measurements, A and B, are performed to obtain a value for the same measurand. Both measurements do not necessarily need to be performed at the same location and the same time. Both measurements consist of the same basic elements: reference standards are used to calibrate an instrument. The measurement is performed and the experimental data are evaluated using some auxiliary data. The rows of lines represent the distribution of the measurement uncertainty of the respective step. The both measurements are, however, evaluated on a mere mean value basis. The value shown by the measurement equipment is taken for granted.

The examples in the preceding chapters, however, have shown that a single value need not be the true one! In measurements A and B (cf. Fig. 1.49) slightly different mean values are obtained. Because an assessment of the complete measurement uncertainty budget is not performed, the magnitude of uncertainty in each step of the measurement process remains unknown. As a result, the values from both measurements are different, but the fact of a difference in the values is all what can be stated. The likely position of the true value, which was aimed upon, remains obscure. Notwithstanding obvious errors in the experimental set-up, the measurement process or the evaluation,

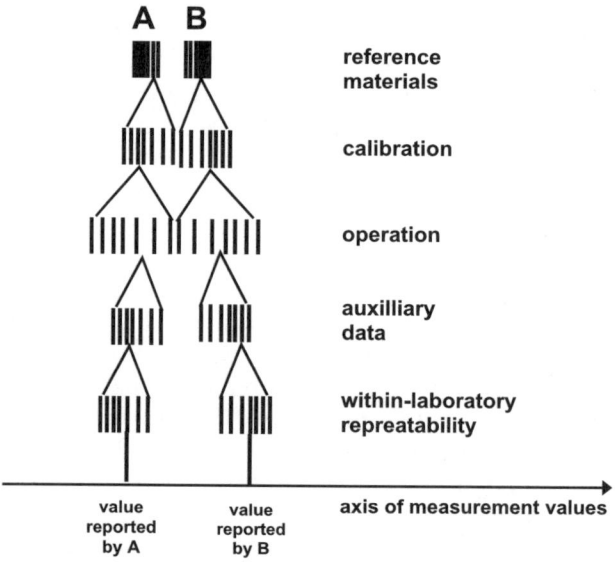

Figure 1.49. A schematic representation of two measurement processes of the same measurand in presence of measurement uncertainty. The true value is aimed on but unknown. Note that this figure is a modification of Fig. 1.5, although without a (known) conventional true value. How far away from the true value may each datum be?

it is impossible to judge the data without detailed information on the influences and effects having caused a deviation.

It may be argued that the both measurement processes in Fig. 1.49 are shown in a way such that the both results separate from each other. Taken the probabilistic background, it is also possible that the obtained mean values at each stage compensate and the final results are much closer. In fact, the set-up shown in Fig. 1.48 is reminiscent of a Galton board where balls form an approximate normal distribution, provided enough balls (measurements) are available to run over a small but sufficient number of steps. In case of a Galton board, however, the balls have only two choices. Therefore, a Galton board is only a crutch to illustrate a complete measurement process. The more complex the measurement process, the less is the probability that two measurement values accidentally coincide.

1.8.6
Overestimation of Measurement Results

In the case of thermodynamic data, a larger number of values for the formation constant of the same species is rarely available. Hence, the difference between various thermodynamic constants may be large, or in a group of a few data one value may be farther away from the other two or three. Removing this supposedly "extraneous" value by some argument is a common "variance reduction" technique. Figure 1.48 illustrates that without an assessment of the measurement uncertainty of each influence quantity, the likely position of the true value cannot be assessed. All variance reduction argument is more or less arbitrary in such a situation. Another situation may arise if measurement uncertainty is evaluated for a limited number of influence quantities, for example repeatability. Repeatability estimates, for example derived on basis of lack-of-fit (cf. Sect. 1.8.2), are sometimes available for a number of thermodynamic data.

An example is given in Table 1.16 for the U(VI) hydrolysis species $(UO_2)_2(OH)_2^{2+}$. Because it is unclear whether the "$\pm$" symbol represents a standard deviation or a confidence region, the individual literature values are considered to represent mean values and standard deviations of normal distributions.

An experiment has been performed where 33 U(VI) solutions in sulfate medium at room temperature have been characterised UV-Vis spectroscopically (Meinrath et al 2006). The respective spectra can be found on the book CD. The data analysis was performed by target factor analysis in combination with a threshold bootstrap scheme using computer-assisted target factor analysis (TB CAT) (Meinrath and Lis 2001). The respective cause-and-effect diagram is given in Fig. 1.50. As has been discussed in detail elsewhere, the majority of effects given in the fishbone part of the diagram can be neglected. Important uncertainties in influence quantities are the free OH^- concentration, the total sulfate concentration and the U(VI) concentration. The ISO Type A un-

Table 1.16. Comparison of some formation constants lg K_{22}^a available in literature for 0.1 M perchlorate solutions and ambient temperatures (Meinrath et al. 2006)

Method	lg K_{22}	Reference
P	-5.95 ± 0.04	1
P	-5.89 ± 0.37	2
P	-5.85 ± 0.02	3
P	-6.09 ± 0.06	4
P	-6.09 ± 0.02	5a
S	-6.28 ± 0.02	5b
S	-5.97 ± 0.06	6
S	-6.14 ± 0.04	7
S	-6.17 ± 0.03	8
S	-5.80 ± 0.20^b	9

[a] For reaction $2 H_2O + 2 UO_2^{2+} \Longleftrightarrow (UO_2)_2(OH)_2^{2+} + 2H^+$
[b] Recalculated from lg $\beta_{22} = 21.76 \pm 0.11$ with lg $K_w = -13.78 \pm 0.02$
P potentiometric titration; S UV-Vis spectroscopy
1 Overwoll P, Lund W, Anal Chim Acta 142 (1982) 153
2 Maya L, Inorg Chem 21 (1982) 2895
3 Vainiotalo A, Mäkitie O, Finn Chem Lett (1981) 102
4 Tsymbal V, Rapport CEA-R-3476, CEA, Saclay (1969)
5a,b Bartusek M, Sommer L, Z Phys Chem 67 (1963) 309
6 Meinrath G, Kato Y, Yoshida Z, J Radioanal Nucl Chem Articles 174 (1993) 299
7 Meinrath G, Schweinberger M, Radiochim Acta 75 (1996) 205
8 Meinrath G, Radiochim Acta 77 (1997) 221
9 Meinrath G, Lis S, Piskula Z, Glatty Z; J Chem Thermodynamics 38 (2006) 1274–1284

certainties are given in the right hand side box. These must be evaluated by computer-intensive statistics. The threshold bootstrap has been implemented to evaluate the spectra.

In Fig. 1.51, a comparison of the resulting probability density of the formation constant lg K_{22} species $(UO_2)_2(OH)_2^{2+}$ with the normal distributions obtained from the literature values given in Table 1.16 is shown. The distribution of the complete measurement uncertainty budget is much wider than the data given in Table 1.16. Upon inspection of the normal distributions representing the literature data, only the curve of Maya (cf. legend of Table 1.16) is comparably distributed and strongly overlaps with the complete measurement uncertainty distribution.

The other normal distributions are much narrower and imply a higher precision; however, in several cases, the normal distribution do not overlap with each other! Thus they imply a large discrepancy among each other. This observation is not uncommon. The most likely interpretation can be directly taken from an Editorial by P. de Bièvre (2003): *Whether chemists like it or not, most of the "error bars" which accompanied results of chemical measurement in the past were too small if they were to intend to be "accuracies". They represent*

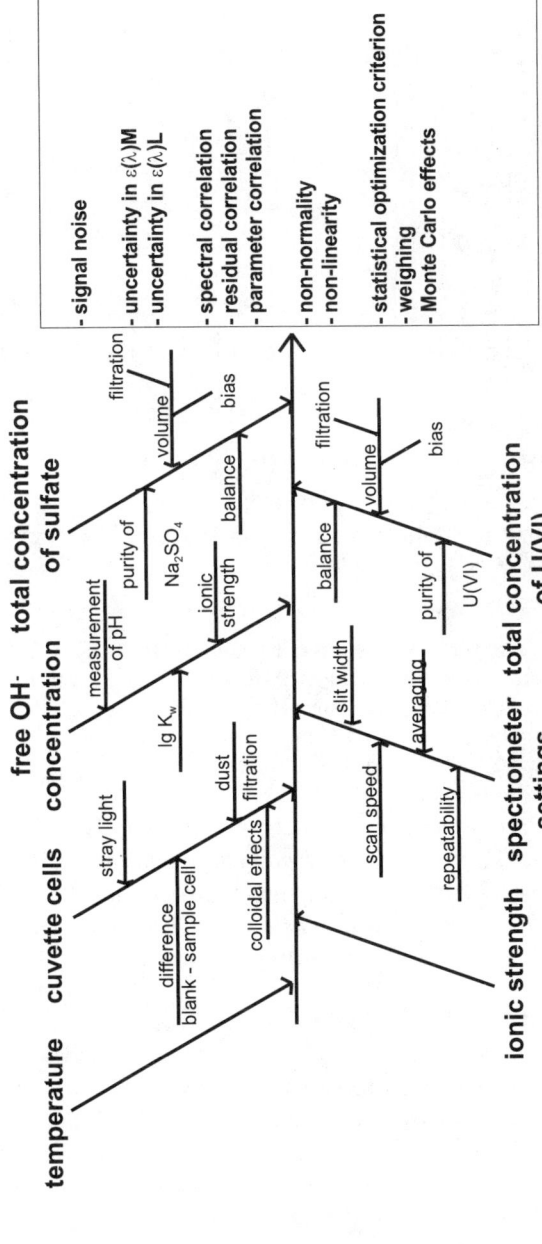

Figure 1.50. Cause-and-effect diagram of a UV-Vis spectroscopic investigation of U(VI) in sulfate medium at pH ~ 3.5

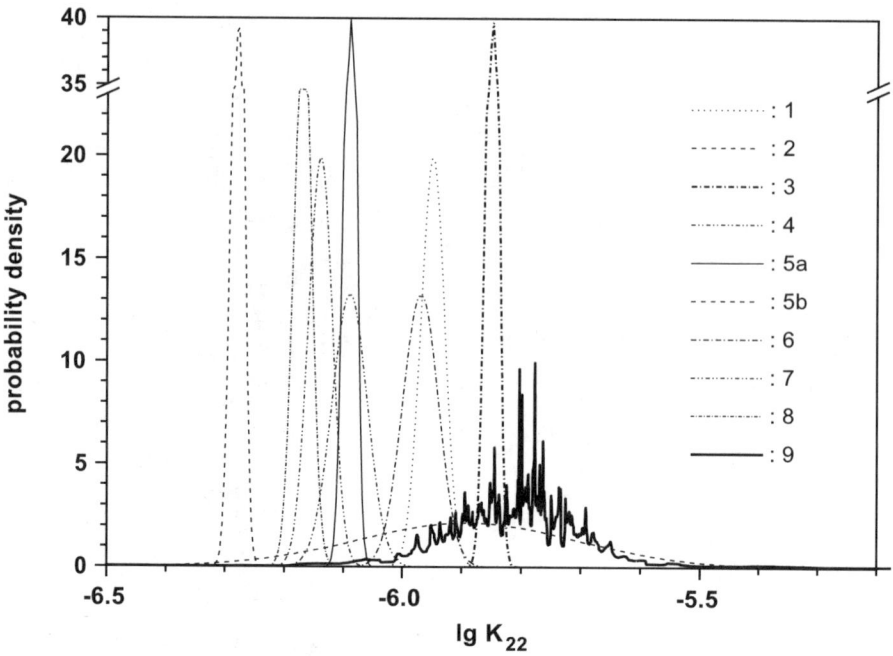

Figure 1.51. Comparison of the values reported in literature for lg K_{22} (cf. Table 1.16) with the results from this study. Literature data is transformed into the probability densities (Normal distributions; *light curves*) for comparison with the probability density of the complete measurement uncertainty (*thick curve*)

repeatabilities or reproducibilities of measurement results, but rarely, if ever, "total" uncertainties. There are several possible explanations for this. One of them is that the interpretation of the definition of measurand as the "quantity subject to measurement", enabled them to do so. The intended or announced measurand was not meant (e.g. concentration), but the actual "quantity subject to measurement" (e.g. a ratio of electric currents in a spectrophotometer). Thus, "uncertainty" in many chemical measurements was reduced to repeatability or reproducibility of measurements of electric currents, generated by ionisation processes, photomultipliers, etc. It was very convenient to "omit" the uncertainty related to the chemical preparation of the samples, since that fell outside the definition "quantity subject to measurement" (the electrical currents). It was more gratifying to be able to present good repeatabilities, since that provided prestige for the measurement laboratory: the "error bars" looked good because they were small. All of which comes to the surface and becomes clearly visible in interlaboratory comparisons (ILCs) where the results of measurements of the same quantity in the same material are displayed with "error bars" representing repeatabilities only. All of a sudden, a number of results show up as being "bad" because their "error bars" do not overlap. In many cases they even do not cover the *reference value of the ILC (when available). Had a more complete "uncertainty" been assessed by way of larger "error bars", the result would not only look all right, it would also be all right. The price to pay was that it did not look so attractive.* There remains nothing to add.

The example of $(UO_2)_2(OH)_2^{2+}$ in U(VI) sulfate solutions does not represent an ILC, neither is the derived value for lg K_{22} claimed to be traceable. However, the lacking overlap of uncertainty ranges reported with thermodynamic data from literature (if available at all) is striking. For most data in literature, where some important details of the measurement process remain unclear, a metrological post-assessment of the complete measurement uncertainty budget is difficult. Such an effort seems, however, to have more merits compared to a visual selection of "good" and "bad" data by review panels that do not have (documented) metrological and statistical experience.

Hence, as long as non-traceable measurement values remain inside the laboratory of determination or within a group P accepting such data on a conventional basis, no other party is affected. Such data may serve as a common basis of discussion. As soon as others are affected by decisions based on these data, the conventional basis will break in case of conflicting interest. Hence, data collections on the basis of reported thermodynamic data have enormous deficits preventing their application in controversial situations. The few examples summarised in this chapter also illustrate that conclusions based upon such data may be misleading.

1.8.7
Summary: The Need for Performance Assessment of Analytical Chemistry Methods

"What can we know? or what can we discern, when error chokes the windows of the mind?"

(J. Davies, 1570–1626)

Measurement methods do have a limited resolution power. Ignoring this limitation may have unforeseeable consequences. In most cases, however, the consequences do not directly contribute to scientific progress. This fact may be illustrated by an example from astronomy.

In 1877, the Italian astronomer G. Schiaparelli reported the observation of straight line structures on planet Mars. He referred to these structures as "canali", meaning channels but being mistranslated into "canals". These canals were not accepted undisputedly (e.g. Evans and Maunder 1903). However, the idea of intelligent life on other planets captured the public and scientific attention for several decades. The subject developed, mainly due to the efforts of the US astronomer P. Lowell. Schiaparelli in 1895 wrote the book "La vita sul pianeta Marte" (Gutenberg Projekt 2006). Within the following 50 years, the idea of artificial channels on Mars was intensively debated (e.g. Antoniadi 1910; Comas Sola 1910). The idea of canals as the product of an intelligent Martian civilization slowly faded in the first decade of the twentieth century. The photos taken by Mariner 4 in 1965 finally gave direct evidence of the absence of large-scale engineered structures on Mars.

The example of Martian canals can, nevertheless, be fruitful for the current discussion on performance assessment of analytical methods in chemistry. It was clear already in 1877 that the structures described by Schiaparelli and others were on the edge of the resolution capacity of the instruments. Observations of the canals were made during the period of closest approaches between Earth and Mars. The list of positive observations is long, despite the ongoing disputes. The "canals" case also shows how expectations influence the observations and their interpretation. Fig. 1.52 shows two Mars drawings by K. Graff, director of the Hamburg observatory, after 1928 the Vienna observatory. His visual observation abilities still are legendary among astronomers.

Graff's drawings (from observations with a 60 cm refractor) show the influence of the "canals" idea in 1901, reported as "clearly seen", which in the course of the years (there are similar drawings until 1928) disappear more and more.

The idea of Martian canals nevertheless has been fruitful, culminating in HG Wells' "War of the Worlds" (Wells 1898) and the "little green Martians" have become an element of Western culture. But the truth is different: there are no canals on Mars.

It may be argued that the Mars canals error is a result of purely visual observation, a highly subjective process. The collection of numerical data has its epistemological importance from their objectivity being obtained by measurement tools instead of human senses. The foregoing discussion, however, shows that this objectivity is a relative one. An experiment is a question to nature. This question can only be asked with a relative accuracy, and the response given by nature rarely answers exactly the intended question. So what information is in the data? To extract the likely answer from a collection of data, the range of possible answers has to be tested. Appropriate tools have been are being developed. The field of science dealing with this task is statistics.

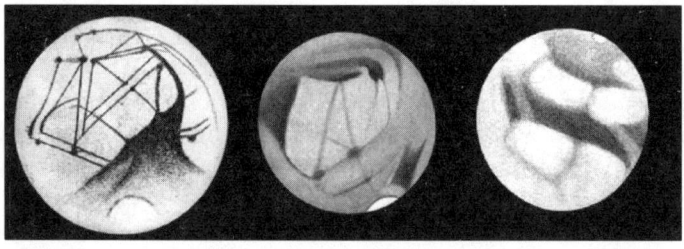

Schiaparelli 1877   **Graff 1901**   **Graff 1909**

Figure 1.52. The "canali" on Mars in a drawing by G. Schiaparelli (*left*) and observations of K. Graff at the Hamburg Observatory. The 1901 drawing is still influenced by the "canali" interpretation, while Graff's 1909 drawing does not any longer indicates straight lines (Schramm 2006)

"Most people are not natural-born statisticians. Left to their own, human beings are not very good at picking out patterns from a sea of noisy data. To put it the other way, we all are too good at picking out non-existent patterns that suit our purposes and prejudices" (Efron and Tibshirani 1993). To reduce the risk of overinterpretation and underestimation of information in a given data set, it is important to understand the performance of an experimental method. Currently, techniques in analytical chemistry are qualified according to overall performance parameters, e.g. limit of detection, recovery. Noise and random influences do not play a role in these performance assessments. It is, however, essential to have a clear understanding of the interpretation limits within a given data structure. Otherwise, misinterpretation is unavoidable.

The following Part II will deal with deficits and criteria for thermodynamic data and the methods used for their determination. Here, homogeneous solution parameters and surface interaction parameters will be used as examples. For a limited number of methods for thermodynamic data determination, cause-and-effect diagrams have been evaluated and applied, e.g. solubility measurements (Meinrath 2001), UV-Vis spectroscopic measurements (Meinrath and Lis 2001, 2002; Meinrath et al. 2004, 2006, 2006a) and potentiometric titration (Kufelnicki et al. 2005; Meinrath et al. 2006b). These reports may serve as a starting base in the discussion of deficits and criteria presented later on.

"Door meten tot weten" (Through measurement to knowledge)

(H. Kammerlingh-Onnes)

This summary must begin with a clarification. Von Weizsäcker (1974) said about Plato: "...the thoughtlessness has not be done to say: I know the answer and I will write it down for you" (Lindner 2002). Likewise, the exact details of a way to better, more accurate and more reliable thermodynamic data is not known to the authors. Ignoring uncertainty, however, is not a means to trustable measurement values, either. It is evident that a more detailed knowledge on the accuracies and precisions of currently available chemical measurement data are urgently needed for decision-making in politically sensitive situations. There is little chance to gain progress if the current status cannot be assessed by hard, objective criteria.

It is one of the successes of a science to become part of daily life, to play an important role in the society and obtain public funding not as a kind of alms but as an investment for future profit. All sciences need to justify public funding. Science is not funded for the progress in the past – there is heavy competition for a society's resources. Sciences have to prove their relevance and their continued ability to contribute to the social and economic stability of the society.

The deficit analysis of thermodynamic has made evident the current state to which the measurement of thermodynamic data currently contributes to knowledge. There are two ways to judge the situation. The first, pessimistic, view is depression. Most thermodynamic data satisfy the criteria to be seen as "rumours". There may even be an impetus to bash the messenger despite the fact that he illustrates and underpins this message with examples and graphs: a defeatist.

The second way is a happy one. Thermodynamic data, for many years, have been determined to characterise the elements and classify them with respect to their properties. The Periodic Table of the elements and the Pearson concept of soft and hard acids and bases are but a few examples. In the pre-1980s, nobody was thinking about computer-based modeling. The years since 1980 have anyway brought forward new insight and new capabilities. Some of the mathematical, numerical and statistical methods used in the deficit analysis have been developed in this period, the computing abilities have rocketed and still increase. There is plenty of new knowledge to gain with more reliable data, new methods of determination and evaluation.

For the time being the following statements can be defended:

- On the basis of existing thermodynamic data, predictive geochemical modeling is not reliable.
- Neither the analytical details concerning the determination of a certain thermodynamic datum nor the details concerning the evaluation of a certain thermodynamic datum are sufficient in almost all cases.
- The values of many (most?) thermodynamic data are random numbers whereby the random number-generating process is unknown.
- It is comparatively simple to demonstrate that some thermodynamic data reported in literature or having passed the scrutiny of a PhD procedure are not the result of a detailed and critical data analysis.

A major reason to these statements is the simple fact that the most basic element in the determination of a value has been ignored: A measurement is a comparison. No comparison can be made with arbitrary accuracy and precision. Hence, the determination of thermodynamic data, as an element of analytical chemistry, is unavoidably associated with uncertainty and bias. These contributions must be quantified. Otherwise, all data analysis will go astray into an unpredictable and uncorrectable direction.

Whatever the best way might be to analyse and to report measurement uncertainty, the chemical sciences have frittered away the time to come forward with an own viable and demonstrated proposal to account for measurement uncertainty of its genuine SI unit, amount of substance. Now, the ISO Guide to the Expression of Uncertainty is the valid document. Its language is the world-wide metrological basis to communicate the reliability and quality of measurement values. EURACHEM/CITAC (1998, 2002, 2004) has shown approaches to comply with these international requirements in the field of analytical chemistry. In some countries, courses in metrology in chemistry are offered within chemistry curricula.

2 Metrology in Chemistry and Geochemical Modeling

2.1
Geochemical Modeling – An Introduction

The previous section was devoted to some fundamental and general aspects of metrology in chemistry. Illustrating examples from chemical measurements have been discussed. Metrology in chemistry is not different from metrology in other fields, e.g. manufacturing or physics. The addition "in chemistry" wants to emphasize that application of metrological principles is a rather new topic in chemistry. Metrological concepts apply wherever comparisons are made, and measurements are comparisons. Comparisons become especially important when made relative to the quantity "amount of substance". The importance results from the complexity of manipulations and transformations often necessary to arrive at a value for the quantity "amount of substance". Determination of values for the quantity pH by two-point and multi-point calibration and the evaluation of formation constants for the formation of chemical species in aqueous solutions by a solubility experiment were shown in Part I to illustrate this complexity and its consequences. A value is influenced by multiple, sometimes purely numerical, effects, for example, non-normality, correlation in influence factors, correlation in residuals, the optimization criterion itself.

From the viewpoint of determining values of chemical quantities, it may be satisfactory to report merely mean values. However, the limitations of such an approach becomes evident if data must be compared which result from different measurements (Hibbert 2006). The question whether the quantity pH of a natural water (for instance water percolating from a source in the environment) has changed over a certain period of time is an illustrative example. The difference in the values does not necessarily indicate a change in the water composition. Only if the observed change is larger than the resolution capacity of the analytical method such a change has been substantiated. With a known resolution capacity, fitness-for-purpose of an analytical method can be established.

In such situations an experimenter therefore requires two pieces of information: a) what is the resolution power of the experimental method and b) what is the required quality of an experimental method. The term "resolution power" refers to the minimum difference between two values which cannot result from random influence factors but must be a property of the two items that are compared. If the observed change between the two measurement values is large, even a poorly resolving method may be satisfactory (e.g. indicator tips in case of pH). If the observed change is small, the question arises whether the observed difference is by chance or significant. Another similar situation arises from compilations of thermodynamic data. Often, several published values are available for a reaction of interest. In the vast majority of cases, the values are different. Notwithstanding obvious misinterpretations and errors, a basis is required on which the published data are judged. If the experimenters reporting the data themselves have not carefully studied the likely variability of their data, which other criterion could be applied, except the belief of the compilers to recognize a "good" value if they see it? A realistic estimate of the uncertainty is even more important if further conclusions and/or decisions are based upon this value. Predictive geochemical modeling is a good example where all three aspects, a) fitness-for-purpose of analytical methods, b) comparability of measurement values over time and space and c) progression of uncertainty of individual measurement values in subsequent application of these values for decision making etc., are relevant issues.

The focus on geochemical modeling may highlight another relevant aspect. In almost all but the most simple settings (for instance by laboratory column experiments) the results of geochemical modeling cannot be verified. Geochemical modeling is often applied because the dimensions of the problem do not allow direct experimentation (e.g. because the time scales are too long), or the accessibility of the zone of interest (e.g. a plume in the subsurface) is poor. Economic considerations are an additional aspect favouring computer simulations.

A decision maker (cf. Fig. 1.14) has to arrive at a conclusion within a limited time. Scientific arguments are only one factor. Other sources, e.g. social, political and economic arguments, may play the prevalent role. Scientific studies may support consensus-finding because scientific arguments are often considered as "objective", less influenced by wishful thinking and vested interests.

Pondering the importance of a reliability criterion for a computed prediction, a broad range of studies might be expected exploring the reliability of geochemical modeling results. In fact, a variety of studies are discussing the possibilities of improving the predictive power, to rank the input quantities according their influence on the modeling output (sensitivity analysis), and to underscore the significance of calculated results. There are, however, only rather few studies available investigating the uncertainty of the output from geochemical modeling calculations themselves. None of those studies aims at the complete measurement uncertainty budget. As it has been put elsewhere: "The number of persons that examine the uncertainties in simulation are few compared to the number of persons who make simulations without uncertainties" (Ödegaard-Jensen 2006).

This finding may seem surprising. Some arguments may hint on possible reasons. First, while statistical hydrology extensively studies the impact of distributed input on the distribution of output (e.g. Fortin et al. 1997), the number of input quantities increases drastically if chemical transformation processes are combined with groundwater flow. For most quantities characterising the chemical reactions, reasonable estimates of uncertainties (not to mention complete input distributions) are unavailable for reasons discussed already in Part I. Second, the numerical burden in geochemical modeling is drastically higher compared to a mere hydrogeological simulation. It should be kept in mind that the systems treated by hydrogeology, involving advection, convection, dispersion, and diffusion, are by no means trivial. Treatment of these processes simultaneously with chemical changes increases the dimensions of a system tremendously. Chemical changes are described by sorption, decay (chemical decay and radioactive decay), generation of substances (e.g. by dissolution of gases or anthropogenic input) and reactions. Already the calculation burden just to generate output from a single input parameter set is considerable. Third, the audience may not be interested in a discussion of uncertainties at all. A decision maker will certainly prefer clear-cut information and tend to avoid a discussion of prediction uncertainties. Likewise, organisations funding related research will prefer projects promising more efficient solutions, compilation of better data bases and faster algorithms, instead of work implementing known methods to arrive at clumsy distributions instead of clear numbers with many decimals. A discussion on how "more efficient", "better" and "faster" might be objectively defined is almost painstakingly avoided. Occasionally qualifiers like "verified" or "validated" can be found, however, without definition of their meaning. Fourth, pondering the almost complete absence of traceable chemical measurement values (e.g. formation constants, solubility products, sorption coefficients for the various sorption models, kinetic data) in literature, the assessment of a complete measurement uncertainty budget for these values is a necessary requirement, not an accomplished one. Therefore, at today's state of metrology in chemistry, a geochemical modeling output satisfying a metrologist's requirements is a goal lying in a distant future. But this insufficient situation does not justify the overinterpretation and underestimation characterizing today's discussion of geochemical modeling output results.

Part II will introduce geochemical modeling. The numerical tasks of groundwater simulation and the fundamental techniques to solve these tasks will be introduced briefly. Groundwater modeling is a very developed field. The descriptions necessarily will be superficial. Geochemical modeling will be introduced as an extension of groundwater modeling. Groundwater modeling is a more recent development; more or less it became technically feasible by the enormous increase in the CPU clock rates and digital storage capacities. Geochemical modeling introduces chemical interactions into groundwater modeling. Thus chemical information obtained from experimental measurements enters the computer codes. The impact of metrology and, at the current state-of-the-art, the resulting deficits of contemporary geochemical modeling will be outlined. Finally a deficit analysis will be given and proposals are made how to improve the situation in the short and middle term.

2.1.1
Groundwater Models

"He who has heard the same thing told by 12 000 eye-witnesses has only 12 000 probabilities, which are equal to one strong probability, which is far from certain."

(Voltaire)

This chapter intends to introduce into the basic concepts of geochemical modeling. There is no intention to provide an exhaustive, systematic outline. For such a purpose appropriate texts are available (Kinzelbach 1986, 1989; Holzbecher 1996; Hardyanto and Merkel 2006). For the following discussion, it is sufficient to distinguish between hydrogeological models and geochemical models. A hydrogeological model is intended to predict groundwater flow. The empty spaces in the subsurface, having a wide range of spatial dimensions with a large variety in their mutual connectivity, can be constantly or temporarily filled with water. Those spaces being constantly filled are termed groundwater zone, the temporarily water-filled spaces comprise the vadose zone. The water forms the mobile phase, while the material forming the walls of the empty spaces is the stationary zone. For the present discussion, the stationary phase is generally the geological formation. The water-filled spaces in the geological formations are aquifers. Water flowing through an aquifer will be physically affected, for instance due to viscosity changes or temperature variabilities. In order to describe the variability in an aquifer's empty spaces, the parameter porosity is defined. The porosity P gives the relative amount of empty space in a defined volume, whereby only water-accessible spaces may be considered. Hence the enormous variability in the natural subsurface is described by a dimensionless figure characterising a given volume.

A key element of any hydrogeological model is the Darcy law, shown in Eq. (2.1):

$$\frac{\Omega}{A} = K_f \frac{\Delta h}{L} \tag{2.1}$$

where Ω is the amount of water flowing and A is the cross-section of the aquifer. The quotient Ω/A is termed the Darcy velocity v. L gives the length of the aquifer over which the flow is considered while h is the height. Hence, the Darcy law assumes a flow driven by gravity. The K_f value is a property of the porous medium and can be described by Eq. (2.2):

$$K_f = \frac{k \delta g}{\mu} \tag{2.2}$$

with k: permeability, δ: density of water, g: constant of gravity, μ: viscosity. Equation (2.2) shows that the parameter K_f directly depends on the permeability k. The hydraulic conductivity K_f is defined by Eq. (2.1). The Darcy velocity v may be decomposed into the three spatial dimension v_x, v_y and v_z, given in Eq. (2.3) in its infinitesimal form for illustration

$$v = -K_f \begin{pmatrix} \frac{\partial h}{\partial x} \\ \frac{\partial h}{\partial y} \\ \frac{\partial h}{\partial z} \end{pmatrix} = -K_f \nabla h .$$

(2.3)

If the aquifer is anisotropic, the scalar K_f is replaced by a tensor $\mathbf{K_f}$. The pressure p of a water column is given by Eq. (2.4):

$$p = h \, \rho \, g$$

(2.4)

With Eq. (2.2) an alternative form of Eq. (2.3) can be obtained in Eq. (2.5):

$$v = -\frac{1}{\mu} k (\nabla p - \delta g)$$

(2.5)

where k is the permeability tensor. Equation (2.5) is valid for the saturated zone and allows a description of the Darcy velocities in a three-dimensional homogeneous section of three-dimensional space.

The flow in an aquifer is furthermore characterized by the conservation of mass. Water is considered as an incompressible medium. If an amount of water is flowing into a saturated volume, the same amount of water has to leave this volume. From a detailed development of this condition Eq. (2.6) results:

$$\delta S P = -\nabla \cdot \delta v + q$$

(2.6)

where S is the saturation (relative amount of water in a porous volume; S is 1 in case of a saturated zone) and q represents any source or sink. Equation (2.6) is one of several alternative forms of the time-dependent differential equation describing the flow of water in a porous medium, where the stationary phase is time-independent. Depending on the boundary conditions, Eq. (2.6) is further modified. Only the movement of water is described by Eq. (2.6). Application of Eq. (2.6) to real problems results in numerical systems where analytical mathematical solutions are not available. Consequently, alternative methods are required to obtain at least approximate solutions. Commonly, spatial discretisation is used.

2.1.2
Discretisation Methods for Space

Instead of continuous equations describing the variables at all points in space, discretisation implies that the variables are defined only at a limited number of points in space. The spatial locations where the variables are defined are called a mesh or a lattice. The mesh can be one-dimensional (a line of intervals), two-dimensional (a mesh) or three-dimensional (a lattice). The various intervals may be equally (equidistant) or irregularly spaced. A mesh or lattice may be made up of the same geometrical shape (triangle, square or cube, tetraeder or prism) or from different geometrical elements.

Four discretisation methods are commonly applied: finite differences, finite elements, finite volumes and compartment models. For the following discussion it is not important to understand the details of the discretisation methods. It is, however, important to understand that no chemical transformation is considered. Exclusively the transport of the mobile water phase is treated.

2.1.2.1
Finite Differences

For applying finite difference methods, the differential quotients in Eq. (2.6) are replaced by difference quotients. Then an approximation for the second order derivatives is given by Eq. (2.7).

$$\frac{\partial^2 f}{\partial x^2} \approx \frac{f(x_1 + \Delta x_1) - 2f(x_1) - f(x_1 - \Delta x_1)}{\Delta x_1^2} \tag{2.7}$$

where x is the space variable and x_1 the location where the function f is evaluated. Note the approximation sign, $\approx$. The degree to which the right side of the approximation corresponds to the left side, termed truncation error, is a field of intense research and a source of a large variation of algorithmic details. Equation (2.7) is valid only in a homogeneous situation where the coefficients contributing to f are independent of x. If the coefficients, e.g. in an inhomogeneous porous system, the approximation becomes more complex:

$$\frac{\partial}{\partial x} K_x \frac{\partial h}{\partial x} = K_x \left(x_1 + \frac{\Delta x_1}{2} \right) \frac{f(x_1 + \Delta x_1) - f(x_1)}{(\Delta x_1)^2}$$
$$- K_x \left(x_1 - \frac{\Delta x_1}{2} \right) \frac{f(x_1) - f(x_1 - \Delta x_1)}{(\Delta x_1)^2} , \tag{2.8}$$

where K_x gives the hydraulic conductivity along the space coordinate, whereby the "space" is one-dimensional. In a one-dimensional situation, the discretisation points can be located at the interval borders or in the center of an interval. In two- and three-dimensional space, the difference equations can be evaluated at the intersections of the mesh (node-centered) or at the centers of each block (block-centered).

Finite differences are applied in many hydraulic simulation codes, despite the fact that the consistency of the numerical formulations are debated (Aziz and Settari 1979). Simulation codes on basis of a finite difference approximations have shown to be convergent if the lattice is sufficiently smooth.

2.1.2.2
Finite Volumes

The key idea of the finite volume approximation is the mass conservation in a space volume. Mass is a conservative variable – at least within the rounding errors of a numerical processor. Finite volumes are often discussed as a generalisation of finite differences, whereby the method can be applied to irregular lattices and even curvilinear boundaries (Narashiman and Witherspoon 1976, 1977, 1978; Narashiman et al. 1978; Pickens et al. 1979; Peyret and Taylor 1985). An important step in finite volumes is the calculation of means. Instead of providing the analogous equations of Eq. (2.8), a short discussion of forming a mean will be given.

Means. A characteristic statistics of a set of data is the mean. Given a set of n values, three different means can be obtained: the arithmetic mean Eq. (2.9)

$$\overline{X}_a = \frac{1}{n} \sum_i x_i ,$$ \hfill (2.9)

the geometric mean Eq. (2.10):

$$\overline{X}_g = \sqrt[n]{\prod_i x_i}$$ \hfill (2.10)

and the harmonic mean Eq. (2.11):

$$\overline{X}_h = \frac{n}{\sum_i \frac{1}{x_i}} .$$ \hfill (2.11)

For the three values $(3, 6, 9)$ the means are $\overline{X}_a = 6$, $\overline{X}_g = 5.451$ and $\overline{X}_h = 4.909$. The arithmetic mean is the most familiar of the mean values while the harmonic mean is occasionally interpreted as "weighted arithmetic mean". This weighing, however, may have considerable influence on the simulation output. Holzbecher (1996) provides an illustrative example underscoring the importance of mean value formation in finite volume approximation. The numerical data is shown in Table 2.1, where the Darcy velocity is calculated for varying hydraulic conductivity K_2 in a two-layer system by a finite volume approximation using different means. While by using the arithmetic mean the overall hydraulic conductivity is almost unaffected by the changing hydraulic conductivity in the bottom layer, this picture is drastically modified if the mean is calculated as harmonic mean.

While the Darcy velocity is apparently constant in case of the arithmetic mean, the Darcy velocity is drastically reduced when K_2 decreased in case of the harmonic mean. This latter observation is physically reasonable: the lower the

Table 2.1. Numerical values for the Darcy velocity in the bottom layer of a two-layer system, calculated for two different methods to obtain a mean. The hydraulic conductivity K_1 in the top layer is fixed while for K_2 different values are applied

K_2	10^{-5}	10^{-6}	10^{-7}	10^{-8}	10^{-9}	10^{-10}
$\overline{X}_a$	$5.5 \cdot 10^{-5}$	$5.05 \cdot 10^{-5}$	$5.005 \cdot 10^{-5}$	$5.0005 \cdot 10^{-5}$	$5 \cdot 10^{-5}$	$5 \cdot 10^{-5}$
$\overline{X}_h$	$1.818 \cdot 10^{-5}$	$1.980 \cdot 10^{-6}$	$1.998 \cdot 10^{-7}$	$2 \cdot 10^{-8}$	$2 \cdot 10^{-9}$	$2 \cdot 10^{-10}$

$K_1 = 10 \cdot 10^{-4}$; X effective Darcy velocity over layers K_1 and K_2

hydraulic conductivity, the less the amount of water is flowing. Furthermore, calculation of a mean may have influence on the propagation of rounding errors etc. Numerical stability is a key issue in modeling complex systems with computers because the limited numerical accuracy in combination with an enormous number of calculations may lead a code astray easily.

2.1.2.3
Finite Elements

Finite element methods decompose the differential Eq. (2.6) into linear combinations of basic functions. To chemists this approach is familiar from quantum chemical calculations where the wave functions ψ are decomposed into linear combinations of known, hydrogen-like wave functions. In a formal manner, Eq. (2.12) describes the decomposition step:

$$f(x, y, z, t) = \sum_{i=1}^{n} \lambda_i(t) f_i(x, y, z) . \tag{2.12}$$

The (time-dependent) coefficients λ_i give the weight of each of the N basis functions. While in case of quantum chemical problems the basis functions often are approximate hydrogen-like orbital functions, the basis functions in a finite element approximation are piece-wise linear or quadratic functions of the type given in Eq. (2.13):

$$f_i(x, y, z) = \begin{cases} \alpha_{ik} + \beta_{ik} + \gamma_{ik} & \text{in neighboring elements} \\ 0 & \text{elsewhere} \end{cases} \tag{2.13}$$

where the three coefficients α, β, γ denote the vertices of a triangular element and k describes the index of the element connecting to the i-th element. The simple form of the basis functions allows fast differentiation.

The solution of the time-dependent flow Eq. (2.6) is thus reduced to the solution of systems of linear equations. In case of finite elements the resulting matrices are also sparse. Efficient numerical algorithms exist to solve these systems of equations. An important point for finite element approximations is the non-conservativity of the resulting mass balances.

2.1.3
Discretisation Methods for Time

The analytical solution of a system of differential equations allows calculation of the state of the system at any time t, if the conditions at a time t_0 are known. In a discrete representation of the system, an analytical solution is not available. The conditions at a time t_i (with $t_i > t_0$) must be obtained by iteration. The time between t_0 and the time of interest t_i is divided into m time intervals, Δt. Starting from t_0, the state of the system at $t_0 + \Delta t$ is evaluated. At the next step, the system conditions at $t_0 + \Delta t$ are used as starting conditions for the calculation of the subsequent time step, at $t_0 + 2\Delta t$. After m iterations the condition at t_i is obtained. This simple iterative scheme is occasionally termed the Euler method.

In order to arrive at the new state at time $t_0 + k\Delta t$ ($k = [1, m]$) the time dependence of the variables must be known. The operator T (the 'recipe') holds the information on the time dependence of the variables. In other words:

$$\frac{\partial f}{\partial t} = T(f) \, . \tag{2.14}$$

Simple solutions can be obtained by using a weighted summation:

$$\frac{f(t + \Delta t) - f(t)}{\Delta t} = \kappa T(f(t)) + (1 - \kappa)T(f(t + \Delta t)) \, . \tag{2.15}$$

If $\kappa = 1$ the forward Euler method, for $\kappa = 0$ the backward Euler method is obtained. If $k = 0.5$, the method is termed the Crank–Nicolson method (Crank and Nicolson 1947).

Another widely used method for time discretisation is the Runge–Kutta approach. Here a time step is divided further into sub-steps. A variant of Runge–Kutta method is the Heun procedure (Hermann 2004). The variety of methods for time discretisation indicates that the methods have implicit limitations while clear criteria to judge the different discretisation methods are lacking. Higher numerical stability usually requires more calculations while each calculation may fail due to computational instability. Therefore, stability and rounding-off errors are important aspects of all implementations (Stoer 1976; Meis and Marcowitz 1978).

2.1.4
Compartment Models

Groundwater models describe the flow of water in a porous medium. The description is based on lumped parameters like the porosity or the hydraulic conductivity. The subsurface is not available for direct study. Any groundwater model must be based on a limited, often rather sparse amount of direct experimental evidence. The flow rate from a well or the pressure in a piezometer are such quantitative information on which a groundwater model is calibrated. Commonly the only validation criterion of a (often complex) model are a limited numbers of experimental information, e.g. flow rates or piezometer levels. A groundwater flow model is a simple calculation stating only that in a given volume of a porous material, the amount of fluid mass that flows in, less the amount of fluid mass that flows out, gives the change in fluid mass in the volume (Voss 1993). The fluid balance is the major physical law involved in a groundwater model. The use of a computer to run a groundwater model merely allows the volume to have complex geometry and to be subdivided into many parts. The simple mass balance concept is then applied to each subvolume (Voss 1996). While state-of-the-art computer techniques allow the construction of very complex models with a large number of subvolumes, a model should not be more complex than the experimental information available for the area of interest requires, and allows. In fact, an appropriate simple model with few volumes does not need to be less descriptive than complex models with a highly sub-divided subsurface space. Compartment models implement these ideas. The subsurface region of interest is subdivided into a small amount of cells. The exchange of water between these cells is described by the respective exchange rate equations. There are some limitations to the cell size in transport models that separate them from compartment models. Some of these criteria, e.g. the Péclet criterion, will briefly discussed below. Compartment models can be very satisfactory and efficient. An example of a compartment model is Ecolego (Avilia et al. 2003).

As early as 1985, van der Hejde et al. (1985) reported a total of almost 400 computer implementations of groundwater models. The overwhelming fraction was intended for predicting future behavior. The steep increase in CPU performance during the past 20 years has further sparked the development of groundwater models. Today an even larger number of computer implementations is available while complex codes which required main-frame computers in the 1980ies are today running on standard desktops calculators. The simple concept of groundwater modeling, however, has not been changed.

2.1.5
Geochemical Models

The fundamental importance of water results in part from its excellent solvent properties for polar substances and the overall scarcity of the resource. Only 2.5% of all water on Earth is fresh water. Water flow does not only transport water but also the dissolved constituents which themselves may constitute a risk for the resource water. The chemistry of substances dissolved in water is a field of intense research (e.g. Stumm and Morgan 1996) and concern (e.g. Rydén et al. 2003). Chemical transformation and interaction can be numerically described at least approximately. Considering the variety of water uses and water consumption, to include at least the most fundamental knowledge about chemical reactions in water is desirable.

2.1.5.1
Diffusion, Dispersion, Sorption and Reaction

The mass balance for a substance i in a volume V_i is given by Eq. (2.16):

$$m_i = V_k\, c_i \tag{2.16}$$

A substance in a volume of water can move independently of the water flow by diffusion in presence of a concentration gradient (1st Fick law):

$$j_i = D_{m,i}\, \nabla c_i \tag{2.17}$$

where j_i is the concentration flow vector and $D_{m,i}$ the molecular diffusion coefficient of substance i. The molecular diffusion coefficient is temperature dependent and change with the total amount of dissolved substances.

A second important process is dispersion (Fig. 2.1). The friction between the mobile phase and the stationary phase cause a velocity profile over a pore. Flow velocities in different volumes may be different due to different porosity. In short, even if a substance i would not show diffusion, the concentration profile of that substance would change with time as a result of water flow. Third, the paths of particles of substance i in a porous medium may be different. For the same macroscopic distance different particles may require different time. This process also results in a modification of a given concentration profile. The dispersion is not a material property and cannot be described by a substance-related constant. The dispersivity α_l (longitudinal dispersivity) is described in analogy to the diffusion by Eq. (2.18):

$$H_l = \alpha_l\, |v|\ . \tag{2.18}$$

The longitudinal dispersion H_l depends on the mean flow tensor v and the dispersivity which itself depends on concentrations and the dimensions of the studied area (Scheidegger 1961; Kinzelbach 1986). Hence, dispersion may be also considered as a lumped-parameter process which has the potential to be used as an adaptable (e.g. fitting) parameter.

Dissolved substances in the mobile phase interact with the stationary phase. This interaction is often described as sorption. A sorbed particle does not flow with the mobile phase but is retarded. In fact, sorption of particles is considered as a major retardation mechanism in the groundwater. The numerical description of a sorption process is therefore often replaced by a lumped-parameter coefficient $R_{f,i}$, the retardation factor of substance i. Sorption is a summary term for all processes where a substance is selectively taken up by

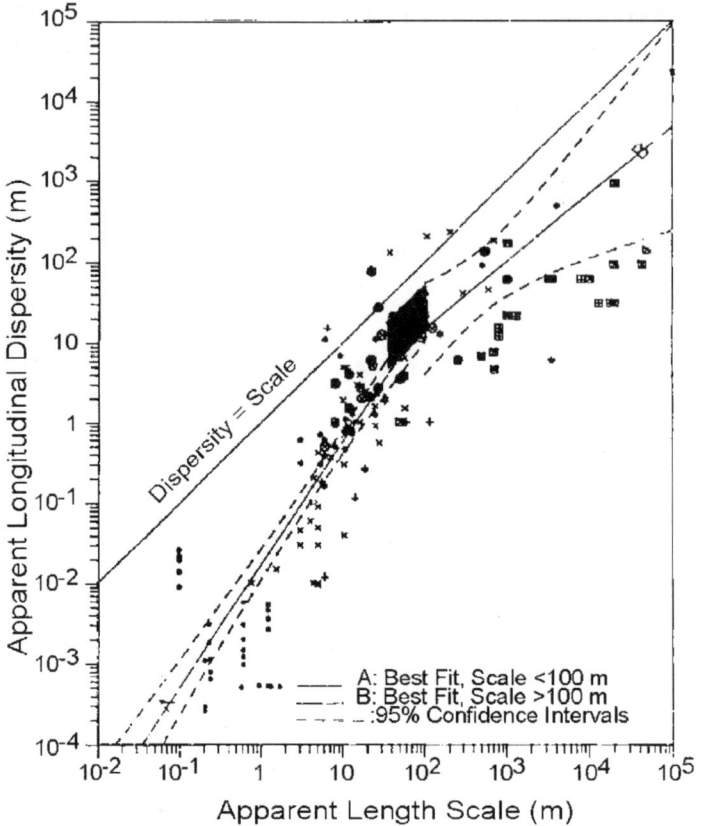

Figure 2.1. Estimated dispersivity as a function of length scale (Yucca 2002) illustrating scale dependence of dispersivity

another substance. The term "sorption" is always applied if the nature of the underlying process is unknown (Römpp 1999).

The simplest way of expressing this interaction is the assumption of a Nernst-like equilibrium between the sorbed substance i on the surface and the substance in the mobile phase:

$$K_D = \frac{c_{i,s}}{c_{i,aq}} . \tag{2.19}$$

Hereby is $c_{i,aq}$ the concentration (amount of substance per volume) of substance i in the mobile phase while $c_{i,s}$ is the concentration (amount of substance per area) on the stationary phase. Equation (2.19) is also given as

$$c_{i,s} = K_D \rho c_{i,aq} \tag{2.20}$$

where is the density of the mobile phase and $c_{i,s}$ is given in units of mass of sorbed substance per area. By using the assumption of a simple, concentration-independent equilibrium according to Eq. (2.20), the retardation factor $R_{f,i}$ is given by:

$$R_{f,i} = 1 + \frac{\delta_s}{P} K_d \qquad (2.21)$$

where δ_s is the density of the stationary phase and P gives the porosity.

Because sorption is the major retardation factor, the simple K_D assumption has been replaced by more complex models (e.g. Westall and Hohl 1980; Bolt and van Riemdijk 1982; Hiemstra et al. 1989; Kosmulski 1997). The consequence of more complex models is the increase of variables. Fitting a complex model with many adaptable parameters to experimental data will always result in an apparently good fit. This fact has been shown in Part I: The fit is apparently good because the model is over-fitted.

A further important aspect of chemistry is reaction. Substances transform into each other. These transformations follow certain rules. The key concept of chemical transformation is the law of mass action. Given a reaction

$$a\, A + b\, B \rightleftarrows c\, C + d\, D \qquad (2.22)$$

the respective equilibrium state is given by:

$$K = \frac{[C]^c\ [D]^d}{[A]^a\ [B]^b}, \qquad (2.23)$$

where K is the reaction constant and A, B, C, and D are chemical substances while a, b, c, and d are the respective stoichiometric coefficients. Note that Eq. (2.22) describes a reversible reaction. In a logarithmic form, Eq. (2.23) is written as:

$$\lg K = c \lg [C] + d \lg [D] - a \lg [A] - b \lg [B]\ . \qquad (2.24)$$

From the second law of thermodynamics comes the following relationship:

$$\Delta G_r = -RT \ln K;, \qquad (2.25)$$

where ΔG is the change in the free Gibbs energy of reaction r, R is the gas constant, T is the absolute temperature in Kelvin and K is the reversible equilibrium constant of this reaction. Now, a direct relationship between a formation constant K and the second law of thermodynamics is available, implying that K itself is a constant of nature. Determining K under laboratory conditions nevertheless allows one to apply the measured/evaluated value under different conditions. The details, however, show that this simple transferability often requires special attention especially if the reaction involves charged species (ions) (Ringbom 1958). Appropriate textbooks are available (e.g. Hartley et al. 1980; Merkel and Planer-Friedrich 2005). In short, the task of estimating the composition of an aqueous solution holding multiple mutually interacting constituents forming a variety of species was tackled by (mean value) computation. Implementing chemical reactions into a groundwater model is a comparatively recent development (Simunek et al. 1995).

2.1.5.2
Advection-Dispersion Equation, Consistency and Stability Criteria

The numerical description of a substance in a flowing medium is described by equations of the following fundamental form:

$$\text{RPS}\frac{\partial \theta}{\partial t} = \nabla D \nabla C - v\nabla C + q , \tag{2.26}$$

where S is the saturation of the porous medium and P gives the porosity. The amount of substance of a constituent is given by C while q represents all sources and sinks. Chemical reactions, radioactive transformations and influences of other processes (e.g. leaks in a pipeline etc.) are generally incorporated into the detailed formulation of q. Equation (2.26) includes second order mixed terms of the type $\frac{\partial}{\partial x}D_{xy}\frac{\partial C}{\partial y}$ and first order terms of the type $v_y\frac{\partial C}{\partial y}$. These terms, too, are solved by spatial and temporal discretisation on the basis of the similar approximations as before. Hereby several approaches to the discretisations for first order and second order terms are available, e.g. the Courant–Isaacson–Rees method (Courant et al. 1952) and the Friedrichs method (Friedrichs 1954) for the first order terms.

Approximation of the infinitesimal and continuous quantities by discretisation generates an error, the discretisation error. The discretisation error is of fundamental interest because it cannot always be estimated directly (e.g. by using Taylor expansions). For some cases, e.g. the Crank–Nicolson method, the numerical errors have been studied (deVahl Davis and Mallinson 1972; van Genuchten and Gray 1978; Bear and Verruijt 1987).

Consistency is an important criterion for a numerical algorithm to converge. Consistency is the property of a code implementation. Consistency implies decreasing truncation errors with increasing (finer) discretisation. However, even consistent algorithms do have their inherent limitations, which are often "forgotten" over the convenient (and often visually attractive) GUIs and colourful graphics representation delivered by modern computer implementations (Voss 1998). The advection-dispersion Eq. (2.26) comprises two elements: the parabolic dispersion term and the hyperbolic advection term. A satisfactory solution is only possible with Lagrangian methods and the random walk algorithm (Häfner et al. 1997). A reason for the limitations of other algorithms, especially finite difference method and finite element method, results from the Péclet number stability criterion. The Péclet number Pe gives the ratio of advective and dispersive fluxes:

$$\text{Pe} = \frac{\Delta x}{\alpha_l} \tag{2.27}$$

(neglecting diffusion), where Δx is the grid spacing and α_l is the longitudinal dispersivity. The stability criterion requires Pe $\leq$ 2. If the Péclet number criterion is not fulfilled, the code will show high numerical dispersion, a large grid orientation effect and, occasionally, oscillations. In fact, the Péclet criterion is a reason for the existence of compartment models. Compartment models might be considered as a special group of transport models where the compartments cells are rather large. The Péclet criterion does not allow that. The large cells would be numerically unstable, resulting in excessive dispersion and oscillations. Thus compartment models need other numerical approaches. Compartment models will not be discussed here, even though they are important occasionally, e.g. in performance assessment of nuclear waste repositories. Examples for compartment models are Amber (Quintessa 2002) and Ecolego (Avilia et al. 2003), which, however, aim at long-term risk assessment from radioactive materials.

A further stability criterion, valid for all discretisation algorithms, is the Courant criterion. The Courant criterion is given by Eq. (2.28)

$$Co = \frac{v\Delta t}{S\Delta x} \le (0.5 \ldots 1) , \tag{2.28}$$

where v gives the velocity field, S the saturation and Δx the grid spacing. Thus, the Courant criterion sets an upper limit to the time step, Δt. In practice, the combination of Courant and Péclet criteria requires small time steps and high discretisation (small volume sizes) to avoid the domination of the simulation output by numerical dispersion. Further stability criteria exist, e.g. the von Neumann criterion. For the discussion here it is sufficient to understand that the approximate solutions for the advection-dispersion Eq. (2.26) are limited by certain criteria requiring a minimum discretisation and limit the magnitude of time-steps. For a practical application of hydrogeological models, these criteria demand a minimum in CPU time to produce results which are not governed by truncation errors and other contributions to numerical dispersion.

The chemical interactions are generally coupled to the transport codes in a separate step: each time step calculates the movement of the dissolved species according to Eq. (2.26). Then, the chemical reactions take place and the next time step is calculated and so forth until the time of interest is reached. There is almost no limit which specific speciation code from the wide variety of available codes is selected (cf. Merkel and Planer-Friedrich 2005), as long as it can interact with the transport code. The speciation code calculates the composition of a solution on basis of specific input information, e.g. the total amounts of elements. On basis of the information on the formation constants (cf. Eq. (2.24)) the composition of the solution in terms of chemical species is calculated. This approach assumes that the solution composition is unaffected by other influences (for instance microbiological activity) and always in thermodynamic equilibrium (attempts to implement kinetic effects will be neglected here). Thus speciation calculations by means of Eq. (2.25) are also approximations.

A geochemical modeling code therefore comprises several components. The hydrogeological model (the flow model) is a basic unit, and the speciation code

Table 2.2. Algorithms to solve linear and non-linear equations and differential equations

Type of system	Algorithm	Reference
Linear		
	Gauss diagonalisation	Nash (1981)
	Cholesky factorisation	Wilkinson (1965)
	Jacobi diagonalisation	Jacobi (1846)
	Singular Value Decomposition	Golub and Reinsch (1970)
	QR decomposition	Francis (1961, 1962)
	Gauss–Seidel elimination	Bronstein and Semendjajew (1979)
	Relaxation methods	Agmon (1954)
	Conjugate gradients	Kammerer and Nashed (1972)
Non-linear		
	Picard iteration	Mathews (1989)
	Levenberg–Marquard method	Marquard (1963)
	Newton–Raphson iteration	Lawson and Hanson (1974)
	Simulated annealing	Cerny (1985)
	Simplex method	Nelder and Mead (1965)
	Metrolopolis algorithm	Metropolis et al. (1953)

is a second basic unit. The speciation code itself consists of two components. The first component is the collection of formation constants K for those species that have been identified to exist (including solid phases, sorbed species, and gases). This component is commonly termed "thermodynamic database". In fact, these files hold an often more or less arbitrary selection of values collated from literature. In some cases these data have been reported for a variety of ionic strengths and are "corrected" by some ionic strength correction method, e.g. Davies equation, extended Debye–Hückel equation etc. Some databases are at least consistent, but only the Joint Expert Speciation System (JESS) data base (May and Murray 1991, 1991a, 1993, 2001) provides a database with convenient access to all thermodynamic data, the method to ensure consistency, and access to the respective literature references. It is available over the World Wide Web (JESS 2006).

The second component is the numerical solver. This component provides the species concentrations from the total element concentrations and the information in the thermodynamic database. The numerical solver needs to consider linear functions only if exclusively complex formation in aqueous solution needs to be considered. As soon as other chemical processes, e.g. solid phase precipitation or dissolution, sorption processes etc., need to be considered the solver must be able to deal with non-linear equations. Non-linear equations are commonly solved by iterative techniques. Table 2.2 lists some relevant algorithms to solve linear and non-linear equations and differential equations.

2.1.5.3
Uncertainty in Modeling

A computer code for geochemical calculations nowadays comes usually very handy and user-friendly. A graphical user interface (GUI) is almost a standard feature. The average user is no longer a specialist in the field, but often has to rely on the code and its versatility for the problem at hand (Klemes 1986; Voss 1998). Errors occurring in such codes cannot be easily recognised or even be traced. The codes even may be of such a complexity that no individual can have the complete survey over the source code. Hence, there must be alternative approaches to study the quality and reliability of these codes, and also to find tools for the influences uncertainty-affected input quantities will have on the simulation output.

Uncertainty is ubiquitous in groundwater modeling and geochemical modeling. When modeling groundwater flow an important source of uncertainty are the starting conditions and the boundary conditions. The leakage rate of a contaminant, the variations in the flow itself and the duration of a contaminant input are often very difficult to assess. The input quantities are a further source of uncertainty. Not only do the thermodynamic input quantities have to be seen with scepticism. Already the very existence of a certain species (see below) may be questionable. Extensive data collections with sorbed species (especially in surface complexation models) introduce a large number of degrees-of-freedom into a numerical system where even the very basic quantity to describe chemical interaction processes with the stationary phase (commonly termed "sorption") is doubtful: the surface area. Four uncertainty sources are commonly distinguished: modeling uncertainty, prediction uncertainty, statistical uncertainty and intrinsic uncertainty.

Modeling uncertainty is due to the unavoidable simplification necessary to describe a complex natural system by functional, mathematical relationships. In some occasions modeling uncertainty is treated by a calibration process on basis of observation data.

Prediction uncertainty accounts for the limited amount of knowledge available at the time of analysis. Effects that are not known, not recognised or not adequately described may cause a deviation of the simulation from the reality. Human errors during data collection, recording and analysis are one example. The processes commonly described by sorption may be seen as a similar example.

Statistical uncertainty comes from the inadequate description of natural variability and its inadequate use within the model. Occasionally the expectation is issued that statistical uncertainty can be minimised by collecting more information. There may be situations where this expectation may be fulfilled. It should therefore be kept in mind that the ultimate goal of a simulation computer model is to model the behavior of nature. If data collection (e.g. by drilling holes) is modifying the study site this approach may jeopardise its own intentions.

Intrinsic uncertainty accounts for the variability in the physical parameters. It does not account to the limited knowledge that can be gained but the fact that a parameter (e.g. a porosity or temperature) can drastically change over a very small distance. From the knowledge of the physical properties at one point only limited conclusions can be drawn on the same properties in some distance.

2.2
Handling of Uncertainty in Geochemical Modeling

"Prediction is very difficult – especially if it is about the future."

(N. Bohr)

2.2.1
Uncertainty and Sensitivity Analysis

2.2.1.1
Measurement Uncertainty, Probability and Risk

The effect of uncertainty of experimental quantities in geochemical modeling is rarely discussed in the chemical literature. Metrology in chemistry is an approach to transfer the experiences with metrological concepts, and the associated enormous benefits, from physics and engineering to chemistry and, thereby to fill an existing gap. Support for metrology in chemistry is also coming from the field of risk analysis (e.g. Helton 1994). The necessity of nuclear waste disposal in geological formations (Ewing et al. 1999) and the regulatory requirements to assess the uncertainty related with the predictive models for nuclear waste repositories (e.g. Eisenberg et al. 1987) demand a reliable basis of uncertainty assessment also for the chemical quantities. Hence, highlighting the benefits that metrology in chemistry can contribute to environmental risk assessment is not of mere academic interest. Metrology in chemistry is developing concepts and protocols for those situations where the data obtained by chemical measurement will affect other people. Those "other" people will have to carry the risks resulting from a decision based on these data (Oreskes et al. 1994).

Risk and probability are closely intertwined notions (Kaplan and Garrick 1981). Probability is also an important concept in metrology in chemistry. Given a set of experimental observations, metrology in chemistry asks: What is the probability that a certain value can be observed (measured) in a subsequent repetition of the experiment at another place and another time? The question may be asked the other way around: What is the risk that the value of a quantity obtained by repetition of the experiment at another time and another location will be found outside a certain range? The concept of the complete measurement uncertainty budget tries to derive an estimate for the respective risk by building on the basic experience that "chance obeys laws".

Probability is often considered as a numerical measure of a state of knowledge, a degree of belief or, taking the reverse, a degree of doubt. It is important to realise that probability is the numerical expression of a subjective quantity. Surprisingly, this subjective "knowledge" can be quantified. The Italian mathematician Cardano, in his treatise "Liber de ludo alea" (Cardano 1526), is considered today as the founding father of probability theory. His small booklet, however, appeared in print only in 1663. Even though Cardano has anticipated many of the fundamental results of Pascal and Fermat, he is rarely given credit for this. Ignorant of the contributions of Cardano, probability theory arose from the discussion of a question of Chevalier de Meré about chances in gambling with dices. De Meré's question to Pascal gave rise to an intense exchange of letters between Pascal and Fermat, in which the fundamental theory of probability was established. As Laplace wrote: "It is remarkable that a science which began with the considerations of play has risen to one of the most important objects of human knowledge". The word probability is considered "as a meaning the state of mind with respect to an assertion, a coming event, or any other matter on which absolute knowledge does not exist" (de Morgan 1838).

There are three implications about probability. The classic view formulated, e.g. by Laplace and de Morgan, holds that the notion refers to a state of mind. None of our knowledge is certain; the degree or strength of our belief as to any proposition is probability. Another view defines probability as an essentially unanalyzable, but intuitively understandable, logical relation between propositions. Therefore, we must have a logical intuition of the probable relations between propositions (Keynes 1921). The third view of probability rests on the statistical concept of relative frequency. One of the most important "frequentists" for chemists is R.A. Fisher, whose "Statistical methods for research workers" (Fisher 1951) and "The design of experiments" (Fisher 1937) for many decades served as a kind of "bible" in practical statistics. A large part of today's statistics appearing (often in a distorted way; Thompson 1994) in chemical communications are based on R.A. Fisher's frequentist approach to experimentation. Most chemists are unaware that Fisher's frequentist approach is only one among several others. Further enlightenment of the meaning of probability can be found in the treatises of Nagel (1939) and Jaynes (2003). In summary, six fundamental facts can be stated regarding uncertainty:

1. Uncertainty is a fact of life, it cannot be avoided.
2. Uncertainty never disappears, although it can be reduced.
3. Uncertainty describes a range of situations from a complete lack of specific knowledge to knowing everything but the exact outcome of an action.

4. Different uncertainties warrant different responses.
5. Not to make decisions because uncertainty exists is foolish.
6. Decision-making ignorant of relevant uncertainties is risky.

2.2.1.2
Geochemical Models

The variety of numerical approaches to solve the mathematical equations arising in groundwater modeling and chemical speciation has given rise to a series of computer codes. Table 2.3 summarises some of these programs. Table 2.3 is by no means exhaustive. There might be several hundred of codes implemented on various levels of complexity. For most of these programs few information about the numerical and implementation details are available. In part the programs are commercial (POLLUTE/MIGRATE and CHEMKIN). For other programs source code is available (e.g. PHAST and TReAC).

A model's ability to describe past behaviour justifies no valid claim that the model is also able to predict future behaviour satisfactorily. The idea behind the models given in Table 2.3 (and most of those not given in Table 2.3) is the close relationship between the behavior of a groundwater system in nature and its approximation in computer code. This claim is commonly underpinned by demonstrating the agreement with a computed, simulated behavior and various test cases; e.g. break-through curves, plume distributions, and simplified situations where analytical solutions are available (Nitzsche 1997). It is a common feature of almost all models that the satisfactory behavior is shown by a single run of the code. It is, however, an irrevocable fact that all experimental work is affected by uncertainty. This point has been discussed in Part I and will not be repeated here. Hence, regardless of the aims and means of the programs, they all suffer the same limitation: uncertainty (Ekberg 2006). The LJUNGSKILE calculations (Ödegaard-Jensen et al. 2004)) given in Chap. 1.7 underscore the consequences of uncertainty in the predictive speciation of a batch solution (without any transport). These results

Table 2.3. Geochemical codes (two- to three-dimensional)

Program name	Reference
CHEMTARD	Bennet et al. (1994)
OS3D/GIMRT	Steefel and Yabusaki (1996)
CHEMKIN	Reaction Design (2006)
POLLUTE/MIGRATE	GAEA (2006)
PHAST	Parkhurst et al. (2002)
TReAC	Nitzsche (2000)
CORE2D	Samper et al. (2000)
MINTRAN	Walter et al. (1994)

are underscored by similar studies (e.g. Ekberg and Lundén-Burö 1997; Meinrath 2000a; Meinrath et al. 2000a; Ödegaard-Jensen et al. 2003; Denison and Garnier-Laplace 2005). Therefore the likely impact of measurement uncertainty in input variables on the variability of output prediction is of considerable interest to both the developers of such programs, their users, the users of the output data and those affected by decisions based on such simulation results.

2.2.2
A Conceptual Model of Geochemical Modeling Codes

It is obvious that for computer codes with dimensions of a geochemical modeling program the evaluation of a cause-and-effect diagram is futile. A groundwater model is usually based on very few experimental data (e.g. a geological characterisation of the geological formation, piezometer pressures in wells and boreholes). Educated guesses for relevant input parameters (e.g. intermediate pressure heads, porosities and permeabilities, retardation factors) are common. The hydrogeological model is usually calibrated (adapted) to the site of interest by varying certain parameters to increase the coincidence of predicted and measured physical quantities (e.g. source flow rates). Hence a groundwater model often is a "black box" for the user.

A chemical speciation code directly or indirectly coupled to the transport code often comprises a combination of algorithms (cf. Table 2.3) to arrive at a numerical solution for a given chemical system. In principle, the algorithms are stable and applied successfully for many years in all areas of science, humanics and technology. However, in chemical systems the range of values may vary over orders of magnitude. The matrices may be nearly singular and/or ill-conditioned. Rounding errors (notorious candidates for producing fatal errors due to rounding and overflow are matrix inversions, combinations of differences of large numbers, and divisions by zero). Usually the user is left without notice over the intrinsic processes in a speciation code – partly because the programs try to handle such errors internally.

The only component of a speciation code directly accessible to the user is the thermodynamic "database". Commonly this "database" is a list of chemical compounds together with their formation constants on the basis of Eq. (2.30). Several such collections are available. A selection is given in Table 2.4. Table 2.4 is not exhaustive. Its purpose is simply to illustrate the variety of efforts to collate the large number of measurement (mean) values of (potentially) relevant thermodynamic data available in literature. Most of the data collection is related to the respective computer programs, for instance EQ3/6, SOLMINEQ, WATEQ4F.

Thus, a geochemical model can be considered as a combination of three elements: a groundwater model, a speciation code and a "database" of chemical

Table 2.4. Selected references to databases for geochemical modeling

Database name	Reference
CODATA	Garvin et al. (1987), Cox et al. (1987)
IUPAC	Academic Software
NIST	Martell et al. (2003)
WATEQ4F	Ball and Nordstrom (1991)
JESS	May and Murray (1991b)
CHEMVAL	Chandratillake et al. (1992)
MINTEQA2	Allison et al. (1991)
SOLMINEQ	Kharaka (1988)
GEMBOCHSV2EQ8	Woolery (1992)
SUPCRT	Johnson et al. (1992)
HATCHES	Cross and Ewart (1991)
NAGRA TDB	Hummel et al. (2006)
EQ3/6	Woolery (1992)
Holland and Powell minerals database	Holland and Powell (1990)

data. The database is often the only component of the geochemical model that can be manipulated to a larger extend by the user. Thereby the format of the input is program-dependent and must be carefully maintained in order to run the program successfully. Figure 2.2 gives a graphical representation of such a geochemical model structure.

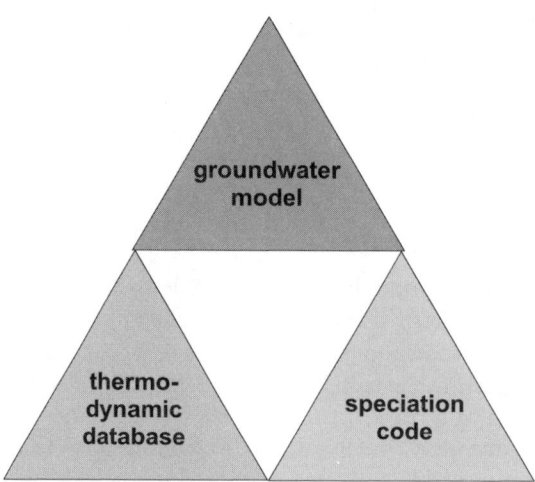

Figure 2.2. A conceptual model for coupled geochemical codes. These codes combine a groundwater model, a speciation code and a database of chemical parameters describing the magnitude relevant processes, e.g. complex formation, solid phase formation and dissolution, adsorption and desorption, formation of colloids etc.

2.2.3
Importance of the Thermodynamic "Database"

The availability of different databases, often independently collated from more or less the same inventory of literature, quickly resulted in the insight that the variety of possible interpretations of the same source data will result in considerable variations in the "recommended" data. Because of the potential relevance of geochemical modeling in decision-making processes (e.g. safety analysis of nuclear waste disposal sites) activities were initiated to harmonise (that is, standardise) the values held in the different data collections. Standardisation, however, is nothing else than the establishment of a convention. The degree to which the values in a database agree with the "true" values of the respective formation constants decides whether a modeling effort provides a dependable answer.

The alternative to standardisation is an assessment of the reasonable variability in a measurement value, and the subsequent attempt to reduce this variability due to a detailed understanding of the sources of the variability. The former way is comparatively easy and may be understood as some type of diplomacy. The latter way is the scientific one, tedious and time-consuming, often costly and frustrating. On basis of the experience of the past 450 years, from Galilei to the twenty-first century, the statement can be made with great confidence that the only successful approach is the latter one.

In Fig. 2.3, a section from the PHREEQC database is shown as an example for the content of a thermodynamic database. The aqueous solution species are characterised by a reaction equation, the respective equilibrium constant K, following the key word "log k" according to Eq. (2.25) and, optionally, a value for the parameter ΔH_R in the van't Hoff isochore Eq. (2.29).

$$\frac{\partial \ln K}{\partial T} = \frac{\Delta H(T)_R^{\circ}}{RT^2} , \tag{2.29}$$

where the superscript $^{\circ}$ denotes a standard state, $\Delta H(T)$ the enthalpy of reaction and T the absolute temperature in Kelvin. In case of PHREEQC, (which is the speciation code in TReAC), an enthalpy of reaction represented by the key word "delta_h" may be given in a polynomial form to account for the temperature dependence of $\Delta H(T)$.

All values in this database are mean values. Hence, from the large number of chemical equilibrium data in literature, often given with a considerable spread, the user must decide for one value in each constant. The example of the U(VI) species $(UO_2)_2(OH)_2^{2+}$ in Table 1.16 is just one illustrative case (Meinrath et al. 2006). By considering the large number of species potentially formed, e.g. in a sample of natural water (possibly also holding organic ligands), it becomes evident that data selection may become a time-consuming task. Due to the absence of hard selection criteria, the resulting collection is necessarily subjective. The

#75 CaOH+
Ca+2+H2O=CaOH++ H+
   logk   -12.598000
   delta_h 14.535000 kcal
#76 CaCO3
CO3-2+Ca+2=CaCO3
   logk   3.153000
   delta_h 1.806000 kcal
#78 CaSO4
SO4-2+Ca+2=CaSO4
   logk   2.309000
   delta_h 1.470000 kcal
#79 CaPO4-
PO4-3+Ca+2=CaPO4-
   logk   6.459000
   delta_h 3.100000 kcal
#80 CaHPO4
PO4-3+Ca+2+H+=CaHPO4
   logk   15.085000
   delta_h -0.230000 kcal
#81 CaH2PO4+
PO4-3+Ca+2+2 H+=CaH2PO4+
   logk   20.961000
   delta_h -1.120000 kcal

Figure 2.3. Example section illustrating the PHREEQC thermodynamic database

resulting data collections may be expected to be quite variable. There have been and still are large standardisation projects under way (e.g. the CHEM-VAL projects of the Commission of the European Communities (Chandratil-lake et al. 1992)). The variability may also be caused by the different aims of a geochemical modeling code. Codes for modeling radioactive transport need different species information than those modeling brine behavior or littoral intrusion of seawater. The Holland and Powell (1990) mineral database mentioned in Table 2.3, for instance, is devoted exclusively to rock-forming minerals and fluids. Therefore it holds thermodynamic information (including uncertainties) on, e.g. corundum and spinel, but not on, e.g. $Fe(OH)_3(s)$ and $UO_2CO_3(s)$. Reference to this database is made because it gives detailed information about the procedures to ensure internal consistency. The uncertainties specified in the database are those required for consistency. There are alternative methods to derive internally consistent thermodynamic data, e.g. by using Bayes estimation (Königsberger 1991; Olbricht et al. 1994). Whatever the method for achieving internal consistency might be, the procedure starts by using experimental data. Due to the general paucity of information concerning the quality of the literature data, it is impossible to assess their reliability. It must be emphasised that there is no (what-so-ever complex) method to reassess the literature data. Major reasons are the general unavailability of the original measurement data and the nature of a scientific publication itself (Medawar 1990). Therefore, metrology in chemistry aims at the development of a framework, protocols and procedures to improve the situation at the only location where improvement can be effective: in the laboratories with the experimenters.

A further variability within databases may result from a different conception of the database. In the example given in Fig. 2.3 the species with number #77 is missing. This species is "CaHCO3+". Experimental data on the behavior of Ca^{2+} in carbonate solutions is commonly unable to distinguish the specific form of carbonate co-ordinating to Ca^{2+}. In almost all cases, an interpretation of the observed effects by coordination of Ca^{2+} with CO_3^{2-} (resulting in a neutral species $CaCO_3°$) or HCO_3^- (resulting in a charged species $CaHCO_3^+$ and a neutral species $Ca(HCO_3)_2°$) is (numerically) satisfactory. To design a decisive experiment would require to assess carefully the measurement uncertainty of the complete data in order to find conditions where the differences in the interpretation due to a single species ($CaCO_3°$) or a two-species ($CaHCO_3^+$ and $Ca(HCO_3)_2°$) model are significant. Hence the existence of the $CaHCO_3^+$ species is considered doubtful by some chemists. If both species, $CaCO_3°$ and $CaHCO_3^-$ are included into the database, the co-ordination of Ca^{2+} by carbonato ligands (CO_3^{2-}, HCO_3^-) is overestimated. Of course, in systems where calcium interactions play a role the difference in the database may also cause differences in the model output. Such differences are often considered "minor" and not discussed further in the presentation of respective simulation results.

Thermodynamic databases may become extensive. The CHEMVAL-2 database holds 1356 aqueous species. Not all species will be relevant in a simulation of some site-specific conditions. However, the modeler will be interested which input is most strongly affecting a modeling result. This information becomes crucial if different models result in different simulation output. Sensitivity analysis is often capable to answer such questions despite the fact that large parts of the geochemical modeling code is a "black box" to the user.

2.2.4
Sensitivity Analysis

In a sensitivity analysis, the magnitude of an individual entry in the complete set of input data is remained fixed while all other entries are varied within a wide range according to some random procedure. A practical implementation may create a random field of entry vectors by randomly sampling the entries from distributions specified for each entry. With this modified input data sets the modeling is repeated until each entry vector with modified entries is applied in a simulation. If the output does not change widely, the one single entry which is kept constant must have some influence on the simulation procedure. On the other hand, if the output value does change widely the single constant entry cannot have importance in obtaining the simulation output. When such a cycle is repeated systematically with each entry value, a distribution of simulation outputs exists. The wider the distribution, the less important is the entry. The relevance of each entry in the input data set on the output result of the simulation program can be assessed by ranking according to the magnitude of the output distribution. Sensitivity analysis is not only a helpful tool in comparing and analysing "black box" modeling output. It is also an element of "good modeling practice" (GMP) (e.g. Stanislaw 1986; Banks 1989; Kleijnen 1995; Schneider et al 2006).

2.2.4.1
Example of a Sensitivity Analysis

An example of a sensitivity analysis is given for the solubility of uranium in a groundwater (Ekberg and Emrèn 1996). The groundwater composition is given in Table 2.5. Uranium may be present in granitic groundwaters and the question arose upon the maximum concentration uranium may have in such groundwaters before precipitation would set in. Hence a geochemical modeling program is applied. In the present case the speciation calculation were performed by PHREEQE (Parkhurst et al. 1980), the FORTAN version

Table 2.5. Groundwater composition of a geological formation (Ekberg and Emrèn 1996)

Element	Concentration/[mol dm^{-3}]
pH	8.1
pe	−4.4
Cl	0.18
Na	0.094
Ca	0.047
S	0.0058
Mg	0.0017
Br	0.0005
Sr	0.0004
K	0.0002
C	0.00016
Si	0.00015
Li	0.00014
F	0.00008
Mn	0.000005
P	0.0000006
N	0.0000018

of PHREEQC. A solid phase of $UO_2(s)$ was added as a solid phase. A variety of chemical species may form in such a solution, e.g. $(UO_2)_2(OH)_2^{2+}$, $CaOH^+$, $U(OH)_3^+$, depending on the pH and pe of the solution. The generation of a random input matrix for the sensitivity analysis is illustrated in Table 2.6. Instead of showing all possible species, only four representative species $UO_2(s)$, $CaOH^+$, UOH^{3+} and UO_2^{2+} are shown and the matrix size is limited to four random values. The species UO^{2+} may form at appropriate pe and pH according to

$$U^{4+} + 2H_2O \Longleftrightarrow UO_2^{2+} + 4H^+ + 2e^- . \tag{2.30}$$

In the standard database of PHREEQE the equilibrium constant of this reaction is given as $\log k = -9.1$.

With the information in Table 2.6, four repetitions of the solubility calculation can be done. While the entries for the three species $UO_2(s)$, $CaOH^+$ and UOH^{3+} vary from run to run, the value for UO_2^{2+} remains constant. From this scheme four values for the solubility of uranium would be calculated. On basis of these four values the mean and the standard deviation could be obtained. The standard deviation value would be the smaller the more important the parameter "UO_2^{2+}" is for the modeling result. The (very limited) sensitivity analysis scheme of Table 2.6 would be completed by keeping successively each of the other parameters unchanged while the remaining three vary. After four cycles (in each cycle the value of one of the input parameters is held constant) for each of the four parameters the respective standard deviation would be available. The smaller the standard deviation the higher the relevance of a parameter for the calculated output. A meaningful sensitivity analysis often comprises hundreds of runs – the fundamental concept, however, remains unchanged.

Table 2.7 summarises the sensitivity analysis results for the species with major impact on the simulation result. By analysing the results, e.g. in terms of a redox diagram of uranium in natural aqueous systems (Fig. 2.3), it seems reasonable that the solubility is mainly affected by the solubility product of the relevant solubility limiting U(VI) solid phase $UO_3 \times 2H_2O$ (Meinrath et al. 1996). On the other hand, the complexation of U(IV) by carbonate has obvi-

Table 2.6. A (highly abbreviated) random input matrix for sensitivity assessment of the parameter entry UO_2^{2+}

Species	1st run	2nd run	3rd run	4th run
$UO_2(s)$	−4.3	−4.8	−5.1	−4.5
$CaOH^+$	−12.2	−12.0	−12.8	−12.9
UOH^{3+}	−0.2	−0.8	−0.5	−0.4
UO_2^{2+}	−9.1	−9.1	−9.1	−9.1

Table 2.7. Distribution of outputs from a sensitivity analysis of solubility of $UO_2(s)$ in a groundwater (Ekberg 2006). The list is limited to the eight most relevant parameters

Species	Variance
$UO_3 \times 2\,H_2O(s)$ (schoepite)	$1.507 \cdot 10^{-3}$
$U(OH)_4$	$1.511 \cdot 10^{-3}$
UO_2^{2+}	$1.582 \cdot 10^{-3}$
H_2S	$1.643 \cdot 10^{-3}$
pe	$1.690 \cdot 10^{-3}$
UO^{2+}	$1.698 \cdot 10^{-3}$
$CaSO_4^{\circ}$	$1.702 \cdot 10^{-3}$
HCO_3^-	$1.710 \cdot 10^{-3}$

ously been suppressed. The amount of C in the groundwater can be reasonable attributed the carbonato species HCO_3^-. The concentration of this species depends, however, on the CO_2 partial pressure. The concentration of C (cf. Table 2.5) is about 1 order of magnitude lower than expected at the given pH. Hence, the tendency to form carbonato species in the groundwater is much lower than assumed in the calculation of Fig. 2.4. The appearance of H_2S seems surprising but the redox diagram indicates that the groundwater is accurately on the S(VI)/S(-II) redox boundary. Due to the different solubilities of $UO_3 \times 2$ H_2O and $UO_2(s)$ the redox conditions are important. From the groundwater data in Table 2.5, this relationship is not readily to be recognised. The sensitivity analysis spurs a more detailed investigation. It is underscored by the equal importance of the redox potential pe.

There are a variety of techniques for efficient sensitivity analysis discussed in literature (Imam and Helton 1988; Ekberg 2006). Not all techniques are suitable under all circumstances. While the general concept of sensitivity analysis is rather straightforward, some fundamental points of concern remain. The technique described in the above example is a Monte Carlo type sensitivity analysis. Alternative techniques are the response surface technique and the Fractional Factorial Design approach. It is common to both alternative techniques that they assume a certain general form of the response (linear or quadratic) of the geochemical model. Geochemical models do not usually follow such a simple, functional trend. Monte Carlo (MC) techniques are suitable for moderately sized problems. The CPU demand increases, however, in a quadratic manner with the problem size. Wherever MC methods may be applied successfully, Latin Hypercube (LHS) approaches (Chap. 1.7) are alternative choices. The more extended the problem is in terms of the amount of input parameter, the more efficient is LHS over MC usually. Finally local approximation by a differential analysis has be applied. It is important to understand that such an approach gives an answer only for the local point under study. In geochemical models the input variation, however, may cover orders of magnitudes. The use

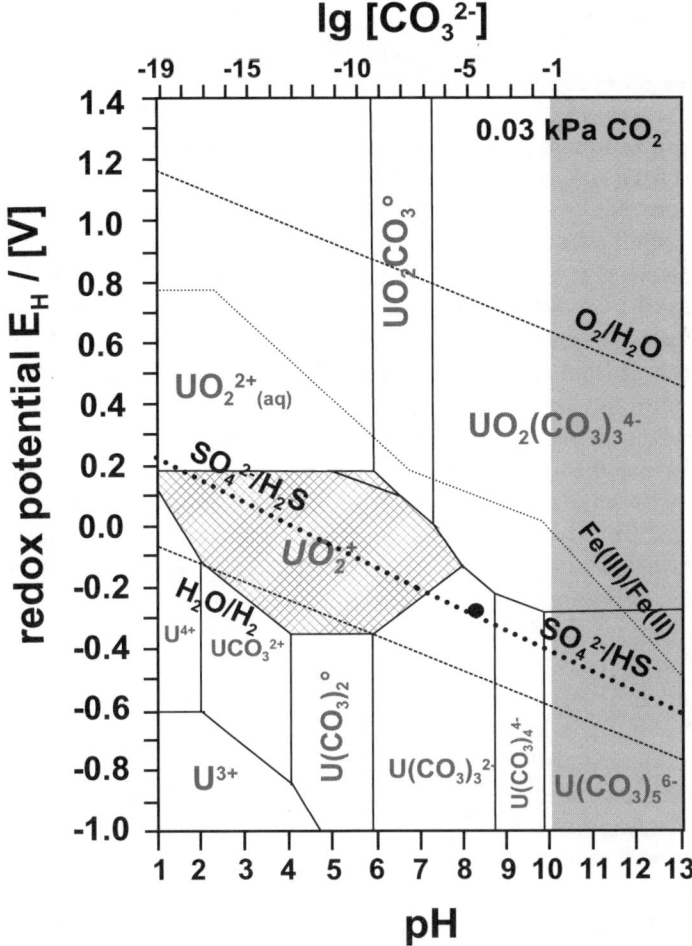

Figure 2.4. Redox diagram of uranium in low ionic strength water in equilibrium with the atmospheric CO_2 partial pressure. The *hatched area* gives the stability field of the questionable U(V) species, UO_2^+. The *black point* (•) indicates the approximate position of the groundwater Table 2.5 close to the redox boundary U(IV)/U(VI)

of locally restricted derivative methods may become misleading here (Imam and Conover 1980).

2.2.5
Uncertainty Analysis

It is important to realise that sensitivity analysis can be an element of uncertainty analysis but sensitivity analysis itself is not a type of uncertainty analysis. The uncertainty in geochemical modeling (both stationary and in combination with transport) is only recently gaining the attention it deserves, especially if it concerns chemical quantities. In the present context of metrological impact assessment for chemical data in complex situations, this short section highlights several important points. First of all, a basis for objective assessment of chemical information in literature is rarely available.

In fact, uncertainty is a psychological phenomenon. Uncertainty is doubt. Nature does not have doubt. Light emitted from a distant sun in one direction of the universe is emitted with the same velocity from another sun in the opposite direction of the universe. If there is some discrepancy in the measured value for the quantity "speed of light", this discrepancy is a result of our limited ability to assess ("measure") its velocity. The accuracy in measuring the speed of light has tremendously increased over the past centuries. Surely the photons generated by the two stars have the same speed. If there are differences in the measurement values it results from our limited abilities to measure the speed of light accurately. From the first attempts (by opening lanterns separated over some distance) it became apparent that this method is unable to give an answer with regard to the speed of light. The "lantern method" was not fit for purpose. In 1857, Kohlrausch and Weber obtained the prestigious Bavarian Maximilian Order of Arts and Science for measuring the velocity of light to $c = 3.1074 \cdot 10^8 \, \mathrm{m \, s^{-1}}$. The good agreement of this value and today's standard value ($c = 299\,792\,458 \, \mathrm{m \, s^{-1}}$) within 3.5% should not be overestimated. Kohlrausch and Weber did not consider at all the systematic errors (even though all disturbing effects were carefully considered). Today's estimate of the uncertainty in the value for c given by Kohlrausch and Weber is 6% (Walcher 1988). The doubt attributed to a measurement value comes from the discrepancies observed for values of the same quantity when measured repeatedly, at different time and at different location. This doubt also refers to the limited predictability of a value to be obtained from a future experiment on basis of the previous experiments. Here, "the frequentists" approach to assessing variability and the Bayesian approach in statistics meet (Efron 1986; Jaynes 2003).

In short, while the term "probability" may be tracked back to antique scriptures (e.g. Cicero 80 BC) the modern notion of probability has risen from gambling. Chevalier de Mére, a French nobleman with an interest in the analysis of observations, once asked the noted mathematician Pascal about the observed difference in certain events when playing with dices. The answer to de Meré's question wasn't simple and Pascal started a discussion with Fermat. Hence, our modern anticipation of probability has risen from gambling and the perception of doubt and risk. In the present context, the crucial point is that "chance obeys laws" (Poincaré 1908).

2.2.5.1
Example of Uncertainty Analysis

In 1999, Hamed and Bedient (1999) concluded: "The impact of parameter uncertainty on achieving remediation/containment goals is important. Failure to account for such uncertainty can dramatically hinder the efficiency of the remediation/containment scheme, creating significant economic ramifications". Further: "The uncertainty of the parameters describing the aquifer material, the chemical, and the leaking source have been the focus of many research efforts. (...) This was motivated by the recognition of the impact that such uncertainty has on the predictive ability of groundwater fate and transport models".

An example of these research efforts has been given by Nitzsche et al. (2000). The study was spurred by solubility calculations where uncertainty in the solubility products was taken into account (Ekberg and Lundén-Burö 1997). If a minor variation in the parameters defining the solubility of a substance has a significant impact on the concentration values calculated in a single speciation experiment, then this impact must also become apparent in a geochemical modeling simulation. The simple reason is that such speciation calculations are repeated multiply during a reactive transport calculation. There is no difference whether the transport step and the speciation step are performed separately for each cell of the mesh (two-dimensional) or the lattice (three-dimensional) or the steps processes are treated by coupled equations. In case of uncorrelated input parameters the complete procedure would correspond to a random sampling experiment. The distribution of the output result, therefore, likely will approximate a normal distribution as a consequence of the Central Limit Theorem. In fact, Ekberg and Lunden-Burö (1997) had interpreted the simulation output by normal distributions with good agreement. A perfectly normal output, as observed under idealised sampling conditions (Meinrath and Kalin 2005) would not be expected because non-random effects, e.g. chemical transformations like redox reactions, should affect the output.

To assess the impact of uncertainty in chemical parameters of a thermodynamic database on the geochemical modeling output, a simplified scenario was assumed. The problem was treated by a Monte Carlo approach with the geochemical code TReAC (Nitzsche 1997). TReAC is based on the standard software MODFLOW (McDonald and Harbaugh 1988) and PHREEQC (Parkhurst 1995). TReAC includes procedures to simulate double porosity aquifers. For sorption, the built-in PHREEQC procedures were used in conjunction with the respective parameters in the PHREEQC standard database.

The geological setting was kept as simple as possible. The complexity of the geological setting would mainly use computing time and, therefore, reduce the number of Monte Carlo steps and, consequently, the significance of the output results. Hence, a sand column was considered with a length of 0.4 m, a cross-section of $70\,cm^2$, effective porosity P = 0.32, vertical plug flow of

$0.14\,\text{cm}\,\text{min}^{-1}$ and a dispersivity $\alpha_l = 1.0\,\text{cm}$. The Courant number was 0.2 and, therefore, well within the Courant criterion Eq. (2.28). Uranium was selected as the element of interest.

The PHREEQC standard database was applied with the modifications summarised in Table 2.8. The uncertainties σ, given as standard deviations of a normal distribution, are much too small under modern understanding of the complete measurement uncertainty budget. Nevertheless, the limited variance caused considerable variation in the simulated properties. For the computer simulations a drinking water enriched in uranium as a consequence of uranium mining was assumed.

Average uranium concentrations are $7.2 \cdot 10^{-5}\,\text{mol}\,\text{dm}^{-3}$ due to saturation with the respective U(VI) solid phase, $UO_2CO_3(s)$. Mean value of the quantity pH is about 6.5. Carbonate concentration is comparatively high with $[HCO_3^-]=1.8 \cdot 10^{-3}\,\text{mol}\,\text{dm}^{-3}$. The water has $7.7 \cdot 10^{-4}\,\text{mol}\,\text{dm}^{-3}\,SO_4^{2-}$, $1 \cdot 10^{-3}\,\text{mol}\,\text{dm}^{-3}\,Ca^{2+}$ and $6.7 \cdot 10^{-4}\,\text{mol}\,\text{dm}^{-3}\,Na^+$. For simulation purposes, the water was numerically equilibrated with $CaCO_3(s)$, and thus showed a CO_2 partial pressure of about 2%.

Figure 2.5 shows a speciation result of a Monte Carlo simulation for such a water (batch). Uranium sulfato species $UO_2SO_4^{\circ}$ has a rather low concentration while the carbonato species dominate the solution. The mot interesting aspect offered by Fig. 2.5 are the overlapping distributions of the species $UO_2CO_3^{\circ}$ and $UO_2(CO_3)_2^{2-}$. The probability density of $UO_2CO_3^{\circ}$ concentration is quite narrow compared to the probability density of $UO_2(CO_3)_2^{2-}$ concentrations. As a consequence there is some probability that $UO_2(CO_3)_2^{2-}$ becomes the prevailing solution species. The total concentration varies due to the varying amounts of solution species. $UO_2CO_3(s)$ dissolves and forms UO_2^{2+}. The solubility product of $UO_2CO_3(s)$ is not varied during the Monte Carlo assessment. However, the small differences in the tendency to form complex species in solution, expressed by the random variation in the formation constants of the

Table 2.8. Formation and reaction constants for U(VI) hydrolysis and carbonato species at $I = 0$ and $25\,^\circ\text{C}$

Reaction	Lg K	σ	Reference
$UO_2^{2+} + H_2O \Longleftrightarrow UO_2OH^+ + H^+$	−5.87	0.08	Choppin and Mathur (1991)
$2\,UO_2^{2+} + 2\,H_2O \Longleftrightarrow (UO_2)_2(OH)_2^{2+}$	−5.93	0.05	Meinrath and Schweinberger (1996)
$3\,UO_2^{2+} + 5\,H_2O \Longleftrightarrow (UO_2)_3(OH)_5^+ + 5\,H^+$	−16.49	0.12	Meinrath (1997)
$UO_2^{2+} + CO_3^{2-} \Longleftrightarrow UO_2CO_3^{\circ}$	10.27	0.05	Meinrath et al. (1993)
$UO_2^{2+} + 2\,CO_3^{2-} \Longleftrightarrow UO_2(CO_3)_2^{2-}$	16.7	0.4	"
$UO_2^{2+} + 2\,CO_3^{2-} \Longleftrightarrow UO_2(CO_3)_3^{4-}$	22.9	0.3	"
$UO_4^{2+}SO_4^{2-} \Longleftrightarrow UO_2SO_4^{\circ}$	2.9	0.1	Burneau et al. (1992)
$UO_2^{2+} + SO_4^{2-} \Longleftrightarrow UO_2(SO_4)_2^{2-}$	3.6	0.1	"

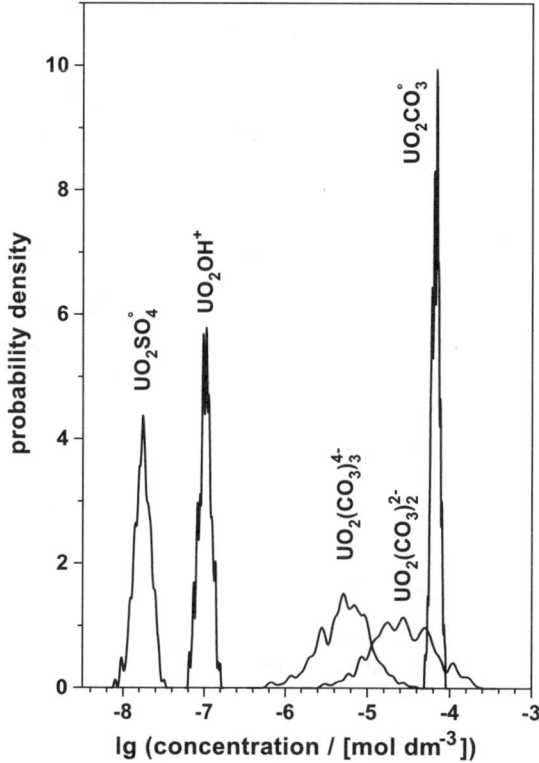

Figure 2.5. Simulated distribution of U(VI) species for a model water used to simulate transport through a sand column

respective species causes small differences in the U(VI) solubility. This variability in the total U(VI) concentration is not caused by sorption processes. Sorption (within the model used in PHREEQC) to quartz occurs mainly with cationic species. The major species under the conditions of the column are, however, neutral to anionic species. The distribution of the U(VI) solubility in a batch solution is shown in Fig. 2.6.

The simulations in Fig. 2.5 and Fig. 2.6 are obtained for a batch situation. If the respective solution is running through a column, a considerable amount of uranium may be sorbed to the surface of the stationary phase. The result is a difference in the break-through concentration. Depending on the relative species concentration in the starting solution, more or less U(VI) is prevailing in solution as a preferentially sorbed neutral $UO_2CO_3^\circ$ species (the role of UO_2SO_4 can safely be neglected here). This concentration reduction does not lead to a major retardation.

Predicting the time when a threshold value is exceeded by a contamination plume at a point of interest is a typical task in pollution assessment. The

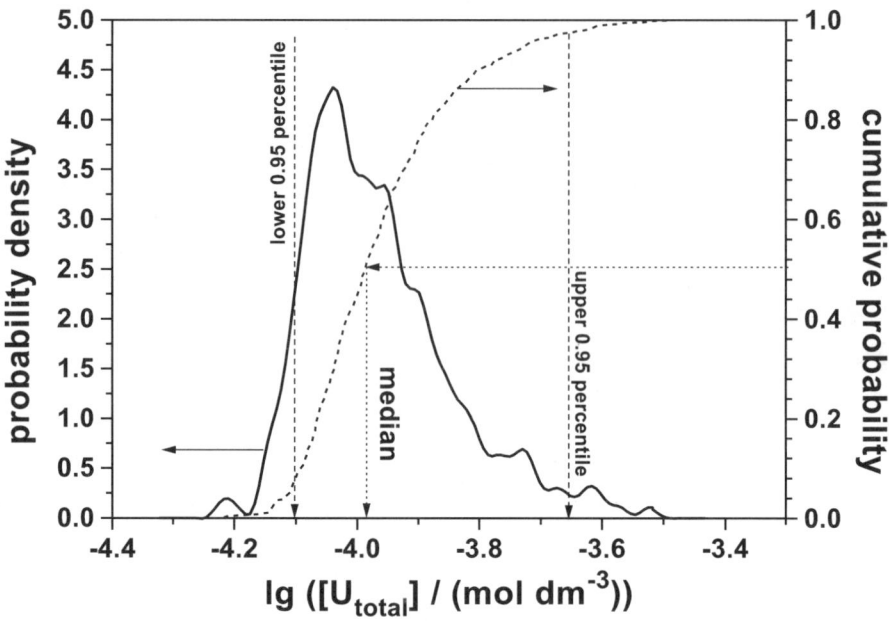

Figure 2.6. Distribution of uranium concentrations in a model water. The results are obtained by Monte Carlo sampling (500 resamplings). Concentration differences result from small differences in the tendency to form complexed U(VI) species

compliance with legal limits is an important issue in the sustainable use of natural resources. Analysing the risk of accidentally exceeding a limiting value is a standard task for risk assessment. Examples are the safe operation of nuclear power plants or the probability that a chemical plant will excessively release potentially hazardous gases, for instance due to an improbable but not impossible combination of individually improbable states in a facility.

In case of the sand column simulation the time was calculated until the break-through curve exceeded an arbitrarily set limit of 10 mg U(VI) dm^{-3} of water. The result is shown in Fig. 2.7. The 95% confidence range of the simulation covers a range from 3.6 days to 9.3 days. This effect is drastic. The demand in CPU time for such a simulation (admittedly limited in its scope) is also considerable: 120 h CPU time at an IBM Risc 6000 workstation. The increase in CPU speed since the year 2000 is considerable and the time demand today would probably be reduced by 50%. Nevertheless, this simplified setting can only emphasize the importance of thermodynamic data in geochemical modeling. Many influencing parameters from the groundwater model were either kept constant or completely eliminated by assuming plug flow in a sand column with a homogeneous porosity. It may be argued that in case of a more realistic hydrogeological setting the uncertainties due to spatial variability in the subsurface as well as uncertainties in the upscaling parameters of the

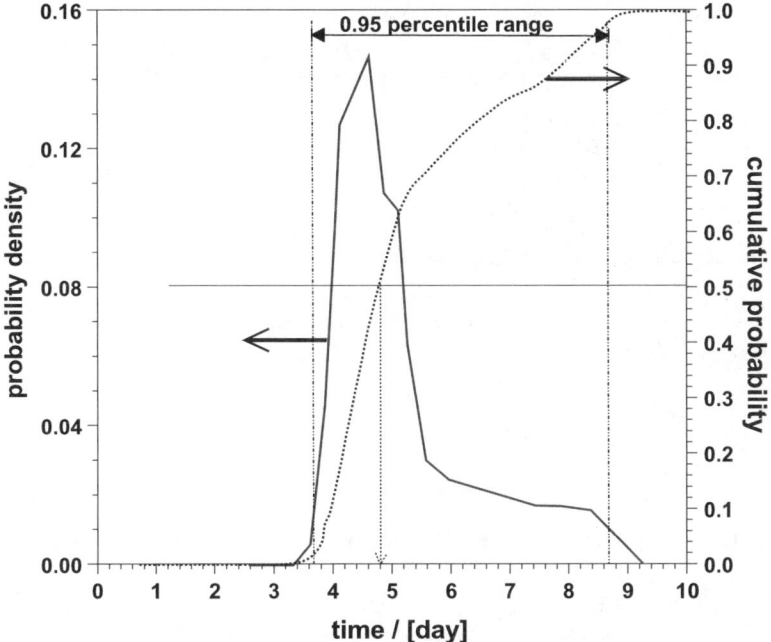

Figure 2.7. Cumulative distribution and probability density for simulated U(VI) concentrations exceeding a threshold value in the column outlet of $10\,\mathrm{mg\,dm^{-3}}$

groundwater model might exceed those from the uncertainties in the chemical database.

A discussion on the relative magnitudes of uncertainties contributing to a geochemical simulation output would require a common basis to quantify uncertainty. Metrology offers a scheme to arrive at such a comparable measure, but only for measurement data. For subjective uncertainties, e.g. guessed distribution curves of porosities and flow fields, as well as adequate contaminant release rates etc. it is difficult to create a similar basis for comparison. A third group of uncertainty contributions are formed by the neglected correlations and dependencies within the parameters. Mathematical expressions need to be balanced dimensionally, because otherwise values for physically meaningless quantities are calculated and carried through the simulation procedure (Ferson 1995). These few examples show that the chemical quantities are only one of many factors affecting the output of a geochemical modeling calculation.

Four types of uncertainties were introduced at the end of Chap. 2.1.5.3: modeling uncertainty, prediction uncertainty, statistical uncertainty and intrinsic uncertainty. The first two types are inherent in our human limitations to understand this world and to describe our insight by models. Therefore, only statistical uncertainty (variability, randomness) and intrinsic uncertainty (incertitude) can be reduced, e.g. due to research and better engineering, by improving the accuracy and precision of measurement techniques or by choosing other engineering options with reduced irreducible uncertainty. In both cases, the current situation must be assessed as accurate as possible – in order to avoid the selection of alternatives which are a) different but not better and b) bring improvement in one side but have adverse side effects.

2.3
Geochemical Modeling and Decision Making

"The universe is like a safe to which there is a combination, but the combination is locked up in the safe."

(de Vries)

2.3.1
Being Confronted with Geochemical Models and Their Results – Some Guidelines

The previous part of this book inquired into some of the consequences metrology in chemistry may have on geochemical modeling. Geochemical modeling is a wide and varying subject. A main characteristics is the variety of fields contributing. A list is given in Table 2.9.

Table 2.9 is admittedly a little arbitrary. The table could be extended by including important contributions of other fields, e.g. geophysics, by presenting a finer separation within the sciences, e.g. splitting "chemistry" in inorganic chemistry, biochemistry, aquatic chemistry. Table 2.9, however, intends to give support to the statement that it is extremely difficult for an individual to have a full insight into the contributions of all the fields listed in Table 2.9. Nevertheless, the output produced by a geochemical transport code is a function of all the contributions from the different fields of expertise. Hence, whoever

Table 2.9. Fields of expertise contributing to geochemical modeling

Field	Example contributions
Rheology	Fluid mechanics, dimensionless characteristics
Physics	Conservation principles, mass transport
Geology	Geological formations, subsurface characterisation
Hydrogeology	Fractures, water saturation, groundwater properties
Hydraulics	Analysis and description of groundwater movement
Chemistry	Thermodynamic data, chemical analysis
Mathematics	Theoretical fundament, formal description
Information sciences	Computer architecture, programming languages
Numerical mathematics	Algorithms
Risk analysis	Probability distributions, uncertainty progression

has to use, to present, or to interpret the output of geochemical models has to cope with subjects where he cannot be an expert. This situation is the same for those forwarding the data and those being recipients of the geochemical models and their output. Those having to make decisions on basis of such data, however, are under most circumstances no professsionals in the field of geochemical modeling. To them, the modeling output must be helpful in taking decisions. Thus, it is a common observation that the model output forwarded by one group of modelers is contradicted by a second group of modelers. The situation easily may become frustrating (Oreskes 1994a).

Groundwater models and geochemical models are widely applied. There are hundreds of codes available with widely varying capabilities and applicability (van der Heijde et al. 1985) and their number is steadily increasing. New codes are often generated by linking existing codes. Geochemical modeling services has become a market where services are offered on a commercial basis. The source code of commercial programs is generally unavailable. The purchaser of model services has to rely on the statements of the presenter about the reliability of a code's output. The purchaser of a modeling service is confronted often with computer-generated "black box" values and a team of consultants working hard to dissipate any doubt that the values might be the best answer to the purchaser's needs. The optical, and sometimes even acoustic, perfection of the presentation makes it difficult to concentrate attention on the quality and validity of the data.

Oreskes et al. (1994) point to the importance of a neutral language in presenting and discussing the relevance of computer-generated simulation output. Claims, such as "validated", "verified" or "confirmed" can only have a meaning within the framework of the simulation approach itself. Judgement terms such as excellent, fair, poor cannot have any absolute meaning but only characterize the relative performance of a model in describing the data it models. In areas, they conclude, "where public policies and public safety are at stake, the burden is on the modeler to demonstrate the degree of correspondence between the model and the material world it seeks to represent and to delineate the limits of that correspondence". The final request is of special importance. It is necessary and helpful to ask for the limits of a simulation tool and its output if this point should not have received due attention during a presentation. There has been some euphoria with respect to computers and their abilities which, seen from a 25 years distance, is difficult to grasp. In the field of geochemical modeling, this euphoria occasionally celebrates some renaissance.

2.3.2
"It's Good! It's Calculated by a Computer!"

There has been a time when the advent of automatic calculation machines was welcomed as the clue to otherwise unreachable information. In a 1984 TIME magazine cover story a computer journal editor is cited: "Put the right kind of software into a computer, and it will do whatever you want it to. There may be limits on what you can do with the machines themselves, but there are no limits on what you can do with software". This statement is a reminiscence of the heady days of the computer euphoria. And it is incorrect. There is a long list of tasks and problems that cannot be solved by a computer-independent of its CPU clock speed, working memory, or architecture. There are seemingly simple problems where no algorithm will provide a computable answer: some of the problems cannot be computed, others cannot be decided (Harel 2000).

In the field of geochemical modeling, the above statement has a bit another notation. There is the expectation by model users and purchasers of modeling services that simulation of a particular field problem by a complex computer model will yield inherently true results. The reasons for this expectations are 1) models are sophisticated technology involving computers; 2) all data that exist for a problem can be included and 3) the model closely represents the physics of subsurface flow. In short: there was (and often still is) the belief that the numerical model can be made to give an accurate representation of reality if only enough detail and data are included (Voss 1998).

In fact, a computer model is at the bottom of a sequence of transformation steps from an observation in nature to the simulation of a process by a (computer) model probably giving rise to that observation. A scheme according to that proposition is shown in Fig. 2.8. We do not understand nature (reality). Seen in an extreme way, reality is a kind of a helpful hypothesis. Wittgenstein's

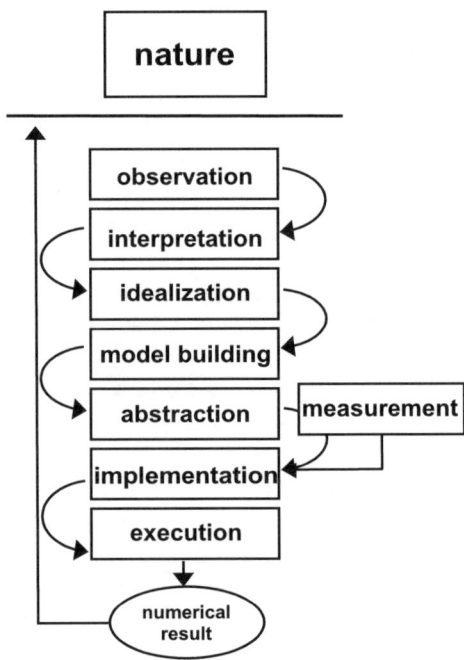

Figure 2.8. Nature, observation and simulation: a successive abstraction

statements "The world is everything that is the case" and "the world is determined by the facts, and by these being *all* the facts" (Wittgenstein 1921) refers to our limits to perceive nature. If the completeness of reality (*all* facts) is reduced to a very limited amount of facts – it is not the reality (the world, nature) any longer. Nature (reality) is observable and conclusions can be drawn. Observation of a phenomenon is usually the first step. It is not quite clear what we actually observe. However, scientific observation is a special kind of observation. Scientific observation is made by measurement. The following statement by Thompson (Lord Kelvin) is well known among metrologists: "I often say that when you can measure what you are speaking about and express it in numbers you know something about it. But if you cannot measure it, when you cannot express it in numbers, your knowledge is of a meagre and unsatisfactory kind". This dictum concludes with the statement that, when you cannot measure then there is no science at all (Thomson 1891). Biologists and geologists have objected this statement. However, scientific observation shall have another quality: it must be reproducible. Singular events may be highly interesting but cannot be subject to scientific investigation. Knowledge obtained from scientific activity may be applied to a singular event. Hypotheses may be raised. But these events cannot be a subject of repeated investigation.

In the present context, Lord Kelvin's statement may be slightly modified: If an observation cannot be expressed by numbers it will be difficult to simulate the observation in a computer. Inquiring into what is measured at all may rise further interesting questions. A key element of the (still) high esteem of science in the general public is its "intention for truth" (von Weizsäcker 1990). But do true values exist at all (de Bièvre 2000; Meinrath 2002; de Bièvre 2006; Fuentes-Arderiu 2006)? If true values do not exist, what is the aim of a scientific measurement? In short, the link between the "real" world and our observation is subject to debate – a debate going for at least 2500 years.

An observation alone is not a basis for science. The observation requires interpretation. It must be set into a context. Within that context the observation is interpreted. A rainbow is nowadays interpreted as consequence of light refraction in droplets. Others interpret it as a divine message. To put an observation into the right context is what "great" scientists have achieved. Fermi, for instance, had irradiated uranium with neutrons before Hahn, Meitner and Strassmann (Hahn 1962). But Fermi's interpretation was a different one – it did not lead him to new, clarifying insight. Fermi had his successes at other occasions, Hahn and Strassmann are remembered as the discoverers of nuclear fission.

The new insight must be put into a model. Science is thinking in terms of models. Often some aspects of the observations cannot be completely accommodated by the interpretation. These observations (or fractions of the observed variance) can be due to random processes (stochastic uncertainty), measurement uncertainty etc. Idealisation is, to some extend, a normal and necessary part of science. It clearly, however, must be set aside from fraud, the willful modification of observations in order to make it fit to a hypothesis.

A model is based on a conceptional scheme. If a relevant process, feature or event is not included into this conceptional scheme, the model resulting from that conceptional notion will not be able to model processes which are influenced by the respective process, feature or event. Idealisation eases model-building. The way of building a model can take any road, from intuition (E. Schrödinger and the wave function) to systematic logical derivation from first principles (P. Debye and the Debye–Hückel law). Less experienced computer-users are not always aware of the fact that modern algorithms fit a model curve to any data provided a sufficient number of variable parameters are included into the numerical description. Model-building is at the heart of science and some guidelines are available. At least at this point of the process, statistics is indispensable (Box 1960; Atkinson and Fedorov 1975; Box and Draper 1987). If numerical data are related to numerical models using numerical criteria: which other field could possibly provide defendable, objective criteria, ready for a later re-examination if not the field of statistics?

A model on paper is an important step for a successful simulation of given observations but commonly the model needs to be transformed from its mathematical form into a computer algorithm and a computer language. Computer languages are not equivalent as the computing machines are not equivalent. A different operating system may require a modified computer code. The step from the paper form to the computer code is most sensitive because up to this point the scientist making the observation and creating the model may be in control of the process. If it turns to the computer implementation, a different breed takes over the lead.

The implementation can be a major hassle. A simple model, neatly expressed in linear equations does not create a major headache, but complex models like groundwater models and geochemical transport models require the simultaneous solution of higher-order differential equations. There are two principal concerns: stability (no crashes due to divisions by zeros, infinite loops, memory overflow, rounding errors, numerical instabilities (quotients of subtractions of numbers with close values are notorious candidates)). Other topics are non-random random number generators and zero determinants in matrix inversions. The numerical operations during the execution of a groundwater modeling code are so numerous that an error analysis is almost impossible. Typical computer codes which numerically solve the partial differential equations, and all of their auxiliary equations, can contain tens to hundreds of thousands of lines of code. Similarly, the inputs and outputs of these codes have high dimensionality, for example, hundreds of input and output quantities (Oberkampf et al. 2004). Therefore, computer codes must be validated, where it is not always clear to what extent a code can be tested in all its facets. Burmester and Anderson (1994) give some general principles that may be helpful in discussion with modeling service providers. It is, furthermore, helpful to remind that all groundwater modeling codes require discretisation; that is: the replacement of smooth curves by individual, distant point locations.

With all that information at hand, facing the output or even the illustrated consequences of a computer model, the reminder might be helpful that people of the civilized world are confronted on an almost daily basis with the output of highly developed, often even governmentally funded computer simulations: weather forecasts.

The implementation also must take into account the measurement data the code should interpret. The algorithms need to account for the conditions of the measurement information. The dimensionality of the implemented physical quantities must be correct. Otherwise the output is garbage. Computer scientists introduced the GIGO principle: garbage in, garbage out. No computer code can forward good results from bad input.

The execution of the code includes the transformation of the input data from the given format into the format acceptable for the computer. In case of the database section in Fig. 2.3, all additions need to follow the list structure. New data must be recalculated to fit within the scheme. In some cases, the numerical values of a new entry are determined in terms of a species already included in the database. An example is carbonato species. If, for instance, a species $U(CO_3)^{2+}$ is to be included into a database of a format given in Fig. 2.3, then the formation constant of the ligand CO_3^{2-} is usually already included into this database. Hence, in most cases the formation constant of species $U(CO_3)^{2+}$ must be corrected to zero ionic strength and recalculated to interpret the formation of $U(CO_3)^{2+}$ correctly with a different formation constant of ligand CO_3^{2-}.

The output of these time-consuming procedures is usually numbers, again. These numbers need to be suitably summarised and presented. The preparation of the summary and the presentation, again, is human activity. The modern computers, however, are graphics machines. With graphical rendering such presentations can be fascinatingly realistic. Without a distant attitude and own interpretation experience it is difficult resist the message. Here, a word from Peter Medawar's (noble laureate for medicine) famous "Advice to a young scientist" (Medawar 1979) might be helpful: "I cannot give any scientist of any age better advice than this: the intensity of the conviction that a hypothesis is true has no bearing on whether it is true or not". In some cases it may be helpful to ponder on the similarity of some geochemical simulation output presentations with computer game graphics.

Hence, whether the complete procedure has any relationship with nature must be determined separately, and with great reluctance. The fact that a simulation program can be made into interpreting a limited number of smooth data points has no meaning. Taking the number of parameters available in such complex codes to adapt the model to the data, good agreement between curves and data is not a proof for the quality of a model but a mere necessity. An interesting discussion has been given by Ewing et al. (1999) with respect to performance assessment of nuclear wastes.

When judging the agreement between simulated data and observations from nature, it must be kept in mind that chemical processes are almost always described by the assumption of thermodynamic equilibrium. Sorption processes are either treated by lumped parameter equations (e.g. K_D values or sorption isotherms). The surface complexation approach is intensively investigated. It is based on the assumption that solution species react with binding sites at surfaces in the same way as happens in homogeneous solution. The resulting values for the interaction parameters (surface complexation constants) are treated as fundamental constants of nature just like the thermodynamic formation constants described by Eq. (2.25). The justification for these assumptions is too often not discussed.

Microbiological processes which govern much of the surface processes in natural aqueous systems (phytobenthos) cannot be included into the modeling at all. These are the major neglected effects in most geochemical modeling studies. Furthermore it must be expected that only a minor part of chemical species in a groundwater is present as a well defined species. Most inorganic material can be filtered off readily (e.g. Palmer 1989). There is, at present, no practicable concept to overcome these limitations.

2.4
Criteria for and Deficits in Chemical Data for Geochemical Modeling

"I cannot give any scientist of any age better advice than this: the intensity of the conviction that a hypothesis is true has no bearing on whether it is true or not. The importance of the strength of our conviction is only to provide a proportionately strong incentive to find out if the hypothesis will stand up to critical evaluation."

(P. Medawar 1979)

2.4.1
Deficit Analysis of Thermodynamic Data

Thermodynamic data are abundantly available in literature and various compilations exist. However, it is almost impossible to judge these data on basis of objective criteria. The statement that data without clear documentation of their determination and uncertainty are "rumours" is certainly valid for thermodynamic data of chemical reactions.

The past and on-going activities to collate available data from literature and to attempt a critical selection process have been mentioned above. The goal is a standardisation: everybody should use the same values for a given reaction. Such a standardisation is nothing else but a convention. The concern with these "critical" databases is the purely subjective procedure of selecting, advocating, rejecting and ignoring data. These concerns may be substantiated by an example from a recently published volume of this database (OECD/NEA 2003). The importance of the OECD/NEA database results from the intention to use the recommended data as input data for nuclear waste disposal in deep geological formations. This collection of "recommended data" includes abstracts of the reviewed manuscripts, among these a manuscript dealing with the interpretation of literature data by statistical, computer-intensive resampling methods (Meinrath et al. 1999a). In the sense of the review (OECD/NEA 2000a), this manuscript (like several others from this author) did not deal at all with the experimental determination of formation constants. The manuscript discussed the advantages of computer-intensive resampling methods in reassessing published literature data. The paragraph of the OECD/NEA collection is picked out because it explicitly states the dilemma faced by all thermodynamic data collections: There is a wealth of data and a multitude of ways to interpret these data but no objective criteria to judge them. No reviewing guideline can assign the literature data what must be assigned to them by the experimentalist(s) as an essential part of the measurement process: quality.

The abstract (OECD/NEA 2003) states: "The estimation of uncertainty of published data has been and still is a problem in the NEA-TDB reviews, because the primary experimental data are rarely available. The reviewers have therefore used both the authors' estimates and their own expert experience on the precision expected of a given experimental method when estimating the uncertainty of equilibrium constants. In systems where one can obtain independent experimental information on speciation by different methods (e.g. potentiometry), one often finds an excellent agreement between the methods, indicating that the uncertainty estimates are reasonable. It should also be pointed out that the uncertainty estimates rarely change the conclusions of predictive geochemical modeling".

This short paragraph highlights several crucial points. First the reviewer(s) claim(s) that different methods result in "excellent agreement" with the subsequent conclusion that the (rather small) uncertainty estimates are meaningful. Are these statements justified? There is, as has been shown at several occasions, no reason to assume "excellent agreement" (cf. Fig. 1.49). For one, the costly and tedious database projects (there are other database collation projects currently going on) would not be necessary at all if the agreement within the literature data would be "excellent". For two, Bond and Hefter (1980) stated: "It is an unfortunate, but only too well known, fact of solution chemistry that different techniques may yield different values of stability constants. Intending investigators therefore need to recognise and understand the known limitations of any method they are considering for use and to bear in mind the possibility of unknown limitations". For three, potentiometry, which is explicitly mentioned in the abstract from OECD/NEA (2003), may serve as a further example. We repeat some of the statements from Filella and May (2005) already cited in Sect. 1.8.1, that the formation constants from potentiometric titrations are "calculated and reported in disparate ways". The result is "difficulty to make valid comparisons between the results from different investigators". Filella and May further state: "There are at least three reasons why this difficulty has become acute in recent years. First, there has been a proliferation of computer programs to determine formation constants. Secondly, many possible methods of calculation, each capable of producing significantly different answers, have become readily available. (...) Thirdly, and perhaps most importantly, formation-constant determination remains necessary but is less fashionable than it used to be. So, results are now often performed by researchers with less experience in this field". Evidently, these authors (P. May is the author of the widely used ESTA codes for analysis of potentiometric data and co-author of the consistent thermodynamic database system JESS) do not conform to the claim of "excellent agreement by different methods". In contrary, both authors further state: "It is well known that the considerable discrepancies between values published for the same chemical system by various authors are a notorious feature of formation constant measurements". An attempt of a metrological assessment of potentiometric titration for the determination of formation constants is reported by Kufelnicki et al. (2005). The study was motivated by the widely varying results on the protonation constants for arsenazo III. The cause-and-effect analysis indicated that the method of potentiometric titration is by far not as precise as assumed on basis of repeatabilities. There are a larger number of uncertainty contributions which become apparent only in interlaboratory comparisons, and by comparing the outcome from independent measurements obtained in different laboratories (Kufelnicki et al. 2005).

The further crucial issue highlighted by the short paragraph from the OECD/NEA review is the statement that "the reviewers have used both the authors' estimates and their own expert experience on the precision expected of a given experimental method when estimating the uncertainty of equilibrium constants". There is, to the knowledge of this author, no record for any of the six reviewers mentioned in OECD/NEA (2003) on the persistent application of any what-so-ever statistical, chemometric or metrological method for a realistic assessment of repeatabilities (there is no expectation that the reviewers should have experience in the evaluation of complete measurement uncertainty budgets) for multi-parameter systems, potentially with a considerable degree of non-linearity and non-normally distributed residuals. Expressed in the other way this paragraph documents the deep conviction of the reviewers that they will recognise a good value if they see it. Whether this conviction is a solid basis for the collation of a database for long-term nuclear waste disposal must be judged by the decision-makers.

The third crucial point highlights the impact of chemical thermodynamic data upon the results of geochemical modeling: "It should also be pointed out that the uncertainty estimates rarely change the conclusions of predictive geochemical modeling". This statement may cause surprise because if there is no impact of chemical input data on the simulation output, why then making the bother and spending years with tedious assessment of thermodynamic data from literature? It would be interesting to know on which information the OECD/NEA reviewers base their assertion. The reports of Hamed and Bedient (1999), Denison and Garnier-Laplace (2005) and the Monte Carlo transport simulation from Nitzsche et al. (2000) are but a few examples in literature telling the opposite evidence, and these reports are only a rather limited selection from literature (e.g. Serkiz et al. 1996; Criscenti et al. 1996; Grauer 1997; Onysko and McNearny 1997; Meinrath et al. 2000).

May and Murray (2001) summarise: "Serious discrepancies between published thermodynamic parameters of chemical reactions are well-known. Since there are many different causes of these problems (such as experimental error, inadequate theory, and carelessness), they can be very difficult to eliminate. The situation is made worse because many thermodynamic data persisting in the scientific literature stem from values that are later corrected or become experimentally superseded. In this regard, a great effort is needed for critical assessment of all the relevant primary measurements and for their transformation into thermodynamically consistent datasets (...) Chemical modeling is thus often compromised".

The OECD/NEA review is singled out here because of the relevance of the field it is directed to and because it also includes published work of the present author. The disposal of nuclear wastes is a considerable problem in many modern societies. The need for disposal is difficult to deny, but the decision on a disposal site is becoming a major headache. Thermodynamic data is reviewed with the eventual goal to underpin the message that long-term safety can be evaluated, e.g. by geochemical modeling. Therefore, the availability of a consistent set of meaningful thermodynamic data bears some importance. Such a database may be implemented on basis of a convention. It should, however, be kept in mind that such a convention can be easily established among a power group (cf. Sect. 1.2) having an interest to close the nuclear fuel cycle (e.g. Commission of the European Communities, US Department of Energy, OECD with its Nuclear Energy Agency, producers of electricity by nuclear power etc.). The conflict of interest, e.g. with the population in the surroundings of the nuclear waste repository and, not to forget, future generations of people having to carry the risk from the waste repository, may render such a convention void. Therefore, relying on a thermodynamic database established by convention, may carry a considerable risk.

As outlined in Part I, the current situation in chemical measurement is not much different from the situation in physical measurements before the advent of a metrological system. The evaluation of the meter reference at the turn from the eighteenth to the nineteenth century resulted in surprising discrepancies of the underlying measurement data obtained during the determination of the Paris meridian length. In those times, nature was considered perfect. Corrections in measurement data to get the (unprecise) measurement results back to (supposed) precision were not uncommon. Because the meter reference, however, was becoming an international standard, the data resulting in its value were scrutinized in much more detail (Delambre 1810). As a result of that scrutiny, existence of measurement uncertainty could not any longer be ignored. In 1805, Legendre (1805) proposed a solution to the problem of judging uncertain data with exact mathematics: the method of least squares, only to learn that in Göttingen, Gauss had used this method for already a decade (Gauss 1809). Hence, the solution to the problem of uncertain data was to accept measurement uncertainty, to study measurement uncertainty and to minimize measurement uncertainty. Without this development modern technology would not have been possible. Hence, the achievements of metrology of physical quantities should be well pondered by chemical analysts. The field of production metrology may teach further lessons (Clapham 1992; Kunzmann et al. 2005). It is essential that a reasonable measurement uncertainty budget is only one relevant factor in a comparable measurement value. The second, metrological traceability, is perhaps even more important (Hibbert 2006). But for most chemical thermodynamic data a metrological traceability doesn't exist. Even worse, the authors are not even aware of its fundamental relevance. Clearly, "improving" repeatability by ignorance of certain published values does not add any value to the quality of the thus obtained "critical values".

2.4.2
Criteria for the Determination and Reporting of Thermodynamic Data

The practical example may underscore the following list of deficits in thermodynamic data:

- Large amount of published data, which at the same time are widely scattered.
- Almost complete lack of documentation to assess the quality of the data; introduces an extreme arbitrariness into all data collections (databases).
- Completely unknown internal consistence within individual data. "Consistent databases" introduce the consistency after the collation of the data.
- Obsolete data and contradictory data (e.g. $Ca(hco_3)^+$ and $caco_3^\circ$; both species interpret the same experimental variance) in existing data collections.
- High competence of reviewers is required; practical experience in the measurement, evaluation and statistical and metrological assessment of data is required. As outlined by Filella and May (2005) this kind of expertise is rare.
- Almost no experience from round-robin tests is available to have a complementary data pool for the assessment of interlaboratory variance (e.g. Via nordtest approach; Magnusson et al. 2004).
- Lack of a realistic estimate for the likely reliability of an entry in the thermodynamic data collection (complete measurement uncertainty budget).
- Often unrealistic values for temperature dependence (temperature effects on formation constants of ionic compounds are far below the resolution ca-

pabilities of the classical methods for the determination of thermodynamic data).

- Lack of tools (protocols, software) to recalculate formation constants and solubility products into reaction enthalpies and vice versa.

Thermodynamic data for geochemical modeling should be determined in the future. The following criteria should be met:

- Traceable analytical measurements, performed and documented according to internationally recognised quality assurance criteria (accreditation, regular participation in proficiency tests, educated and trained personnel, certified calibration material, where available).
- Documentation of the measurements and evaluations according to internationally agreed quality assurance protocols (e.g. Good laboratory practice; OECD 1999).
- Adequately trained personnel for scientific measurements. In these days, the laboratory work is performed mainly by novices to the field (e.g. PhD fellows) (Chalmers 1993; Filella and May 2005).
- Generation of a cause-and-effect diagram for the complete measurement and data evaluation process.
- Annotated list of type b evaluation uncertainties.
- Explicit statement of the empirical distribution function of an evaluated thermodynamic value.
- Documentation of numerical and statistical procedures applied in the evaluation of experimental data. The data evaluation process must include a discussion of influential data points and their influence on the final values.
- Documentation of laboratory notes (e.g. as xerox copies).

2.4.3
Comments

The above lists are understood, from today's point of view, as illusory. The amount of certified reference materials necessary for that purpose would exceed the production capacities. There is little expectation to re-measure the bulk of thermodynamic data in literature. These lists can, however, serve three purposes:

1. Giving a point of reference to which today's quality of thermodynamic data can be compared and indicate a direction to proceed,
2. Specifying the minimum requirements to overcome the unsatisfactory situation outlined, for instance, by Bond and Hefter (1980) and Filella and May (2005),
3. Give an alternative to the highly subjective compilation of standardised "databases" which present at best conventions (cf. Part I for a definition of a convention).

Metrology in chemistry does not require the re-measurement of all data but the statement of a reasonable estimate of measurement uncertainty. The majority of uncertainty estimates (it is assumed that the figure following the "±" in some thermodynamic data compilations and literatures represents a standard deviation, thus indicating normally distributed variables) are much too small. The consequences are mutually discrepant literature values and overestimation of data reliability (Ellison et al. 1997; de Bièvre 2002; Golze 2003).

For existing data, the major deficit of existing thermodynamic data is the complete lack of trust in their reliability. Even though the Guide to the Expression of Uncertainty had been issued over a decade ago, the attempts to implement its concepts into thermodynamic data determination have been spurious. It is, however, unrealistic to expect that the public will accept values (chemical "information") that cannot be related to anything else than the deep conviction of a small group of scientists.

Here, trust is the key word. Thermodynamic data are a result of analytical measurements (determination of the quantity "amount of substance"). Analytical results are based decisions in areas such as economics, science, trade, health care, environmental protection, law and sports. In conjunction with the globalisation of these fields, the range of decisions and the demand for comparable and reliable results are increasing (Zschunke 1998).

2.4.4
Deficit Analysis of Surface Interaction Parameters

Sorption data play an important role in geochemical modeling. Sorption is considered as the main retardation mechanism for contaminants. In nuclear waste performance assessment sorption to technical barriers is considered an additional safety feature. In case of accidental release, sorption to the geosphere formation is expected to reduce the radiation doses experienced by the biosphere. In fact, the determination of sorption data (K_D values and surface complexation constants) has always had an imprint share in the governmentally funded nuclear safety research.

Sorption has not been discussed in the preceding sections of Part II. A reason for this neglect is the complexity of fitting and calculation that enters the evaluation of sorption data. In almost all available literature on the determination of sorption parameters these relevant auxiliary informations were obtained by least-square mean value fitting. The determination of sorption parameters, therefore, is an extended exercise with basically the same elements as the determination of thermodynamic data. Speciation calculations are a crucial element in determination of sorption parameters, independent of whether the experimental observations are interpreted by the simple K_D concept or multi-parameter models (e.g. surface complexation models). The reasons are, first, the dependence of the surface processes from the species composition in solution, and second, the amount of substance bound to a surface must be estimated by the difference of the amount of substance initially added to the system and the amount of substance found analytically after equilibration. Hence, much of the discussion would have been a repetition of Part I. However, criteria and deficit analysis of chemical data for geochemical modeling would be incomplete without considering sorption.

The limited ability of analytical chemistry to quantify amount of substance has been discussed in some detail in Chap. 1.8. The results of a IMEP ILC (IMEP-14) shown in Fig. 2.9 intend to demonstrate these limitations (Papadakis et al. 2004). Clearly, each laboratory could take the stand that their value is correct (e.g. by insisting on their years of expertise in the field) and declare other values for less reliable or faulty. However, there is no reason to do so except the fact that a reference value exists.

Even with respect to this reference value the stand can be taken that the reference value might be erroneous, anyway. Then, progress and communication about measurement values is not possible. If, however, the reference values are accepted and the results reported by all 239 laboratories from 43 countries, spread over five continents are presented as shown in Fig. 2.9, a similar pattern is observed for almost all elements and also most previous IMEP comparisons (www.imap.ws).

Hence with respect to sorption studies, where the amounts of a substance need to be determined before and after an equilibration experiment, the primary concern is not with the single laboratory's values but the limitations in repeating the experiment elsewhere. If repeatability cannot be ensured (demonstrated, e.g. by participation in a proficiency test like IMEP or local round-robin studies organised by national metrological bodies), the resulting values will likewise be enormously scattered. It is clear that no "expert experience" will be able to improve this scatter without the risk of introducing an unspecified amount of subjective bias, as is the case in thermodynamic data.

The term "sorption" is often used in describing the interaction of solute components with a surface, is not sufficiently precise. Sorption is commonly used for phenomena if the detailed process is unknown (Römpp 1999). Therefore, the term "surface interaction parameters" will be used.

The main deficits of surface interaction parameters are:

- Lack of transportability; the applicability of the parameters is commonly restricted to the conditions of their determination. These conditions are commonly laboratory conditions and far away from the subsurface situation;
- There is almost no tool for quality assurance. It is generally not even specified what the "quality" the data need to have to be "fit for purpose";
- The interpretation of measured solubility data is often restricted to some kind of curve fitting. There is generally no reference to the respective numerical and statistical issues that might affect such a process. Fitted parameters are accepted "as given" from the fitting tool;

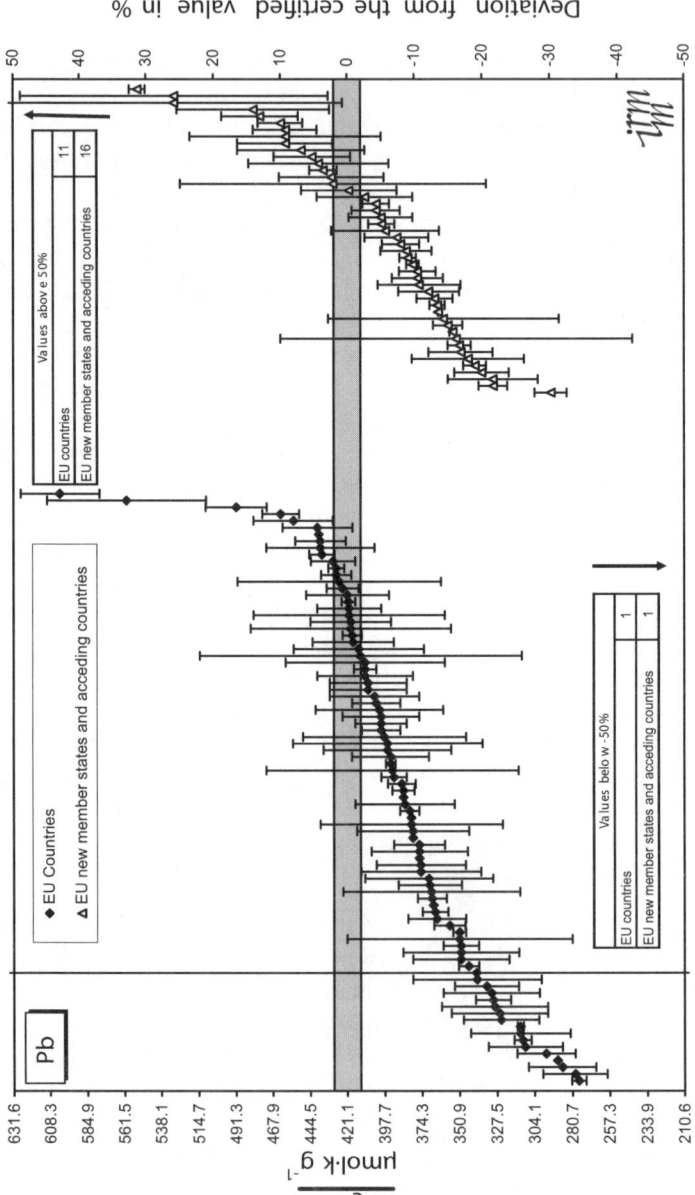

Figure 2.9. Results reported by 239 laboratories participating in IMEP-14 (Papadakis et al. 2004). Shown are the values reported for lead in sediment. The distribution of results for the other elements (As, B, Cd, Cr, Cu, Fe, Hg, Ni, Se, U, Zn) are comparable to the results given for Pb

- Lack of a complete assessment of additional influence factors, e.g. surface area, variability of surface conditions in the experimentation vessel, interface conditions, density of surface sites;
- A variety of theoretical approaches, often in combination with assumptions about the solid-solution interface, which cannot be verified on basis of the measurement data;
- Often considerable scatter in the data that only allow to establish a general trend but are interpreted by multi-parameter models.

2.4.5
Example: Surface Complexation

As an illustration determination of surface interaction parameter (surface complexation constants) values obtained for UO_2^{2+} sorption on aluminium oxide will be discussed (Jakobsson 1999). While Jakobsson (1999) reports on the evaluation of mean values, the following discussion will focus on limitations in such an evaluation due to measurement uncertainties of various kinds. The evaluation presented by Jakobsson (1999) is a typical example for an evaluation of surface interaction constants, however presented in a more complete way than possible in a journal article.

Two kinds of surface sites are discussed: protonated surface sites with a dissociation constant K_a and interaction of dissolved uranium species with these deprotonated surface sites. The surface interaction parameters are considered to be adequately described by equilibrium parameters, K_S. In addition, interaction of the deprotonated surface sites with cations of the background electrolyte (here Na^+) need to be taken into account.

The surface site acidity constants K_a are given by Eq. (2.31). In this definition of the K_a value the surface area is already included:

$$K_a = \frac{(c_{tot} - c_{aq})}{c_{aq}} \cdot \frac{V}{A} .$$

(2.31)

The parameters c_{tot} and c_{aq} designate the total concentration of the sorbed component and the amount of the substance in the aqueous phase after equilibration. V gives the aqueous phase volume and A the surface area of the stationary phase.

The experimental data obtained from potentiometric titrations are given in Fig. 2.10. As the final conclusion from the experimental determination of U(VI) concentrations at $NaNO_3$ and $NaCl$ solutions with different ionic strengths, two surface sites were reported with $pK_{a1} = 7.2 \pm 0.6$ and $pK_{a2} = 11.2 \pm 0.4$. The relevant reactions are:

$$K_{a1} :\equiv SOH + H^+ \Longleftrightarrow SOH_2^+$$

(2.32a)

$$K_{a2} :\equiv SOH \Longleftrightarrow \equiv SO^- + H^+ .$$

(2.32b)

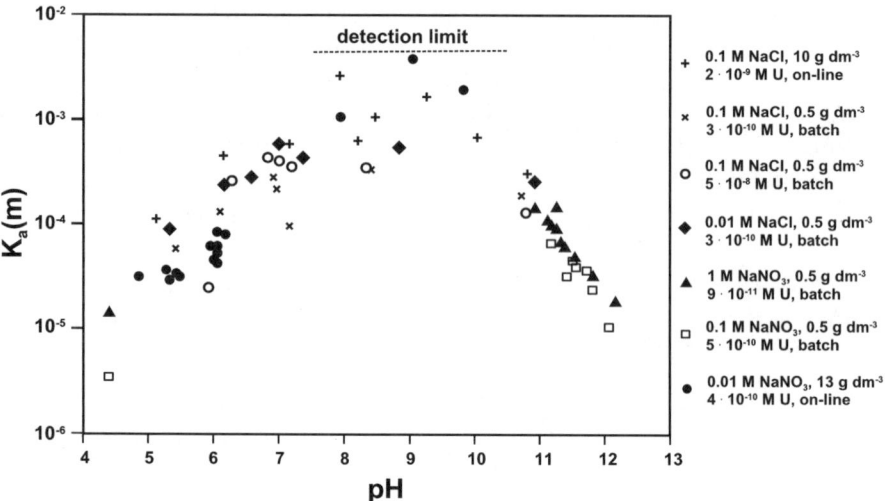

Figure 2.10. UO_2^{2+} sorption onto aluminium oxide as a function of pH for different ionic strengths, UO_2^{2+} and solid phase concentrations (according to Jakobsson 1999)

These surface sites and their respective surface acidity constants K_{a1} and K_{a2} were determined experimentally. The model relies on the assumption that K_{a1} and K_{a2} are constants according to the equilibria:

$$K_{a1} = \frac{\{\equiv SOH_2^+\}}{\{\equiv SOH\}\{H^+\}} \tag{2.33a}$$

$$K_{a2} = \frac{\{\equiv SO^-\}\{H^+\}}{\{\equiv SOH.\}} \tag{2.33b}$$

Furthermore, the interaction of the electrolyte cation, Na^+, with the surface site needs to be taken into account:

$$K_{Na^+} :\equiv SO^- + Na^+ \Longleftrightarrow \equiv SONa . \tag{2.34}$$

The three values enter the evaluation of the interaction parameters of U(VI) with the both sites as mean values. Data interpretation was made on basis of a triple layer surface charge model. The model was chosen because this model allowed the "best numerical interpretation". As a result, two surface interaction parameters for uranium(VI) were derived:

$$\equiv SOH + UO_2^{2+} \Longleftrightarrow SOUO_2^+ : \lg K_{s1}(UO_2^{2+}) = -3.1 \tag{2.35a}$$

$$2 \equiv SOH + UO_2^{2+} \Longleftrightarrow (SO)_2UO_2 : \lg K_{s2}(UO_2^{2+}) = 3.4 . \tag{2.35b}$$

This short summary illustrates the multiplicity of parameters entering a determination of surface complexation constants. Each parameter is either determined independently by some, often complicated, measurement process (e.g. the surface acidity constants), taken from literature (e.g. K_{Na+}), or obtained by least-squares curve fitting. The ancillary data enter the evaluation without their (often considerable) uncertainty.

A small sample of influence factors will be discussed using the concentration determination of U(VI) in NaCl and $NaNO_3$ electrolyte solutions holding $(5 \cdot 10^{-8} - 9 \cdot 10^{-11})$ mol dm^{-3} U(VI) and 0.5 g dm^{-3} to 13 g dm^{-3} Al_2O_3 as example to illustrate some relevant influences. This discussion intends to provide a rationale for the above given deficits and the criteria given below. Most of the discussion points affect other studies of surface interaction parameters and may serve as a guideline for assessing modeling output where surface interaction parameters (e.g. sorption databases) are involved.

Figure 2.9 illustrates that there is no absolute way to determine values for amount of substances. This inability is not a fault of the researcher but an essential element of chemical measurements – their sensitivity to random contributions by influence factors. Without a careful analysis of the detailed procedures, e.g. by participation in proficiency tests, this variability cannot be quantified within a laboratory. The EURACHEM/CITAC bottom-up approach is a protocol to estimate the magnitude of influence. In the study under discussion, radiotracers were used. No information on relevant influence factors, e.g. counting statistics (liquid scintillation counting), is given. Hence, measurement uncertainty with respect to the determination of the metal ion before and after the experiment is unavailable.

Determination of pH was made on basis of calibration materials assessed by Gran titrations. Gran titrations, too, are not free from uncertainties. These uncertainties are commonly not discussed, but of course present (e.g. Burden and Euler 1975). In the graph (cf. Fig. 2.10) only the measured concentration mean values are given forming a saddle point at about pH 9.

The two surface complexation constants (cf. Eqs (2.35b)) cannot be read directly from the experimental data. The both values depend, among numerous other parameters, on the species composition of the aqueous solution. A computer program is necessary to calculate the probable composition of the solution at each pH. These calculations require further ancillary information, e.g. the volume of solution in the experimental vessel, the pH in solution, the amount of sorbed uranium, the concentration of titre solution and its added volume etc.

The concentration of initial and sorbed uranium in solution cannot be determined independently. The amount of sorbed uranium, m_{sorb} depends on the initial amount of uranium, c_{tot}, the uranium solution concentration after equilibration, c_{aq} (cf. Eq. (2.31)), and the volume of the vessel. All three parameters cannot be quantified with arbitrary accuracy, especially at the low total uranium concentration level range $(5 \cdot 10^{-8} - 9 \cdot 10^{-11})$ mol dm^{-3}. Some consequences have been discussed previously. If a standard deviation of 10% is assumed for the assessment of both concentrations, c_{tot} and c_{aq}, the following relationship holds for a solution, where 50% of the initial uranium has disappeared from solution:

$$K_a = \frac{(5 \cdot 10^{-8} \pm 5 \cdot 10^{-9})\,\mathrm{mol\,dm^{-3}} - (2.5 \cdot 10^{-8} \pm 2.5 \cdot 10^{-9})\,\mathrm{mol\,dm^{-3}}}{(2.5 \cdot 10^{-8} \pm 2.5 \cdot 10^{-9})\,\mathrm{mol\,dm^{-3}}} \frac{V}{A}.$$

(2.36)

For the sake of simplicity, it is assumed that the volume V and surface area A are known without uncertainty. The amount of sorbed substance is therefore (cf. Chap. 1.3):

$$c_{\mathrm{sorb}} = (2.5 \cdot 10^{-8} \pm 5.6 \cdot 10^{-9})\,\mathrm{mol\,dm^{-3}}$$

(2.37)

and for K_a follows:

$$K_a = (1.0 \pm 0.25)\frac{V}{A}.$$

(2.38)

Hence, the uncertainty of 10% in the entry quantities to Eq. (2.31) results in a 25% uncertainty of K_a, conditional that there are no other uncertainties contributing, e.g. in the quantity A. The surface area of Al_2O_3 was determined by BET and a surface area of $143\,\mathrm{m^2\,g^{-1}}$ was given. The surface area is often determined by a series of BET measurements. Values for surface area by BET measurements depend, e.g. on the gas used, the pre-treatment of the material, the variability of the size distribution and porosity of materials etc. It should be explicitly mentioned that these uncertainties are mentioned in Jakobsson (1999). As in almost all similar studies, the impact of these uncertainties on the numerically evaluated quantities is not assessed.

Error progression rules indicate that the uncertainties in the amounts of substance in solution and those sampled from solution play an important part in the determination of K_a values. However, a 1:1 distribution between solution and solid results in a value $K_a = 7 \cdot 10^{-6}\,\mathrm{m}$. Higher K_a values (cf. Fig. 2.10) require higher values in c_{sorb}. Simple error progression, however, indicates that the overall uncertainty will rise with increasing sorption. The relationship is illustrated in Fig. 2.11 for four overall uncertainties in concentration determination: 5%, 10%, 15% and 20%. The ability to evaluate a meaningful estimate may fade at locer sorption ratios.

Considering the current great efforts in determination of surface interaction parameters, especially those aiming at the application in performance assessment studies for nuclear waste disposal, it may be surprising that such an analysis is not routinely accompanying all reports of the respective parameters. A similar analysis, using K_D values of iron as an example, has been given previously (Meinrath et al. 2004a). It should be noted that Eq. (2.31) gives a measurand (K_a) in a closed mathematical form. Therefore, such an analysis could be made within the framework of the bottom-up approach by EURACHEM/CITAC (2002).

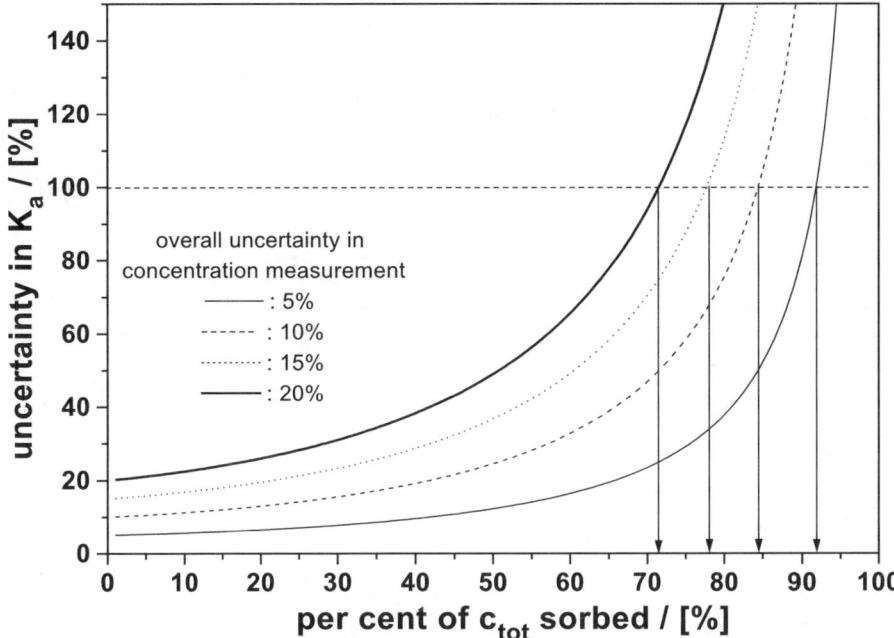

Figure 2.11. Effect of the uncertainty in the determination of amount of substance values upon the uncertainty of the quantity K_a, evaluated by progression-of-error

Determination of K_a values is only a first step in the evaluation of surface complexation constants. The model of surface complexation assumes that there are sites on the surface that bind solution species in the same way as ligands in solution bind to, say, cations. A general equation is given in Eq. (2.39)

$$\equiv SOH_n^r + X^{z+} + yH_2O \Longleftrightarrow \equiv SOH_cX(OH)_y^{(r+z-y)} + (n - c + y)H^+ , \quad (2.39)$$

where $\equiv SOH_n^r$ is a surface site, X^{z+} is a cation of charge z^+ and $\equiv SOH_c X(OH)_y^{(r+z-y)}$ is a surface species (Hiemstra et al. 1989).

Considerable efforts have been taken to provide evidence that such species even exists. There are several studies where structural data, e.g. by EXAFS or IR spectroscopy, have been interpreted in terms of such species.

A formation constant of a surface species depends, of course, on the concentration of X^{z+}. In an hydrolysed solution, its value is therefore determined conditional on the solution species. The tendency to form such solution species is expressed by thermodynamic formation constants (cf. Eq. (2.25)). Consequently, the value of a surface complexation constant crucially depends on the number of species, their composition and the respective formation constants. The data in Fig. 2.12 were interpreted by the thermodynamic formation constants in Table 2.10.

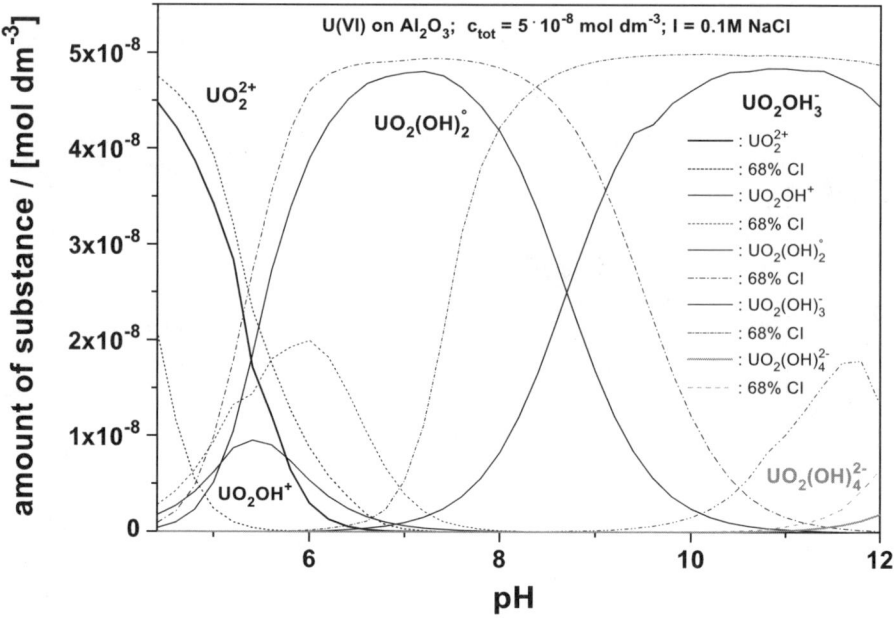

Figure 2.12. LJUNGSKILE speciation simulating solution conditions of a sorption study of U(VI) on Al_2O_3 under a nitrogen atmosphere (Jakobsson 1999). The U(VI) species included into the simulation are given in Table 2.10. (Uncertainty limits give 68% confidence levels). At given confidence level all hydrolysed species may also have zero concentration

Table 2.10. Thermodynamic formation constants for U(VI) hydrolysis species. A few values had been adjusted in Jakobsson (1999) to achieve a good fit of the data

Reaction	Formation constant $(I = 0, 25 °C)$	"Adjusted"
$UO_2^{2+} + H_2O \Longleftrightarrow UO_2(OH)^+ + H^+$	-5.2 ± 0.3	
$UO_2^{2+} + 2 H_2O \Longleftrightarrow UO_2(OH)_2^o + 2 H^+$	-10.3	-12
$UO_2^{2+} + 3 H_2O \Longleftrightarrow UO_2(OH)_3^- + 3 H^+$	-19.2 ± 0.4	-22.4
$UO_2^{2+} + 4 H_2O \Longleftrightarrow UO_2(OH)_4^{2-} + 4 H^+$	-33 ± 2	-35.7
$2 UO_2^{2+} + H_2O \Longleftrightarrow (UO_2)2(OH)^{3+} + H^+$	-2.7 ± 1.0	
$2 UO_2^{2+} + 2 H_2O \Longleftrightarrow (UO_2)_2(OH)_2^{2+} + 2 H^+$	-5.62 ± 0.04	
$3 UO_2^{2+} + 4 H_2O \Longleftrightarrow (UO_2)_3(OH)_4^{2+} + 4 H^+$	-11.9 ± 0.3	
$3 UO_2^{2+} + 5 H_2O \Longleftrightarrow (UO_2)_3(OH)_5^+ + 5 H^+$	-15.55 ± 0.12	
$3 UO_2^{2+} + 7 H_2O \Longleftrightarrow (UO_2)_3(OH)_7^- + 7 H^+$	-31 ± 2	
$4 UO_2^{2+} + 7 H_2O \Longleftrightarrow (UO_2)_4(OH)_7^+ + 7 H^+$	-21.9 ± 1.0	

In almost all surface complexation studies the composition of the aqueous solutions is assessed by some appropriate formation constants. The reason for these calculations is the generally very low species concentrations that prevent a more direct speciation. The species given in Table 2.10 are taken from a review within the OECD/NEA database series (OECD/NEA 1991). For most species in Table 2.10 no direct experimental evidence exists. Species such as $UO_2(OH)_3^-$ and $UO_2(OH)_4^{2-}$ have been inferred from least-squares curve fitting to potentiometric titration studies. The meaning of the figures following the "$\pm$" symbol in Table 2.10 is not clear. At the concentration levels of a sorption experiment (below the solubility limit of the least soluble solid phase under given conditions), an experimental verification of the solution composition, e.g. by emission spectroscopy, is not possible (Meinrath 1997a).

Figure 2.12 presents a LJUNGSKILE calculation on basis of these formation constants using the figures following the "$\pm$" as a standard deviation. For the species $UO_2(OH)_2^\circ$, an upper limit only is given in the original reference ($\lg K \le -12.3$ at $I = 0$). The formation of this species is considered by a uniform distribution in the range $\lg K_{12} = -12.3 - 14.3$.

From the ten species given, four species only play a role: UO_2OH^+, $UO_2(OH)_2^\circ$, $UO_2(OH)_3^-$ and $UO_2(OH)_4^{2-}$, next to UO_2^{2+} itself. The latter two species are highly questionable and "came into existence" only by neglect of CO_2 contamination. U(VI) carbonato species can be detected spectroscopically already at free CO_3^{2-} concentrations of $10^{-11}\,mol\,dm^{-3}$ (Meinrath 1997a; Meinrath et al. 1996a). CO_2 is ubiquitous (e.g. adsorbed to vessel and tools) and reacts readily with neutral to alkaline solutions. Even a nitrogen box cannot prevent CO_2 contamination, which is, in addition, almost impossible to detect without considerable efforts using instrumental analysis (e.g. mass spectrometry). In fact, carbonate contamination of NaOH was shown, e.g. by Jakobsson (1999), at levels detectable in Gran plots.

The uncertainties for each species illustrate the need for probabilistic speciation. All species can be either not present all or be the dominating species. This observation is most evident for $UO_2(OH)_2^\circ$. Within the 90% confidence interval (CI), this species may be either not present or dominating, over the complete pH range of the study.

Nevertheless, the second, third and fourth hydrolysis constant of U(VI) had to be drastically modified by Jakobsson (1999) to achieve a numerically satisfactory interpretation to the data. No information is given about the criteria of the numerical data interpretation. Figure 2.13 gives an interpretation by a LJUNGSKILE speciation. Two main effects can be recognised: hydrolysis of UO_2^{2+} sets in at higher pH and the prevailing species above pH 6 is the neutral $UO_2(OH)_2^\circ$.

Of course, "adjustment" of formation constants is an unacceptable procedure. Either the formation constants in Table 2.10 are incorrect, or the premises of the study need to be reconsidered. Figures 2.12 and 2.13 underscore the ne-

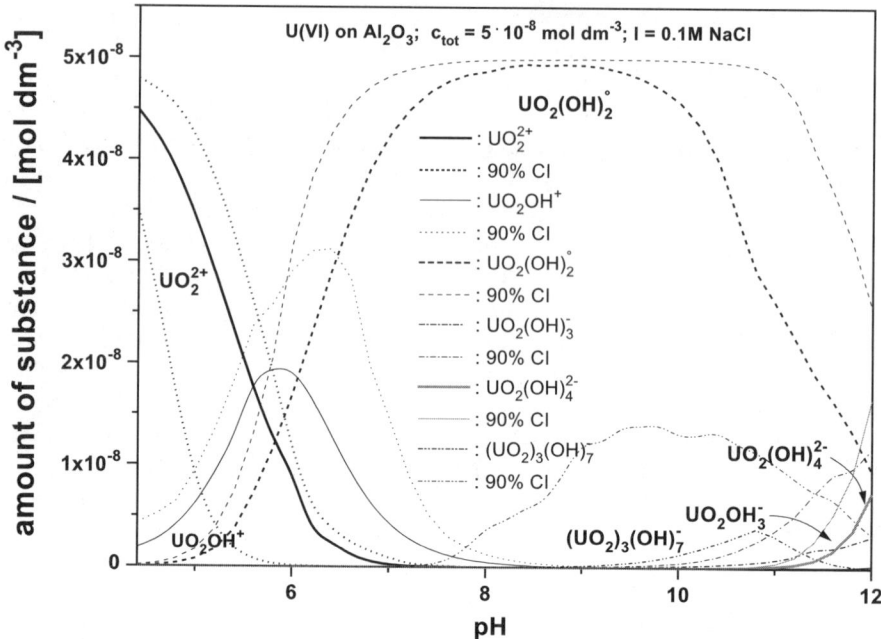

Figure 2.13. LJUNGSKILE speciation simulating solution conditions of a sorption study of U(VI) on Al_2O_3 under a nitrogen atmosphere (Jakobsson 1999). The U(VI) species included into the simulation are given in Table 2.10. For species $UO_2(OH)_2$ and $UO_2(OH)_3^-$ and $UO_2(OH)_4^{2-}$ the "adjusted" values are used. At given confidence level all hydrolysed species may also have zero amount concentration

cessity to evaluate uncertainties when assessing solution conditions by numerical methods (Meinrath et al. 2000, 2000a; Denison and Garnier-Laplace 2004; Ödegaard-Jensen et al. 2004). Instead of presenting a clear picture of the solution composition, Figs. 2.12 and 2.13 communicate the inability of currently compiled databases to allow more than a rough statement about some probable species that might be present in a given solution. In fact, several species might form in solution being either prevailing, or almost absent.

Electrostatic models of the solid-aqueous solution interface. Surface complexation studies almost exclusively report mean value-based data evaluation. In addition to the often considerable amount of solution species, one or several surface species need to be included into the interpretation model (cf. Eqs. (2.32a)–(2.35b)). Furthermore, several models are available to account for the solid-solution interface. Examples are the constant capacitance model (Stumm et al. 1976), the diffuse layer model (Dzombak and Morel 1990), and the triple layer model (Blesa et al. 1984). These models make various assumptions on the effect of charges and ion distribution in the microscopic vicinity of the surface. These dimensions are experimentally almost inaccessible.

Surface complexation models require a number of adjustable parameters. The most important parameters are surface site density and capacitance densities. Surface site densities give an estimate on the number of $\equiv$SOH groups per surface area. Several experimental methods are given to estimate these numbers, e.g. crystal dimension calculations, potentiometric titration, tritium exchange. Determinations of surface site density vary by an order of magnitude depending on the method used. The ability of the surface complexation models to describe anion adsorption using inner-sphere and outer-sphere complexes is sensitively dependent on the value of the surface site density (Goldberg 1991). The same might hold true for cation adsorption. The relative amount of inner-sphere and outer-sphere complex formation is an additional adjustable model parameter.

Capacitance density C relates to surface charges and the surface potentials of the different layers in the various interface models. In most studies using the triple layer model (TLM), C_1 is used as an adjustable parameter, while C_2 is set to 0.2 C/m$_2$.

In the fitting exercise on basis of a triple layer model $C_1 = 1.1\,\mathrm{C\,m^{-1}}$ and $C_2 = 0.2\,\mathrm{C\,m^{-1}}$ were used for the data Fig. 2.10. Figure 2.14 shows the basic structure of a triple-layer model. The data was fitted with FITEQL (Westall 1982). A good

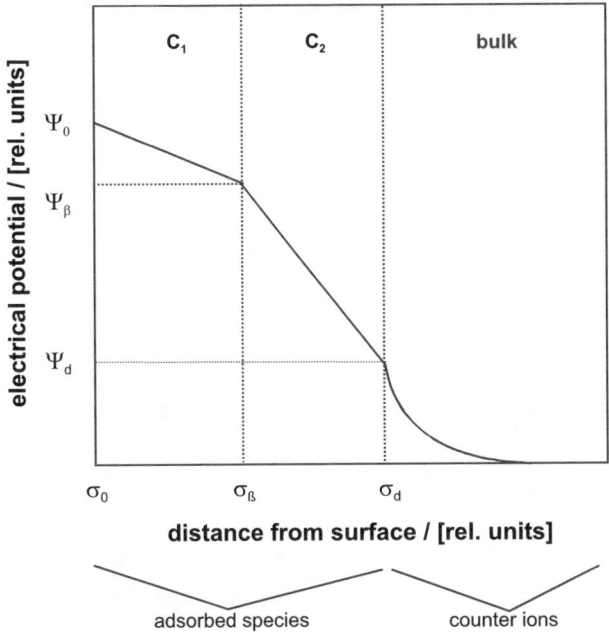

Figure 2.14. Schematic representation of the surface-solution interface in the triple layer model. C_1 and C_2 are the different compartments, σ_0, σ_β and σ_d give the charge at the surface and the compartment boundaries, respectively, while Ψ gives the respective electrical potential

fit was obtained with a model taking two surface complexes into account. Westall and Hohl (1980) state that the adjustable parameters of a surface complexation model are strongly correlated. This statement is especially true for the acidity constants of a surface species ($\equiv$SOH, $\equiv$SOH$_2^+$), surface site density and capacitance. These factors describe surface charging.

This discussion gives an impression of the many factors affecting the values obtained from fitting certain data in terms of a surface complexation model. There are surprisingly few studies discussing the influence at least a few of the uncertainty contributions. In contrary: the many adjustable parameters are generally not seen as fitting parameters which increase the degrees of freedom, but the mean value-least squares fitting results for these parameters were regularly considered as valuable information.

2.4.6
Criteria for Determining and Reporting Data for Surface Interaction Parameters

Determination of values for surface interaction parameters by multi-dimensional curve-fitting computer codes is currently a widespread activity. To be meaningful, the following criteria should be met:

- A definition for "sorption" should be given which unambiguously indicates which effects are included into the interpretation,
- A detailed description of the abilities and limitations to recognise and distinguish between different surface processes should be given,
- Possible unaccounted processes should be mentioned,
- The relevant influence quantities should be listed,
- As a minimum requirement, progression-of-error analysis or Monte Carlo studies should be presented for the major interdependent parameters,
- Analytical measurements must be performed and documented according to internationally accepted criteria (e.g. Analytical measurement committee 1995, Eurachem/Citac 2002)
- Cause-and-effect diagrams for the complete measurement and data evaluation process must be presented,
- Uncertainties from type b evaluations have to be listed and their evaluation documented,
- Dependencies and correlations among the data and the parameters of a surface interaction parameter evaluation must be given.

There is no doubt that the study of surface complexation parameters is a kind of fundamental study which may highlight certain aspects of surface-solution interactions, providing further understanding of the retardation capabilities of geological materials and soils. However, by considering the large amount of parameters and measurement values entering such an evaluation, some concern may be directed to the effect of limiting resolution capabilities of analytical methods and computational tools. "Error progression", "correlation", "estimate" and "quality assurance" are terms that rarely appear in surface complexation study reports. Applying such data without an appropriate estimate of uncertainty, e.g. in so-called "sorption databases", to environmental problems may increase the inherent risk instead of improving the prediction reliability (e.g. Lerche 2000).

Science works with models. Many of these models are deterministic. The deterministic character of these models is, often, caused by the very large number of influence quantities. Hence, the determinism is built on the probabilistic law of large numbers. For instance, the term "statistical thermodynamics" is an excellent reminder. Doubt and uncertainty (two faces of the same medal) are an intrinsic part of science. The field of risk analysis is currently evaluating the protocols and techniques to make uncertainty assessment a helpful tool in decision-making.

A clear presentation of uncertainties (that is, doubt) accompanying a value of a quantity relevant for decision making is also a quality criterion of a study per se. In a study dealing with the credibility of science and scientists in the public, the physicist Maier-Leibnitz gave criteria to assess expert statements. Two relevant criteria are (Maier-Leibnitz 1989):

- A credible expert will present a critical analysis of her/his own statements and present well-founded, justified error limits,
- A credible expert will present all relevant opinions and analyses, even if they are not in favour of his own point of view.

Since the discussions of the Greek philosophers, the best way to evaluate an idea or a theory is to ask questions. Part II of the book intended to highlight some of the aspects of geochemical modeling to which critical questions may be applied with a good chance to stir a discussion. In no situation, a mere mean value figure should be accepted. A model "may resonate" with nature but it is never the "real thing". The burden is on the modeler to demonstrate the degree of correspondence between the model and the material world it seeks to represent and to delineate the limits of that correspondence (Oreskes et al. 1994).

3 Metrological Principles Applied to Geohydraulic Data

"A truth passes through three steps: first it is ridiculed, second violently opposed. Third it is accepted as self-evident."

(A. Schopenhauer)

Metrological concepts apply to all kind of measurements because all measurements are comparisons. Consequently the values from such comparisons are only meaningful if they are related to common standards and references. In this fundamental aspect a laboratory measurement is not different from a field measurement. In the field, however, control over the influence quantities is much more difficult. Nevertheless, a value obtained by application of a geochemical model is dependent on these values. The uncertainty related to such values will affect the reliability of a model simulation, and therefore has to be assessed.

The third part of this treatise deals with the determination of permeabilities. Permeability directly relates to the Darcy law and the hydraulic conductivity in a porous medium. Its reliability (or, reverse, uncertainty) has an important influence on the complete geochemical modeling approach. The hydraulic conductivity is a primary factor influencing the amount of water transported in a subsurface volume, including the substances dissolved into this water volume.

Discussion of permeability supplements the discussion of geochemical aspects, e.g. thermodynamic data and sorption data. Uncertainty in these parameters is the general topic. That these three parameters cannot be described by a neat, closed physical formula is the common aspect. For thermodynamic data describing complex formation in homogeneous solution a theoretical fundament is available that characterises these parameters as fundamental constants of nature under given conditions. Such a basis is not available for surface interaction parameters. However, at least results from some experimental methods are interpreted in favour of certain microscopic processes to be described by a chemical theory. In case of permeability and geohydraulic data, however, only an empirical and idealised relationship is available: the Darcy law. The field of research is almost inaccessible: a rock body usually extending over large spatial scale. Small samples only can be obtained and must serve as pars pro toto.

The value (range of values, distribution of values) may depend on the method applied, the quality of equipment used, the ancillary assumptions taken, and the uncertainties and error sources considered. Therefore, some efforts have been made to summarise the properties of different experimental methods in tables and graphs. There is, however, no claim to have succeeded with this goal. The aim of this treatise is not to create a reference textbook but to put together information for those being confronted with the results of geochemical modeling. These individuals can be, for instance, co-workers in an environmental office of a region, staff members of a national waste disposal agency, political decision-makers and citizens facing the construction of some potentially hazardous facility within their neighbourhood.

Because geochemical modeling is a multi-faculty subject there is probably no single person having the survey on all its aspects. Those involved in the field, either by determining chemical thermodynamic parameters, surface interaction parameters, writing code or compiling data for application in a geochemical modeling code will obtain a more profound insight into the activity they participate.

The concepts outlined here, to our knowledge for the first time in this depth for metrological application in geohydraulics, will be illustrated by two application examples. The first example deals with permeability determinations in rocks with low permeability, the second with an uncertainty analysis in a geohydraulic bore-hole test in rocks with low permeability. Both examples are related to salt rock. Salt rock is one of the primary target host formations for high-level radioactive waste in Germany.

3.1
A Brief Summary in Geohydraulics

"Now there is one outstandingly important fact regarding Spaceship Earth, and that is that no instruction book came with it."

(B. Fuller)

3.1.1
Permeability and Metrology

The term "geohydraulic processes" and/or "geohydraulic data" relates to the movement of a fluid in coarse or solid rocks. The terms also refer to all parameters relevant for these processes. Individual processes involved in geohydraulic flow are advection, convection, dispersion and diffusion. Besides the gradients of the respective potentials, the essential parameters in flow through coarse rocks are the effective pore volume and the capillary interconnections among pores. These physical parameters are describing the fundamentals of permeability in soils and rocks. In coarse rocks, a fluid is flowing in the porous matrix, which is usually treated as a homogeneously distributed volume of pores. In a solid rock, flow of a fluid is depending on the distribution and the connection of permeable volumes (e.g. fractures) and on the pore volume width. Besides these single porosity models, a limited number of rocks (for instance sandstone) are described by a more complex model, the so-called double porosity model. A schematic representation of the double porosity model is given in Fig. 3.1. These rocks contain matrix pores and fractures. The effective permeability, so called in-situ-permeability, of a rock is a function of the water content and the suction. Suction is caused by capillary action of the fluid, water.

Geohydraulic data are essential input parameters to geohydrological models. Geohydraulic models are largely based on the Darcy law (cf. Chap. 2.1). Part III of this treatise will discuss a frame for a quality assessment of geohydraulic data. As is the case with chemical data, geohydraulic parameters are largely given as mean values despite the fact that a considerable amount of randomness is affecting an evaluated mean value.

The problem in the use of quality management concepts in the field of geohydraulic data is the transfer of laboratory data onto a rock body under realistic hydrogeological conditions. A number of communications deal with the treatment of this randomness, mainly by applying statistical approaches. These experiences will be included in the concept of a metrological approach to quality assurance of geohydrological data. Figure 3.2 shows graphically the problem discussed in this contribution. Figure 3.2 can be seen as an abbreviated version of Fig. 2.8. In this part, the attention is directed to the variability reasonable associated with a parameter obtained by measurement from a small section of the modeled subsurface volume.

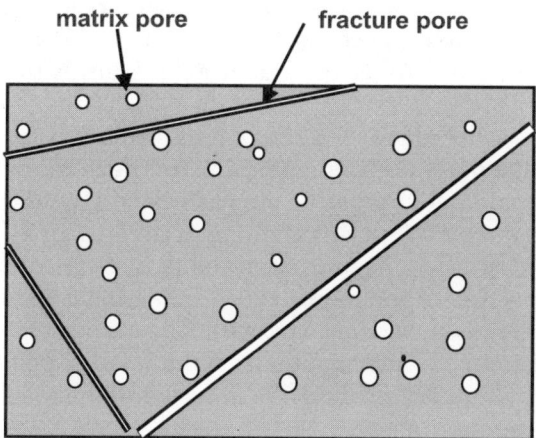

Figure 3.1. Schematic representation of the double porosity model featuring a porous matrix intersected by fractures. A single porosity medium is one in which either matrix or fracture porosity are present, but not both

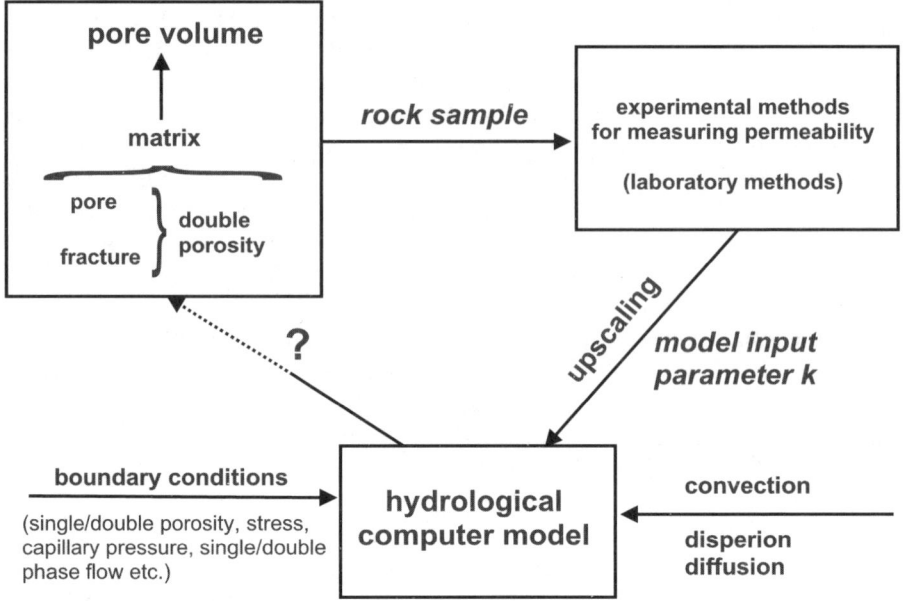

Figure 3.2. Framework for quality assurance of geohydrological data. The crucial, and generally open, question addresses the relationship between the geological formation and the constructed computer model

The relevant parameters which can be measured both in the laboratory and in the field are the effective porosity, the permeability coefficient, the dispersion coefficient, and the diffusion coefficient. Special problems result in the case of occurrence of double-phase or multi-phase flow, since these parameters have to be quantified taking into account the changes in the principal stress parameters (pressure, capillary action, porous and fracture flow) inside the rock formation.

The permeability k describes a material property of a porous medium at the time of investigation. Effects of the pores and pore radius distribution, moisture contents of the matrix or temporal changes of the permeability by the plasticity of the material due to extreme pressures cannot be accounted for by the permeability. In geohydrological modeling, the permeability is commonly obtained from a (very small) fraction of the model three-dimensional subsurface volume. The rock formation is considered as a material with constant geohydraulic characteristics, whereby the information from the small sample fraction (commonly studied by laboratory methods) is upscaled to the total subsurface space under consideration.

Due to the importance of geohydraulic data in rocks for geohydrological modeling several national (DIN) and international (ISO) standards are available setting a framework for permeability determinations in saturated rocks by laboratory methods. For field determination of in-situ geohydraulic properties of rocks, no standards are given. A possible reason for this lack of standards for field methods is the large variety of possible conditions encountered in the field, as well as the current predominance of laboratory determination of geohydraulic data. This preference of laboratory determination is partly caused by this lack of guiding standards, as well as the multiplicity of influence factors which seem to be easier to control in a laboratory environment.

The metrological bottom-up procedures for complex situations, outlined in Part I of this treatise, are suitable for a standardisation of field determination of geohydraulic parameters. In principle, criteria set by a standard shall comply with the fundamental, internationally agreed requirements for quality assurance. So the GUM gives three fundamentals which are applicable to every measurement process, including determination of geohydrological parameters. These fundamentals are uncertainty, fitness-for-purpose and traceability.

Uncertainty is a parameter which characterises the measurement. The reciprocal of uncertainty is reliability. Uncertainty corresponds to the doubt which has to be assigned reasonably to a value of a measurement concerning the true value of the measured parameter. This definition of uncertainty assumes that the parameter in fact has a "true value". In case of a geological sample it is not always certain that the permeability k, which is built on certain model assumptions about that sample, has something like a true value. It is important to note that it is a model assumption that there is a property "permeability" which can be quantified by measurement. Uncertainty is expressed by the dispersion of the values, which could be assigned to the measured variable (GUM § 2.2.3).

Fitness-for-purpose is a characteristic of data, which is generated by a measurement. The fitness-for-purpose permits the user of the data to make technically correct decisions for a stated purpose. Hence, it is not essential that a property of "permeability" exists. It is, however, important that the value used for that property is used correctly within the model framework. Furthermore, the purpose of the value must be known before its measurement and the measurement must be of a quality which is adequate for the purpose. In the other way, there is no use to construct a model requiring a quality for which a measurement procedure does not exist.

Traceability is the property of a result of a measurement whereby it can be related to stated references, usually national and international standards, through an unbroken chain of comparisons all having stated uncertainty. This definition does not differ from other measurements and is taken from VIM (1994). Traceability is the essential element in ensuring comparability of measurement values. This definition illustrates the importance of the availability of standards; the lack of geohydraulic standards underscores the lack of comparability of permeability values reported in literature. It may be argued that the permeability of a rock formation will have a much larger variability than a small fraction of material can indicate; thus comparability of measurement

values of permeability obtained from different samples of the rock formation is of minor relevance. Then the question arises, why permeabilities aren't
estimated. In fact, a variety of methods have been developed to determine permeability values under a broad range of circumstances, both in the laboratory
and in the field.

Because permeability is an important model parameter, the assignment of
a reasonable value (or range of values) has a considerable influence on the
model output and the quality of the simulation output. In a discussion of
model output, the focus will quickly turn to the method of its determination.
A variety of methods exist and the discussion will be eased if objective criteria
for a comparison of the various determination methods exist.

Such a discussion is considerably facilitated by cause-and-effect analysis
because essential influence factors affecting the values obtained by a given
experimental method will become obvious and can be associated with numerical values. Thus, the process of method selection and its performance gains
in transparency which can be communicated and critically scrutinised, e.g. in
a situation of conflicting interest. Cause-and-effect analysis also provides an
objective criterion for ranking influence factors. As an essential condition for
defendable measurement values the goal of the investigation must be clearly
defined beforehand. It may be advantageous to perform a sensitivity analysis
which will guide the subsequent uncertainty analysis. The definition of quality criteria for the measurement values to be obtained from a measurement
procedure will be helpful to assess the success of a measurement campaign
which is, in most circumstances, an expensive and time-consuming activity
which cannot – in contrast to some academic laboratory exercises-repeated
arbitrarily.

Analysing the main requirements of the GUM with respect to geohydraulic
data, the conclusion seems justified that these requirements do not prevent its
application. This conclusion is not surprising because the GUM was established
as a general document relating to all fields of commerce, technology and
science.

Geohydraulic measurements are mainly performed to obtain values for permeability. It is common to obtain such data in the laboratory. The respective
data are then transferred to the rock formation, for instance in case of saturated
permeability tests (cf. Fig. 3.2). Important influence quantities are, e.g. sample
preparation, the measurement process in the laboratory, and the intrinsic uncertainty associated with the numerical method which is used for transfer of
laboratory data to the rock formation. In some rarer cases, permeability data
are determined by field studies, for instance by pump tests. In these cases, a variety of influence factors contribute to the variability of a measurement value:
the sampling, the specific measurement procedure, and the numerical method
of data analysis. As a general rule can be stated that laboratory tests must be
supported by field studies, while field studies are difficult to be appropriately
simulated in a laboratory environment.

The transferability of laboratory tests to the rock formation is acceptable only with appropriate consideration of the rock formation's physical boundary conditions. These conditions, too, cannot be determined with arbitrary accuracy. Already the variability within a given rock formation prevents the uncritical use of laboratory and field data obtained with a small sample or within a limited section of the formation. If a measurement value intended for representing a larger subsurface volume in a computer simulation can be obtained only in the laboratory it is quite difficult to be defended. Data representative for a rock body can only be determined by performing in-situ tests. Trace material and/or isotope investigations (for instance noble gas dating and stable isotope analyses) should be generally considered as supplementary methods for validation and calibration of geohydraulic data – together with other geohydraulic methods.

The uncertainty of the physical parameters in subsurface contaminant transport problems is ubiquitous. This is manifested in the basic heterogeneity of the aquifer formation and the uncertainty related to the chemical, physical and biological properties of the contaminant being released and transported. Furthermore, there is a great deal of uncertainty regarding the leaking source dimensions, concentration, leaking rate and duration. These parameters vary largely from one site to another and also exhibit great spatial variability within the same site (EPA 1999).

There are different sources of uncertainty that the hydrogeologist has to account for. Some of these are summarized in the following. Modeling uncertainty arises due to using a simplistic relationship to describe the actual behaviour of a physical system and the idealization down to operational mathematical expressions. This can be quantified either by comparisons with other, more involved models that provide a more accurate representation of reality, or by comparisons with collected data from the field. Modeling uncertainty can be formally treated in a reliability context by introducing a random variable that describes the ratio between the actual and predicted model response and output.

Prediction uncertainty means that the reliability estimate depends on the state of knowledge that is available to the engineer at the time of analysis. Various factors could affect the model response which are not included in the analysis simply due to lack of knowledge. As the state of knowledge increases, our assessment of the reliability is refined, usually combined with an increased reliability index (Melchers 1987).

Human factors are the total of those errors that arise during collection, recording, and analysis of data. At field sites data are collected, statistical estimators (mean and higher order moments) are obtained, and a probability density function (PDF) is chosen to represent the distribution of each input random variable. Since collected data are usually inadequate and noisy, those PDFs are bound to be biased. This is often termed statistical or information uncertainty (Dettinger and Wilson, 1981). One way of alleviating this problem is to consider the parameters such as the mean and variance themselves as random variables to estimate how uncertainty of the statistical parameters propagates to the model response. Another solution is to collect more data and use a Bayesian approach to update the information (Melchers 1987).

The last type of uncertainty is that resulting from the inherent randomness of the medium variables under consideration. This is quite evident in the soil formations, for which properties such as hydraulic conductivity can span many orders of magnitude at the same site (Bakr et al. 1978; Freeze and Cherry 1979). This type of uncertainty is irreducible, and is often referred to as the inherent, intrinsic, or physical uncertainty (Dettinger and Wilson 1981; Melchers 1987). Although the current research focuses on addressing the physical uncertainty, the approach is equally applicable to other types of uncertainty with the necessary modifications of the formulation.

3.1.1.1
Permeability of Soils

The fundamental equation for the saturated flow in the soil is Darcy law:

$$Q = K_f * A * \frac{\Delta h}{l} \quad \text{or} \quad q = \frac{Q}{A} = K_f * i \tag{3.1}$$

or

$$K_f = \frac{Q/A}{\Delta h/L} = \frac{q}{i} = \text{const} \tag{3.2}$$

with

Q: flow rate
q: effective velocity of ground water flow
A: cross section
Δh: hydraulic difference (potential), difference of the piezometric pressure
L: length
i: hydraulic gradient
K_f: hydraulic conductivity

The hydraulic conductivity K_f is a combined material property of porous a medium and the mobile phase, which is described by Eq. (3.3):

$$K_f = k * \frac{g \, \delta}{\eta} = k * \frac{g}{\nu} \tag{3.3}$$

with

k: (intrinsic/specific) permeability of the porous media
g: gravity of the earth
δ: density of the fluid

η: dynamic viscosity of the fluid
ν: kinematic viscosity of the fluid.

Forces which cause a directed movement of the water in the soil zone are:

- downward directed gravity (gravitation potential),
- the capillary suction (matrix potential), directed upward.

If meteoritic precipitation enters a soil system, an initial saturation of the upper soil zones takes place. Further precipitation causes a movement of the humid front into larger depths. Downward from surface, three zones are commonly distinguished: saturation zone, transport zone and humidification zone (Hölting 1996). The downward directed water transport (seep) depends on the hydraulic conductivity and is a function of soil zone saturation. If the pore water meets a layer of lower hydraulic conductivity, the seeping movement slows down. If the amount of precipitation is larger than the amount capable to percolate, surface runoff is taking place.

Water flow in the unsaturated soil zone can be described with the help of the Richards equation, which concerns a three-dimensional representation of the water movement into the orthogonal directions x, y and z depending on the capillary suction.

$$P * \frac{\partial \theta}{\partial t} = \frac{\partial}{\partial x} k_x(h) * \frac{\partial h}{\partial x} + \frac{\partial}{\partial y} k_y(h) * \frac{\partial h}{\partial y} + \frac{\partial}{\partial z} k_z(h) * \frac{\partial h}{\partial z} \qquad (3.4)$$

with

P:	porosity $[-]$
θ:	saturation/$[\mathrm{m}^3\,\mathrm{m}^{-3}]$
t:	time $[\mathrm{s}]$
h:	suction $[\mathrm{m}^{-1}]$
x, y, z:	space coordinates $[-]$
k_x, k_y, k_z:	permeability of the soil depending on coordinates x, y, z and suction $[\mathrm{m}^2]$.

This equation applies under the condition that no sources, sinks, and/or discharges mark the system. Further assumptions are constant rock porosity, incompressibility of the mobile phase and gas flow being always substantially faster than the mobile phase. If this is not the case (which may occur under certain conditions), then the Richards equation cannot correctly determine the processes in the soil. Furthermore, the Richards equation does not consider macroporous structures (fractures).

The measurement input data necessary for the Richards equation can take place on the one hand via experimental investigations and on the other hand by modeling on the basis of measured soil parameters. In the past few years the approximation of van Genuchten-Mualem (Mualem 1976; van Genuchten 1980) has found widespread application. This approximation describes the relationship between water content and suction and/or of hydraulic conductivity using a uniform parameter set:

$$\theta(h) = \theta_r + \frac{\theta_s - \theta_r}{\left[1 + |\alpha \cdot h|^n\right]^m} \tag{3.5}$$

$$m = 1 - \frac{1}{n} \tag{3.6}$$

$$K_f(h) = K_s \cdot S_e'(1 - (1 - S_e^{1/m})^m)^2 \tag{3.7}$$

$$S_e(h) = \frac{\theta(h) - \theta_r}{\theta_s - \theta_r} \tag{3.8}$$

with

$\theta(h)$:   water content as function of the suction $[m^3/m^3]$
θ_r: residual water content $[m^3/m^3]$
θ_s: saturation water content $[m^3/m^3]$
$K_f(h)$: hydraulic conductivity as function of the suction $[m\,s^{-1}]$
K_s: saturation hydraulic conductivity $[m\,s^{-1}]$
S_e: effective water saturation $(0 \leq S_e \leq 1)$ $[-]$

and the van Genuchten parameters (empirical parameters):

α: (m^{-1})
n: dimensionless
m: $m = 1 - 1/n$; dimensionless
l: dimensionless.

Van Genuchten parameters are empirical parameters, which do not have a defined physical meaning. These parameters have an influence on the shape of the curve, as well as the distribution of values for suction. Equations (3.5) and (3.7) for $\theta(h)$ and $K_f(h)$ apply to the unsaturated conditions. Under saturation conditions, Eqs. (3.9) and (3.10), respectively, hold:

$$\theta(h) = \theta_s \tag{3.9}$$

$$K_f(h) = K_s \,. \tag{3.10}$$

Table 3.1 gives an overview of the relations between suction and pore volume as function of the water content derived from it. The amount of water that any soil can retain and hold readily available to plants is determined by the size distribution of individual pores, especially the proportion of micro pores. In the porous structure of the soil there are two forces trying to remove the water:

1. Upwards: transpiration of plants and soil surface.
2. Downwards: gravity.

The smaller the pores the higher the force by which the water is held. These forces are expressed in bar or centimeters water column, and the term pF is

Table 3.1. Overview of the relations between suction and pore size and the water binding forms derived from it (AG Bodenkunde 2005)

Range of suction HPa (cm water column)	pF-value (1 g cm water column)	Pore diameter in μm	Pore type	Soil water type	Storage capacity type
<60	<1.8	>50	Wide coarse pores	Fast moving pore water	Air capacity and storage capacity for damming wetness
60–32 000	1.8–2.5	50–10	Wide coarse pores	Slow moving pore water	Usable field capacity
32 000– 1 520 000	2.5–4.2	10–0.2	Middle pores	Plant available irreducible water	
Permanent wilting point					
> 1 520 000	> 4.2	< 0.2	Fine pores	Not plant available irreducible water	Connate water

used. As with pH, the p stands for the negative decadic logarithm, while F gives the corresponding suction in units of cm water column. Thus a suction of 2 bar corresponds to a value of pF 3.3. The relationship between moisture content and pF value is strongly dependent on soil type. If the pF value is plotted vs. moisture content two important points relating to the pF value (i.e. suction value) exist:

1. The permanent wilting point (PWP) gives the maximum suction that plants can exert to withdraw water from a porous medium. The PWP is generally taken as pF 4.2 (or 16 bar).
2. The field capacity (FC) is the moisture content of the soil when, after saturation, the soil is allowed to drain freely for 1–2 days. Generally a pF 2.0 corresponds to moisture content at FC. This means, therefore, that for the plant, water is available only in the suction range between 0.1 and 16 bar i.e. pF 2.0 and pF 4.2. Water held at lower suctions will drain very quickly and that held at higher suctions is available to the plant.

Soil zones with respect to geohydraulic characteristics, e.g. water saturation and hydraulic conductivity, are schematically given in Fig. 3.3. The "air access point" is given by the pressure required to overcome the capillary effects. Only if this pressure exceeds capillary suction air is able to enter the soil or rock formation. This suction pressure, expressed in units of the height of a water column, corresponds to the height of the capillary seam and/or the height

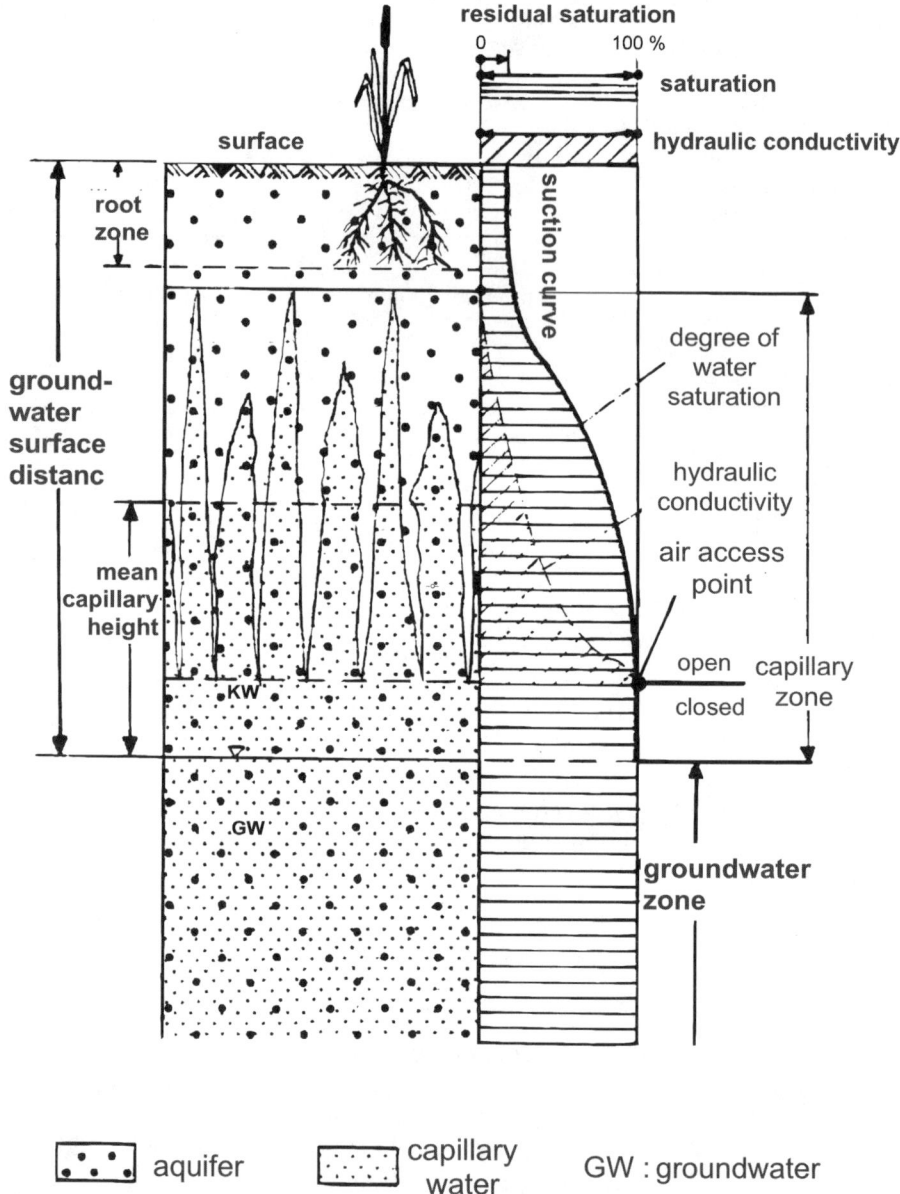

Figure 3.3. Illustration of soil zones with major characteristic geohydraulic properties

of the closed capillary zone. Figure 3.3 illustrates that the "air access point" is significantly above the groundwater level, which corresponds to the hydrostatic pressure level. This implies that there is a water-saturated range above the groundwater level. This range is termed capillary seam or closed capillary zone.

A permeability coefficient (k) of a rock is defined for a rock versus the fluid water at a temperature of $10\,^{\circ}$C in units of m s^{-1}. The definition requires distilled water as the fluid medium although this is obvious that the characteristics of the distilled water may change considerably during the experiment.

3.1.1.2
Permeability of Rocks

At latest since the 1970s the permeability is realized as an important quantity characterising a material, but to achieve at least some comparability with other permeability values a statement of all relevant boundary conditions is necessary.

The Darcy Law describes the movement of the underground water with sufficient accuracy. It is generally applicable to:

- the saturated (water-saturated pores without temporal variance) and unsaturated zone
- stationary and instationary flow
- aquifers and aquitards
- homogeneous and inhomogeneous systems
- isotropic and anisotropic systems
- matrix and related aquifers.

The Darcy law is nevertheless an approximation and a simplification. It has, of course, theoretical and practical limitations which are expressed both in an upper and a lower limit of its validity. Since the Darcy Law is linear equation, there are generally two situations which can be critical for its validity because the flow process cannot be linear:

- very low k values and small gradients,
- very high k values and large gradients.

With a slight extension Darcy law can be transferred into a nonlinear form. Thus the principal description is maintained and linear constraints are removed:

$$V = -K_f \left(\frac{dh}{dl} \right)^m . \tag{3.11}$$

The linear Darcy law is given if $m = 1$, otherwise V is non-linear. For the case of large k values in connection with large gradients it can come to the fact that the flow movement is no longer laminar, but becomes turbulent so that the validity of the linear Darcy law is no longer given. The criterion for a laminar or turbulent flow is the dimensionless Reynolds number, R, which was developed first for flow in tubes and/or open channels.

$$R = \frac{\delta \cdot v \cdot d}{\eta} \tag{3.12}$$

with:

δ: density of the fluid $[\text{kg m}^{-3}]$
v: flow speed $[\text{m s}^{-1}]$
d: tube diameter $[\text{m}]$
v: viscosity in $[\text{kg s}^{-1}\,\text{m}^{-1}]$.

If $R > 220$, turbulent flow in tubes and/or fissures/fractures of the diameters of a tube begins (Streeter 2000). For porous aquifers, Reynolds numbers between 60 (Schneebeli 1955) and 600 (Hubbert 1956) are reported. For Reynolds numbers between $R = 1-10$, flow in a porous aquifer is assumed to be laminar.

In case of low k values and low gradients, a "permeability threshold" was discussed. Below that threshold value fluid movement should be assumed as zero. Swartzendruber (1962), Ludewig (1965), Bolt and Groeneveld (1969), Luckner (1976), Häfner (1985) and Schildknecht (1987) argue for the existence of such a lower limit. Suggestions for its formal, mathematical description have been published. At present, the discussion on the existence of such a permeability threshold is still open. This discussion has, however, not much practical importance because for very low permeabilities (and consequently very low movement of the mobile phase), the importance of convection decreases while diffusion processes becomes the relevant process. At permeabilities in the order of $k = 10^{-9}$ to $10^{-10}\,\text{m s}^{-1}$, diffusion is the determining transport process.

Darcy law is an approximation. In practice, deviation from Darcy law may be interpreted due to several phenomena, which are of important in the preparation and evaluation of hydraulic conductivity tests (Bruck 1999):

- Slippage phenomenon (in case using gas as fluid)
- Turbulent flow
- Mechanical changes of the solid matrix
- Reactions with the pore walls.

In rocks with low permeability test times can be very large due to the very small porosities available for transport of water. For this reason, gases or commonly applied as fluid medium in laboratory tests (e.g. H_2) for the determination of the permeabilities (Wagner 1998). These are commonly rocks with a permeability below approximately $k = 10^{-12}\,\text{m}^2$. During such studies, the so-called slippage phenomenon can be observed in permeability measurements using gases, if the pore diameter of the material has the adequate size for gas molecules (Bruck 1999). At very low pressures, mean free path of gas molecules depends on the temperature and the gas. A proportion of the gas particles is adsorbed onto the pore walls forming a single layer cover or even a multi-layer cover. This cover exerts a lower friction on the gas flow than the rock. Formation of an adsorbed gas cover is reducing the effective pore radius but this effect overcompensated due to friction reduction. As a consequence, measured permeabilities result in an overestimation of a sample's permeability, so-called Klinkenberg effect (Bruck 1999). The Klinkenberg effect describes the difference between the flow of a gas and a liquid through a reservoir, affecting permeability. Gas flow is less impeded by grain surfaces than liquid flow.

Hydraulic conductivity of a rock body can be described by several parameters, summarized in Table 3.2 (GLA 1994). The hydraulic conductivity (k_f value) is defined as a proportionality factor in Darcy law. It corresponds to the flow through a standard area of the rock body under a standard geohydraulic gradient. The transmissivity (T) is similarly defined, however related to a standard column in the aquifer. Theoretically, it can be determined by integration or summation over aquifer thickness (H) from the hydraulic conductivity. Both parameters depend on the fluid characteristics, i.e. on the density and the kinematic viscosity. In order to receive permeability independent of the fluid characteristics, which describes only the rock characteristics, there have to be devided the parameters T and/or k_f by the fluid characteristics (g/v), where g gives the gravitational constant. Thus, the permeability (k) results from the hydraulic conductivity and the transmissibility (T^{ast}) is derived from the transmissivity.

In case of rock formations with a permeability primarily based on fractures, permeability must be differentiated between the rock permeability and the permeability of a rock body. In contrast to the determination of the rock permeability in the laboratory at defined samples under comparable test con-

Table 3.2. Parameters describing the water transport in a rock body (GLA 1994)

Parameters	Symbol	Dimension	Unit	Equation	Relation to other parameters	Characteristics
Hydraulic conductivity	k_f	3D	m/s	$k_f = \frac{q}{J} = \frac{Q}{A \cdot J}$	$k_f = \frac{\varrho \cdot g}{\mu} \cdot k = \frac{g}{v} \cdot k$	Fluid and rock
Permeability	k	3D	m²	$k = \frac{q \cdot \mu}{J \cdot \varrho \cdot g}$	$k = \frac{k_f \cdot \mu}{\varrho \cdot g} = \frac{k_f \cdot v}{g}$	Rock
Transmissivity	T	2D	m²/s	$T = \frac{q \cdot H}{J} = \frac{Q}{B \cdot J}$	$T = \int_{0}^{H} k_f dh$	Fluid and rock
Transmissibility	T*	2D	m³	$T^* = \frac{q \cdot H \cdot \mu}{J \cdot \varrho \cdot g}$	$T^* = \int_{0}^{H} k dh = \frac{\mu}{\varrho * g} \cdot T$	Rock

Q [m³ s^{-1}] flow passage
g [m s −] gravity constant
v [m²s^{-1}] kinematic fluid viscosity
A [m²] area
μ [Pa s] dynamic fluid viscosity
ϱ [kgm^{-3}] fluid density
H [m] aquifer thickness
B [m] cross-section width

Table 3.3. Typical values for porosity and permeability. The values give relevant orders of magnitudes. Porosity and permeability are depending on site conditions

Rock type		Porosity/[%]	Permeability/[m s^{-1}]
Unconsolidated	Sand	25	$10 \cdot 10^{-4}$
sediments	Gravel	30	$10 \cdot 10^{-2}$
	Clay	50	$10 \cdot 10^{-12}$
Consolidated basalt,	Weathered	15	$10 \cdot 10^{-6}$
granite, sandstone,	Fissured/fractured	5	$10 \cdot 10^{-8}$
limestone	Massive	1	$10 \cdot 10^{-10}$

ditions there are no standardised procedures for the determination of the permeability of the rock body.

Table 3.3 summarizes some typical values for porosity P and permeability k in various unconsolidated and consolidated rocks. These values can only give a rough indication. Their values depend largely on local conditions.

3.1.2
Transfer of Laboratory-Determined Permeability Data to the Aquifer

A special problem represents the transfer of the laboratory permeability data (rock permeability) on the permeability of a rock body, a problem, which is discussed in detail in Neuzil (1994). Commonly permeability evaluation methods are based on rather simple pore models (Bruck 1999). In these pore models, only idealised pores with smooth walls are considered, a pore radius distributions is assumed and pathways between pores available for fluid transport are allowed. These methods are mainly applicable to coarse rocks. In recent years more complex geohydraulic characteristics of a solid rock aquifer could be treated, mainly due to the availability of appropriate computer simulation methods. Especially the distribution of fissures and fractures in the rock body can be described by stochastic models. Therefore, further models for the description of porous materials were developed. Three models, the dusty gas model, the lattice models and the fractal model, will briefly be summarised (Bruck 1999):

Dusty gas model. The "dusty gas model" is a theory that describes the transport of gases through porous media. It is so called because it treats the porous medium as a component of the gas mixture, consisting of giant molecules, like dust in a gas. Transport equations for the gas through the porous media are derived by applying the kinetic theory of gases to this "supermixture" of free gas and fixed in space solid molecules. The model has been independently developed at least four times, starting with J.C. Maxwell in 1860; and can be adapted for modeling other phenomena such as aerosol motion (by allowing the solid molecules to move). Dusty gas models include several modes of flow, the "free molecule flow" (Knudsen flow) which is relevant only for gases, bulk flow, continuum flow and, as a rather special flow type, surface flow/surface diffusion. In the latter flow type, the transport is assumed to occur along pore surfaces.

Lattice model. The use of computer-generated porous 3D fields with defined porous structures has become an intense research field in many areas where porous materials play a role. Simulating geological materials has only a comparatively small fraction. An important share in lattice models is held by the lattice Boltzmann models, where the fluid medium is located on nodes of a discrete lattice. Each "fluid particle" has its own direction of movement by which it is transferred to a neighbouring node (movement step). There, particles on a given node collide and attain new velocities and directions.

Fractal model. The distribution of pores and lattice points may vary with the dimensional magnitude considered. Hence, the structure of a computer-generated lattice has to be defined from macroscopic to microscopic scale which is, at least, a tedious process. Fractal models start from the observation that the shape of the inner surface of rock pores follows a self-similar rule. Thus, the theory of fractals is applicable. The SHEMAT software (SHEMAT 2006) is a example of a (freeware) fractal simulation tool (Pape et al. 1999).

The method of neural networks, a model-free self-modeling method of data analysis, has also proposed as a suitable tool to assign rock body permeabilities on basis of experimentally obtained drill core sample permeabilities. The method, proposed by Trappe et al. (1995) can be considered as a modification of a lattice model. In principle, the first step in this method is the installation of logs within the boreholes and, in parallel, the determination of the corresponding permeabilities in an appropriate sample of the drill core. Thus, for each borehole a measured value for the permeability is available. In the first step, the so-called training phase, the neural network is calibrated to generate a relationship between the experimentally determined permeabilities and the borehole location and log properties. In the second phase the log data from new boreholes are associated with permeability on basis of the model-free relationship generated by a neural network. Thus, permeability is assigned on basis of a non-mathematical model generated by the neural network. A similar method is the use of noble gas dating procedures (Osenbrück 1996; Lippmann 1998, Rübel 1999). A comparable procedure is published by Morshed and Kaluarachichi (1998), where neural networks were used for extensive calculations of migration processes of chemical compounds in the aquifer. Neural networks have been popular in many fields of science and technology. A reason for this popularity is the relationship between the neural network and neuron processes in the human brain. The advantage of a neural network to generate a non-mathematical relationship between a cause and an effect is, at the same time, its main disadvantage. There is no clear relationship between the measured effect (log data) and the assigned property (permeability).

3.1.3
Models for the Three-Dimensional Description of Permeabilities

The permeability of a fissured/fractured aquifer or a pore aquifer depends on parameters such as homogeneity and isotropy. Fissured/fractured aquifers are almost always characterised by an extreme anisotropy, because pathways preferentially directed in space exist. In addition, pore aquifers are often anisotropic because due to the sedimentation process a special structure and connection of the pores is formed. It is a frequent observation that aquifers show a higher permeability in flow direction than orthogonal to it.

Three major approaches to the determination of inhomogeneities and anisotropies in a rock formation are available:

- Spatially highly dissolved determination of the permeability;
- Simulation by a genetic model simulating the processes of pore and fissure/fracture formation;
- Simulation by a stochastic model e.g. On basis of the fuzzy logic, where the available knowledge on porosity, pore size and fracture distribution is represented by distribution functions.

Sensitivity analyses show that the spatial inhomogeneities and anisotropies play an important role particularly in the transport modeling of chemical compounds, but are less important in purely geohydrological simulation. Modern standard groundwater models do not have yet appropriate tools for the automatic generation of parameter distributions in a model space. Laboratory and in in-situ methods are available for the determination of geohydraulic data, whereby in each case these tests can be used under stationary and instationary conditions. The following summary gives a survey on internationally applied methods for permeability measurements. In the last years several new applications were developed, mainly for the field of determination of geohydraulic data in low permeable rocks. All methods have their own bunch of uncertainties in the sampling, the measurement and the data evaluation step. Furthermore, the tools for spatial transfer of the laboratory determined geohydraulic data have made progress due to the advancement in computer technology (Mazurek et al. 1998).

3.2
Measurement of Geohydraulic Parameters in Laboratory and Field

"In nature there are neither rewards nor punishments; there are consequences."

(R.G. Ingersoll)

3.2.1
Sample Quality and its Influence on Permeability Data

Chapter 3.1 illustrated permeability as an important but rather complex parameter. It is a material property but only accessible with in part severe restrictions. Porosity is a property of the pore space inside a rock sample. There is, however, no simple relationship between porosity and permeability. A review is available from Nelson (1994). The sampling procedure, necessary to obtain a small fraction from the rock formation under consideration into the laboratory, takes its own influence on the value evaluated for a permeability k.

The methodologies for sampling are occasionally standardised, commonly on a national level. An example is the German standard DIN 4022, Part III, which applies to coarse rocks. Standard DIN 4022 requires a rock sample being obtained from a drill hole drilled with a diamond drill and a double core pipe. A further variant of the generation of "undisturbed" samples from the rock is the "large drilling procedure", i.e. several boreholes are drilled over a large surface. From the drill cores appropriate sections are removed in the laboratory for the intended measurements. Cutting drill core sections in a laboratory commonly has the advantage that the complete drill core serves as a transport protection for the respective sample block.

The permeability of a rock and/or a soil sample is closely linked with sample structure and the rock/soil mechanic parameters (Schreiner and Kreysing 1998). These parameters are subject however to an inevitable change as a result of the sampling process. Obtaining a sample from a rock drill core, for instance, requires cutting, often with diamond saws. The disturbance of the natural equilibrium in the underground due to the sampling affects the mechanical and geohydraulic characteristics of the rocks, where structure, compactness, porosity, saturation, and permeability influence each other. The permeability reacts most sensitively to disturbances induced by sampling (Schreiner and Kreysing 1998).

Documentation of the sampling procedure and test conditions are important elements of quality assurance and quality control. The availability of this documentation is essential to ensure comparability and reproducibility of the measurement results. Unfortunately, such documentation is rarely available in practice.

Apart from the formation of additional junctions and fissures/fractures during the sampling, observed permeability becomes usually smaller with increasing disturbance in the structure and pores of the samples (Schreiner and Kreysing 1998). Even an undisturbed sample changes after the extraction from the rock due to the relaxation from underground stress. Deformations, changes of porosity and migrations of the pore water are consequences. Due to the relaxation of the hydrostatic pressure during sampling the dissolved pore water gases escape, whereby the saturation point decreases (Schreiner and

Kreysing 1998). The same effects occur after temperature increases. Cooling water, used during cutting processes or for transport stabilisation, condenses at the sample surface and in the larger pores. Sampling and sample preparation is heavily disturbing the water balance of a soil/rock Subsequent measures in a laboratory to restore rock body conditions can remediate the situation only in part (Schreiner and Kreysing 1998).

According to a study of Gilbert (1992) evaluating published data from respective literature sources reported particularly high variation coefficents for the permeability of incompletely saturated samples (see Table 3.4):

Isolating a small sample from a larger rock body involves drastic measures: drilling by diamond heads, pouring of cooling water, grinding of sample faces etc. The term "undisturbed sample" must therefore be considered as a euphemism. As a consequence, laboratory tests have very limited validity and field studies are more or less compulsory for validating the laboratory results. With the exception of soils and coarse rocks, normative standards are not available. This lack of standard procedures naturally results in widely varying results for permeabilities assigned to a given rock formation. This variation is further increased by a broader range of experimental methods.

Lack of standard procedures concerns both the test conditions (e.g. pressures) and the preparation/pretreatment of the samples. A standardised sampling procedure would further ease comparability is missing with the sampling, a fundamental requirement, for comparability of measurement results. Uncertainty analyses and/or extensive statistical analysis of measurement procedures are available so far only rarely.

Multiplicity of influences on permeability, for instance of salt rocks (extent of geopressure and its influencing period, loosening, e.g. by the collecting and preparation of a sample, grain size, foreign components such as anhydrite etc.) often results in significant variations of the sample properties. A comparisons of experimental permeabilities obtained in different laboratories on samples from the same salt rock formation is only possible with difficulties, or simply

Table 3.4. Variation coefficients V of soil mechanical values in samples (after Harr 1987, in Gilbert 1992)

Parameter	Variation coefficent V (%)	Reference
Specific weight	3	(Hammit 1966)
Porosity	10	(Schultze 1972)
Saturation index	10	(Fredlund and Dahlmann 1972)
Water content (clay)	13	(Fredlund and Dahlmann 1972)
Cohesion	40	(Fredlund and Dahlmann 1972)
Hydraulic conductivity	240 (80% sat.); 90 (100% sat.)	(Nielsen et al. 1973)

$V = S/M$; S = standard deviation; M = mean value

impossible. As a general observation, the permeability values of samples which were extracted at different sampling sessions commonly differ widely. Therefore, documentation of boundary conditions is essential. Transferring results from laboratory tests to the subsurface rock formation must account on these boundary conditions. Uncertainties affecting the experimental determination of permeability values also contribute to the overall uncertainty in a value for the permeability of a rock formation.

Table 3.5 reflects some consequences of sampling (Fein et al. 1996), using permeability measurements of salt detritus as example. Published data are collected with the intention to create a basis for the determination of the porosity-permeability-function in salt.

Table 3.5 illustrates the dilemma of the rock salt permeability measurements: even though the rock is a quite uniform substance (rock salt), the measurement

Table 3.5. Summary of experimental data the permeabilty of rock salt detrius (Fein et al. 1996)

Reference	Number of values	Fluid	Remarks
Pusch et al. (1986)	12	Gas	Compacted dry with 22 000 °C, insufficient correction of the Klinkenberg effect, turbulence effects, Fuller distribution to $D = 1\,mm$
Walter et al. (1994)	10		Compacted moist and break, salt detritus sieved above $D = 16\,mm$, insufficient correction of the Klinkenberg effect
Liedtke	6	Brine	Compacted moist, salt detritus sieved above $D = 20\,mm$
Liedtke	7	Brine	Compacted dry, $D \leq 8\,mm$
Spiers et al.	7	Brine	Mono grain material mainly, compacted moist and slowly, dried with trichlorethane, mainly free of breaked material
Albrecht and Langer, Suckow and Sonntag	3	Brine	Not compacted salt detritus, $D \leq 5\,mm$
Suckow and Sonntag	3	Brine	Not compacted salt detritus, $D \leq 32\,mm$
IT Corporation	3	Argon	Dry compacted salt detritus, $D \leq 10\,mm$
IT Corporation	6	Argon	Avery Island detritus, $D \leq 10\,mm$, dry compacted
Liedtke	2	Brine	Moist compacted
Liedtke	6	Brine	Moist compacted, D below 20 mm
Walter	4	Gas	Old sample of in situ moistened compacted salty backfill, no breaking
Liedtke, Gommlich and Yamaranci	1	Gas	Very rough granulation (separation), porosity and permeability only inaccurately known

D grain diameter

procedures vary widely. Whatever the numerical results of a permeability determination might be, the permeability values must be considered to have a considerable random component.

The randomness is introduced already by the sampling and sample selection step. Some laboratories used the whole sample of the salt detritus, others sieved the samples and used only separate charges. Due to the preparation these samples exhibit usually different mineralogical compositions and different initial densities, moisture contents and compactions. The differences in the sample preparation can lead also in case of the same porosity to different permeabilities (Fein et al. 1996).

Comparability of values is further compromised by the differences in particle size distribution which in several cases not even has been determined. Commonly either the maximum grain size, occasionally also the non-conformity index and the curvature number was indicated. Fundamental for permeability, however, is the fine grain proportion (Fein et al. 1996).

Comparing the evaporative rock salt formation process of a natural rock salt with compacted rock salt detritus, a different permeability behaviour must be expected despite the same mineral composition (Bruck 1999). The applied compaction pressure has a decisive influence on the resulting material's density and/or porosity. Borgmeier and Weber (1992) report on respective investigations where dried and humid (air exposed) core samples were compared. These samples were squeezed "with a certain pressure". The influence of the humidity on the reduction of permeability under geopressure is substantial, while in extremely dried cores practically no reduction of the permeability could be observed (Bruck 1999).

In undisturbed rock salt the permeability is very small due to geopressure with typical values in the order of magnitude $k < 10^{-20} \, \text{m}^2$ (Schulze 1998). With deformation, a micro fracturing and consequently no permeability increase occurs, as long as the state of stress remains below the dilatancy limit (Schulze 1998). Dilatancy is a rheological phenomenon of viscous fluids, as rock salt, to set sold under the influence of pressure. In case of rock salt, the dilatancy limit gives the pressure where this phenomenon sets in. Exceeding the dilatancy limit will cause damage due to cracks, fissures and fractures. The dilatancy limit may already be exceeded during the extraction of samples from the underground rock because this process is connected with a more or less rapid stress relaxation. This procedure leads to structural loosening, to the forming of flow paths and thus to a permeability which does not exist in the underground rock (Bruck 1999). The simulation of a geopressure in the laboratory may result in a slow closure of these artificially induced pathways due to the inelastic flow of rock salt. Permeabilities obtained from a rock salt sample therefore may be time-dependent (Bruck 1999). Samples from rock without viscous flow property will remain disturbed.

3.2.2
Pore Volume Measurement

3.2.2.1
Basics of Pore Volume Measurement

A major challenge in the determination of rock permeabilities is the extreme sensitivity of the permeability on minor changes in the pore structure of a rock. This dependence is a major reason that permeability is among the rock properties most difficult to predict (Mavko and Nur 1997). It is defined as the ratio of pore area (cavity volume) to the total volume of a rock. This definition applies equally to coarse and solid rocks.

$$P = \frac{V_p}{V_{tot}} = 1 - \frac{V_s}{V_{tot}} \qquad (3.13)$$

with

P: porosity
V_p: pore volume (m^3)
V_s: volume of solid material (m^3)
V_{tot}: total volume of a porous material (m^3).

Measurement of geohydraulic characteristics of a rock in the laboratory often carries the hope to obtain more detailed and reliable information on the pore structure. However, pore volume measurements are as uncertain as permeability measurements. While porosity indicates only an average of pore volume percentage in a rock, the internal structure of the pores must be described by additional parameters such as tortuosity, constrictivity and pore radius distribution. Thus, the experimental information about the pore structure and/or about fissure and fracture structure are affected by various influence quantities.

Porosity described by Eq. (3.13) covers the entire pore volume in a material. It is also refereed to as total porosity, P_{tot}, and total cavity volume V_{tot}. Porosity covers the continuous pore channels, the dead end pores and the pore area locked in itself (gas inclusions, fluid inclusions). In a typical rock (e.g. granite), fluid inclusions are the only cavities which are refereed to as pores. The cavity volume is formed by fissures/fractures, fissures, fractures and junctures. If this text refers to fractures or fissures/fractures, the other cavity-forming structures, especially fissures/fractures and junctures, are included.

The cavity of a porous material and/or a fissured/fractured rocks is always filled with a fluid. In the groundwater zone (also called saturated zone) the fluid is generally water. In the case of a substantial contamination with an organic liquid (e.g. petrol) or in a petrol deposit also an organic liquid can fill the cavity partly or completely. In the unsaturated zone (the range above the groundwater level), two fluids (water and air) are present in the cavities. Rock surfaces are always covered with a thin film of water. Water has a higher wetability than air. Only within a limited range close to the surface this water film can evaporate. At elevated temperatures (arid climate), air-filled structures can exist. Within the contact zone of individual rock particles, water is accumulated in dead end pores and in pores with small pore radii due to surface tension and capillary forces. The term "reducible water porosity" refers to such pores. The remaining, gas-filled pore area is called drainable pore volume.

The gas-filled pores, in return, are not available for water. Thus, these pores reduce the effective pore volume. In addition, while water is an almost incompressible fluid, air is easily compressible. Due to the high compressibility of the gas-filled pore volume, the effective pore volume of such rocks is also pressure-dependent. The distribution of the aqueous and gaseous phase in the pore volume are determined by:

- The surface tension of the fluids;
- The fissure/fracture structure (particle size and form, pore radii, pore form etc.);
- The hydrostatic pressure.

Porosity P, as an integral parameter, is not directly related to hydraulic conductivity, k_f despite the fact that all porous rocks with hydraulically connected pores shows hydraulic conductivity. There is, however, no simple relationship between the number and size of existing pores and the observed permeability. Therefore, values of permeability are difficult to predict. Laboratory studies have shown that permeability depends on a somewhat longer list of parameters: porosity, pore size and shape, clay content, stress, pore pressure, fluid type, saturation – an almost overwhelming complexity. In spite of this, the general behaviour can often be expressed successfully using the remarkably simple Kozeny–Carman relation (Kozeny 1927; Carman 1937; Scheidegger 1974).

$$k = G\frac{P^3}{S^2} \qquad\qquad (3.14)$$

where k is the permeability, P the porosity, S is the specific surface area (pore surface area per volume of rock), and G is a geometric factor.

Solid rocks have to be differentiated into rocks with pore area (e.g. sandstone) and rocks with small and/or almost no pore area but cavities in the form of fissures/fractures (e.g. granite). The size of the pore volume of a porous material is determined by:

- The particle size distribution, the particle shape and the compaction;
- The kind of the binding between the individual grains and;
- The petrostatic pressure of the overlaying stratigrafic layers (geopressure).

The size of the pore volume of a solid rock without pores is determined by:

- Number of the joints, fissures, fractures and junctures;
- Opening width of the joints, fissures, fractures and junctures;
- Fissure/fracture filling.

Beside porous rocks (and sediments) and rocks with a certain fissure/fracture cavity volume there are rocks with double porosity (cf. Fig. 3.1). The porosity of these rocks is determined by pores and fissure/fracture porosity. Important examples are fissured/fractured sandstones.

Double-porosity aquifers pose high requirements (NTB 2001-03) to computer models for flow and transport of chemical compounds. The demands are due to the heterogeneity and spatial variability of the hydrogeological structure. Facing this problem special models are needed which can account for the relevant processes and the interactions between them. In the idealised fissures/fractures so-called flow channels are formed during the flow process, whose opening width is variable. The opening width can be described by a stochastic model using regional uniform statistical parameters. The required information on the fissures/fractures is: direction, length, amount, width, cross-linking, water passage availability and roughness. Although all these parameters have influence on the rock permeability, simple simulations tools use only some of these parameters.

The determination of the various parameters necessary for modeling permeability is often performed by modeling using geohydraulic fissure/fracture network models (e.g. NAPSAC; Hartley 1998; SHEMAT 2006). For this purpose, the numeric model simulates stationary flow inside the pore/fissure/fracture network in different directions. The permeability components are computed for each spatial direction and expressed as permeability tensor. By repeated simulation of the flow with different implementations of the pore/fissure/fracture network confidence intervals for the equivalent permeability are obtained. Hence, on the geohydrological level probabilistic modeling is already a standard procedure. The probabilistic techniques are commonly of the Monte Carlo type. Thus, the computer-intensive statistical approaches introduced in Part I of this treatise may be comparatively new for the description of variability in chemical thermodynamic parameters. However, probabilistic modeling is definitely not a new aspect for hydrogeological parameters, e.g. permeability.

A further advantage of the procedure is the possibility of transferring geohydraulic characteristics not only from local scale to regional scale, but also in reverse from the equivalent permeability to the transmissivity of the water-containing systems in the local scale.

Nevertheless the transfer of modeling results to the rock body is difficult to justify, because the results are based on statistical data (fissure/fracture frequency, fissure/fracture width, etc.) which were measured and calculated on samples whose relevance can easily be questioned: drill cores, boreholes, shafts and pits in mines. In all these cases, stress relaxation of rocks is a common effect, causing changes of the fissure/fracture distribution and structure. So far, no procedure for experimental assessment of fissure/fracture distribution in situ is known.

In summary, numerous uncertainties are associated with porosity measurement. These uncertainties affect, in turn, recharge, hydraulic conductivity, fracture porosity, dispersivity and matrix diffusion. Models for the transfer of sample rock porosity to rock formation porosity has to take these uncertainties into account.

3.2.2.2
Pore Volume Measurement Methods

A variety of different experimental methods for porosity determination are available. Only one single method is based on grain density (DIN 18124, 1997). The remaining method measures fluid transport through sample material. The methods differ in the sample preparation procedures, in sample size, in the fluid and in the method to generate a pressure gradient over the sample. Flow measurements determine, at least in principle, the effective porosity (inaccessible pores and fissures/fractures are ignored) because only accessible pores/fissures/fractures are available for the fluid medium to move over through the sample along the pressure gradient (Rietzschel 1996).

The following, often experimental methods for permeability determination will be summarized briefly:

- Pyknometer
- Lift and soaking test
- Gas porosimetry (Bousaid 1968)
- Carbon tetrachloride method
- Formation resistivity measurement
- Mercury porosimetry.

Pyknometer. Pyknometer methods are gravimetric methods. Their success is based on the high performance of weighing. Weights can be routinely determined with high accuracy. Traceability chains and appropriate calibration services for weights are available in all industrialized countries relating to the BIPM kilogram standard. A rock sample is emerged into a suitable fluid in the pyknometer. The mass of the fluid, the sample mass and the mass of the ensemble after addition of the rock sample is determined by weighing. From these data the amount of fluid inside the rock sample can be obtained. Multiple variants of the pyknometer are in use, e.g. air comparison pyknometer, capillary pyknometer.

The lift and soaking test determines the drainable pore volume. The rock sample is allowed to soak in a suitable fluid. The accessible pores will be filled with the fluid. In a second step, the amount of fluid inside the rock is determined by applying an external pressure to remove the fluid from the pores/fissures/fractures. An advantage of this procedure is determination of porosities at samples with irregular geometry and independent of size. A major limitation are rocks which components possibly interfering with the fluid (for instance water and salt). The lift and soak method is only applicable pore volume determinations of solid rocks. Application to rocks holding, e.g. dissolvable materials [salt, Fe(II) components, carbonates etc.] is misleading (Rietzschel 1996).

Gas porosimetry. Porous rocks are accessible to gases. By removing water from a rock sample in vacuum and contacting this "empty" sample with a known volume of gas, the amount of gas entering the rock volume can be measured.

Carbon tetrachloride method. The carbon tetrachloride method uses the different buoyancy of rock bodies in different fluids. The buoyancy of a sample in water is determined and set into the relationship to the buoyancy in carbon tetrachloride. The advantage consists of the fact that carbon tetrachloride shows a high wettability.

Formation resistivity measurement. This is an electrical measurement. There are two methods of the formation resistivity measurement: a) an electric flow is sent in the rock formation and the resistance against the electrical current is measured; b) an electric current is induced in the formation and the formation resistance is measured. Again the sample with and without pore filling is examined.

Mercury porosimetry. Mercury porosimetry is widely considered the best suitable method for the quantitative determination of the porous structures of solids. This method provides information on pore sizes distribution, pore volume, the apparent and true density of most porous materials. Mercury porosimetry is independently of rock type and sample shape. The method is based on intrusion of liquid mercury into the pores. Due to the high surface tension of hydrophobic mercury, an external pressure is required to push mercury into a capillary or pore. Mercury porosimetry is based on the capillary law governing liquid penetration into small spaces. This law, in the case of a non-wetting liquid like mercury, is expressed by the Washburn equation:

$$D = \frac{1}{p} 4\gamma \cos\phi \qquad (3.15)$$

where D is the pore diameter, p is the applied pressure, γ the surface tension of mercury and ϕ the contact angle between the mercury and the sample, all in consistent units. The volume of mercury penetrating the pores is measured directly as a function of applied pressure. This p-V information serves as a unique characterization of pore structure. Application of mercury porosimetry in the nuclear waste disposal is given by Hellmuth (1995).

3.2.2.3
Discussion and Deficit Analysis of Porosity Determination

During the discussion of total porosity determination according to DIN 18124 (1997) the following deficits became apparent:

- Large inaccuracies in the determination of the total volume V_{tot} are possible (more precise measurements can be made, in principle, by buoyancy measurement).
- The total porosity determined from the grain density contains both the accessible and inaccessible pore volume; because the inaccessible pore space doesn't contribute to permeability, its inclusion into the total porosity may cause an overestimation of hydraulic conductivity.
- The small sample quantity for the determination of the grain density in the pyknometer test imposes additional statistical uncertainties because only a small number of grains fit into a pyknometer volume. Thus the sample is often not representative for the bulk it has been sampled from.
- Large uncertainties are involved in transferring grain density from small sample sizes to an inhomogeneous natural material, determined at a very small quantity.
- The methods discussed above are not suitable for solid rocks. An accurate determination of the porosity of solid rocks in the laboratory is almost impossible because the sample volume must generally be considered as too small.

3.2.3
Laboratory Experiments for the Measurement of Permeability

In case of low permeable rocks the pore volume distribution and the hydraulic conductivity are inaccessible to direct measurement. Therefore, the permeability is determined directly. Generally, the samples are subjected to high pressures. Either the rock sample is pressurised in total or pressure gradients over the sample are established. Thus, most laboratory permeameter tests for determination of gas permeabilities submits a sample to various stress conditions which unusually differ from the situation encountered in the natural rock body. The tests involve stationary and instationary conditions (Borgmeier 1992; Boebe 1994). Many variations of the laboratory permeameter determination method are known, where the triaxial permeameter is more or less the standard application (Gloth 1980).

The range of application for the methods is indicated to $10^{-8} \geq k \geq 10^{-21}\,m^2$ by Autio et al. (1998). This corresponds to about $10^{-12} \geq K_f \geq 10^{-14}\,m/s$ depending on the rock type and fluid. Applications in combination with mercury porosimetry were used for the determination of permeabilities in granites, gneisses, tonalites, basalts, granodiorites, and marble (Hellmuth et al. 1995).

3.2.3.1
Overview on Laboratory Methods for Permeability Measurement

Following briefly discussed procedures for the determination of the permeability with gaseous fluids are usually used in practice:

- Lab permeation test (i.e. triaxial permeameter, mini-permeameter);
- Chamber methods;
- Perfusion of rock samples with gaseous fluids;
- Gas chromatography;
- ^{14}C-PMMA method and Helium gas method.

3.2.3.2
Description of Methods

Laboratory permeation tests (triaxial permeameter). The principle of the triaxial permeameter method is the establishment of a steady-state flow condition in a cylindrical rock specimen. The standard procedure of a permeability measurement with triaxial permeameter is the following: a pressure gradient is maintained across the sample with one end exposed to the ambient pressure and the opposite end at the test drive pressure. A radial confining pressure is maintained around the specimen. The effluent is collected and volume flow rate is determined. A triaxial permeameter set-up is shown in Fig. 3.4. It consists of the pressure unit with strain gauge, load transducer, the pressure housing holding the test specimen, the pressure sensors for radial and axial pressures, and the collectors and sensors for fluid flow and volume determination.

The triaxial permeameter method has developed into a variety of modifications (Christiansen and Howarth 1995), concerning both the apparatus and the technique. In addition to the standard procedure (stationary measurement) given above, instationary methods exist (Bruck 1999). Triaxial permeameter are reported in literature to quantify permeabilities as low as $1 \cdot 10^{-18}$ to 10^{-21} m^2 (Rietzschel 1996). Table 3.6 gives an overview on triaxial permeameter methods and their application limits.

Chamber methods. Permeability measurement by chamber methods comprises a group of instationary procedures (Bruck 1999). Chamber methods can be divided into single chamber and double chamber methods (Fig. 3.5). The permeability determination by a single chamber method represents an instationary procedure, with only one chamber before (the fluid is pressed through the rock sample) or behind (the fluid is sucked through the rock sample) the sample.

The double chamber procedure is based on the measurement of the time-dependent pressure between two chambers, one chamber on each side of the sample with different initial pressure up to the pressure balance (Bruck 1999).

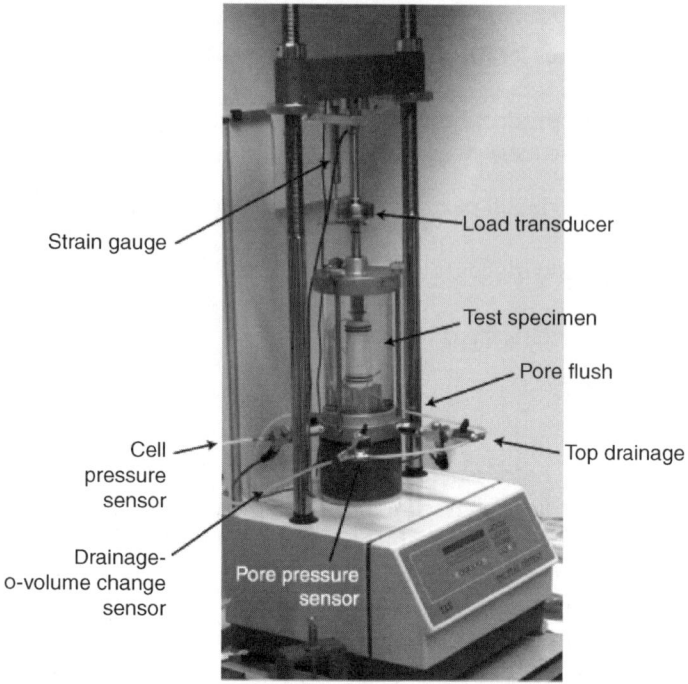

Figure 3.4. Triaxial test laboratory set-up

The evaluation of the permeability and porosity is based on the flow differential equation, whereby the fluid stream(s) from and/or into the pressurized chamber give the boundary conditions. The double chamber method allows to modify the pressure gradient over the sample in order to attain and control stationary conditions.

The practical realisation of chamber tests also has several modifications. The so-called "pulse decay test" represents a special form of the instationary measurements. The investigated sample is pressurized and relaxed in cycles (Koh 1969). Cycle frequency and pressure amplitude at one end of the sample are compared with those at the other side of the sample. The observed frequency differences, phase shifts and/or amplitude changes permit detailed conclusions on the permeability and porosity behaviour inside the sample. The closing pressure test registers a pressure increase at the end of a sample specimen after a stationary pressure gradient over the sample has been established. For this purpose, a valve is installed at the sample end. After a stationary pressure equilibrium is established the valve is closed and the time-dependent increase of the pressure at the sample exit is registered.

Perfusion of rock samples with gaseous fluids. As instationary medium for the perfusion tests different gases are mentioned in the literature: air (Bruck 1998),

Table 3.6. Triaxial permeameter methods for permeability determination

Method	Application limit	Procedure	Remarks	Literature
Standard triaxial permeation test	up to $10^{-12}\,m^2$	DIN 18 130 T 1		DIN 18 130 T 1
Compression test	up to $10^{-12}\,m^2$	Determination in the compression permeability equipment with high confining pressures		DIN 18 137
Closing pressure test		Controlled flow of rock samples over valve at the sample end, after equilibration of the stationary pressure gradient locking of the pressure valve and registration of the timedependent increase of pressure at the sample exit, analytic evaluation	Long-term experiments with low permeability samples because stationary starting situation must be waited for	Pusch 1986

hydrogen (Wagner et al. 1998), nitrogen (Wittke 99), argon (Fein et al. 1996) and helium (Christiansen and Howarth 1995). The pressures applied during this procedure are reported between 1 bar (Wagner et al. 1998) and 20 bar, in some extreme cases even up to 32 000 bar (Christiansen and Howarth 1995). With high pressure equipment permeabilities below $10^{-21}\,m^2$ can be determined.

Gas chromatography. Permeability tests with using a gas chromatography as pressure source is a special application of Darcy law (Bruck 1999). In this test, the cylindrical sample is rinsed at the sides with different gases and different pressures. After establishment of a stationary pressure gradient, the portion of the gas in the sample is determined and calculated into a permeability (Pahl 1995).With this method extremely low permeabilities can be measured. Depending on sample size and gas, $k \leq 10^{-24}\,m^2$. The procedure was developed further by Fröhlich et al. (1995), so that also diffusion coefficients of compacted salt detritus samples (no pressure differences between sample entrance and outflow] could be determined. Under assumption of equivalent pore radii permeability values were assigned to the rock salt sample (Fröhlich et al. 1995). This procedure allows one to obtain qualitative information on the pore structure of a compacted rock sample.

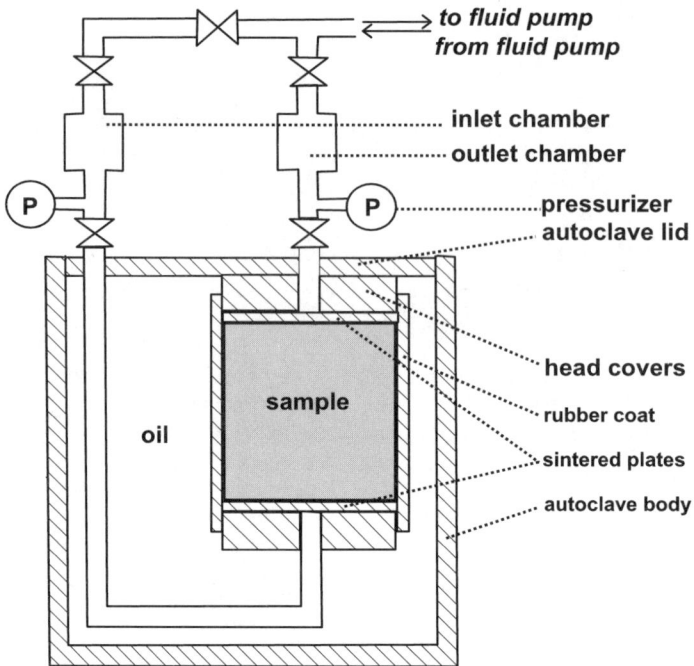

Figure 3.5. Graph of the double chamber test method

[14]C-PMMA-method and helium gas method. The ^{14}C-PMMA-method (PMMA: polymethyl methacrylate) and helium gas method are two rather recent, and therefore rarely applied, methods. Both methods belong to the group of flow tests, whereby with the ^{14}C-PMMA-method the retardation coefficient is determined by surface integration processes. With the ^{14}C-PMMA-method, the rock sample is impregnated with polymethyl metacrylate marked with [14]C. A determination of the diffusion rates is realised (Hellmuth et al. 1995; Autio et al. 1998) using the [14]C ß-decay as a signal source.

PMMA is characterised by a small molecular weight, optically clear, and insensitive to extreme variations in temperature. The ^{14}C-PMMA-method is usually coupled with the helium gas method, where the flow rate of helium gas is determined in a sample saturated with nitrogen (Autio et al. 1998).

3.2.3.3
Deficit Analysis of Laboratory Rock Permeability Measurement Methods

As already outlined previously, a fundamental problem in all laboratory methods for the measurement of permeability is the requirement of so-called "undisturbed samples". Since the extraction of a sample from a soil or rock is an enormous disturbance both under the aspect of rock physics and water chemistry, the extraction of an "undisturbed sample" is a contradiction in itself. Especially in case of soils and rock samples with low permeabilities, the disturbances from sampling (which themselves are very difficult, if not impossible, to reproduce) and the mounting of the sample in the laboratory will always cause deviations from the natural settings.

Conventional laboratory measuring procedures for the determination of the permeability require stationary flow conditions (temporally constant flow rate with constant pressure). Measurements on materials with low permeability therefore require test periods up to several months. It is understood that these time scales pose enormous difficulties to warrant stable test conditions for the determination of smallest flow rates over this period. Beyond that often no stationary flow conditions are present. An objective and provable criterion for stationary conditions does not exist.

A major problem with laboratory permeability tests using gaseous fluids is the choice of the gas pressures themselves. These are commonly set according to the experience of the experimenter because there are no criteria, e.g. for the starting pressures. An advantage of the triaxial permeameter in this respect is the application of radial symmetric pressures that in the ideal situation will correspond to the conditions in the rock before sampling. The documentation of the applied pressure curves is essential to achieve at least a minimum in comparability of the derived permeability values. A modern addition to the triaxial measurement is the use of a rubber coat to shield the rock sample. With a rubber coat, the applied gas pressure at the axial end of the sample cannot escape laterally during the test period (Christiansen and Howarth 1995). Thus, the influence factor of laterally escaping fluid is reduced. A further influence factor is gas adsorption on solid surfaces inside the equipment.

Typical problems of the single chamber method measurements for samples with very low permeabilities and porosities arise due to the absorption of the initial pressure (Bruck 1999). This problem was reduced with the introduction of the double chamber procedure. A further advantage of the double chamber method is the possibility to create well defined conditions on both sides of the sample. An important, but likewise difficult to prove assumption of chamber methods is the complete absence of lateral flow paths.

There is no international standard scale of permeability test methods. Therefore, published results vary widely in the sample sizes, applied pressures, pressure ranges, and experiment duration. Therefore comparability is limited. A further influence factor is the applied gaseous fluids. Generally, no data on sorption of the fluids during the test are commonly available. This error source is only poorly discussed in literature. Hellmuth et al. (1995) reported sorption studies interpreted by BET isotherms for gas sorption during flow tests in granites, tonalites and basalts. For the fluids nitrogen and krypton, the tendency for sorption was small. High sorption was observed, however, for butane.

Table 3.7. Laboratory methods for the determination of the permeability in small-permeable rocks: streaming with gaseous fluids

Method	Application limit	Procedure	Remarks	Reference
Triaxial permeameter	$10^{-15} \geq k \geq 10^{-21}\,\mathrm{m}^2$	Under stationary or non-stationary test conditions determination of the gas permeability using various inner and outer pressures after a pressure input	Many variations in permeameter determination methods are in practice (Gloth 80)	Borgmeier and Weber 1992; Boebe 1994
Single chamber method	Up to $10^{-21}\,\mathrm{m}^2$	Application of a chamber before the sample and production of a pressure jump, application of pressure at the sample entrance and measurement of the time-dependent volume stream at the sample exit and/or the decrease of pressure in the entrance chamber	Determination of permeability and porosity	Wallick and Aronofski 1954; Bruck 1999
Double chamber method	Up to $10^{-24}\,\mathrm{m}^2$	Measurement of the time-dependent pressures in chambers before and behind the sample with different initial pressure up to the pressure equilibrium, then computation of the permeability and porosity of the flow differential equation	Determination of permeability and porosity	Finsterle and Persoff 1997; Bruck 1998, 1999
Gaseous fluid perfusion	At ambient pressure conditions: $k \geq 10^{-18}\,\mathrm{m}^2$ with high pressure equipment $k \geq 10^{-21}\,\mathrm{m}^2$	Flow under stationary or non-stationary test conditions	Calculation of permeability	Stationary test conditions: Bamberg and Häfner 1981; non-stationary test conditions: Pusch et al. 1986
Gas chromatography	$k \geq 10^{-24}\,\mathrm{m}^2$	After application of a stationary pressure gradient the gas portion which has flowed through the sample is measured and converted into a permeability	Procedure was developed further by Fröhlich (1995) so that also permeability values can be determined	Pahl 1995

Table 3.7. continued

Method	Application limit	Procedure	Remarks	Reference
^{14}C-PMMA-method	$10^{-8} \geq k$ $\geq 10^{-21}\,\mathrm{m}^2$ (corresponds depending on the rock type to $10^{-12} \geq k$ $\geq 10^{-14}\,\mathrm{m/s}$)	Impregnation of the rock samples with ^{14}C which is doted with polymethyl methacrylate and determination of the diffusion rates	Rarely applied	Autio et al. 1998
He gas method	$10^{-8} \geq k$ $\geq 10^{-21}\,\mathrm{m}^2$ (corresponds depending on the rock type to $10^{-12} \geq k$ $\geq 10^{-14}\,\mathrm{m/s}$)	Measurement of the flow speed of helium gas in a nitrogen saturated sample	Rarely applied	Autio et al. 1998

Permeability measurement by gas chromatography requires a large machine and takes considerable temporal expenditure. This is a main reason for the limited application this method has found so far. Its advantages are the low permeabilities which can be resolved in principle. With the extreme pressure applied, rock-specific gas sorption and, eventually, liquefaction inside rock pores and fractures can falsify the result. For the ^{14}C-PMMA method and the helium gas method, there are up-to-date no investigations available as a basis for uncertainty assessment. For this reason, these methods will not be discussed further.

The orientation of the sample in the rock can be of fundamental importance, especially if anisotropy comes into play. Spatial orientation can be determined principally in the laboratory however a careful sampling documentation is necessary. Each mechanical action in the underground, however, disturbs equilibria, e.g. by relaxation of the pore water pressure during drilling, by displacing and consolidating the material (Schreiner and Kreysing 1998), and by escape of water-dissolved gases.

3.2.4
Contribution of Geophysical Methods to Permeability Measurements

3.2.4.1
Overview on Geophysical Methods

The direct determination of values characterising the geohydraulic conditions in a rock body characteristics using classical methods of geophysics is possible only rarely. Table 3.8 summarises some geophysical methods together with the major characteristics with respect to permeability determination.

Direct access to water content and permeability is offered by nuclear magnetic resonance (NMR) (Kenyon 1992). In the field, geophysical procedures are commonly applied as supplements to geohydraulic tests (cf. Table 3.8). Examples of the application of these procedures are given by Hubbard and Rubin (2000) and Lindner and Pretzschner (1998).

The distribution of electrical and dielectric characteristics can be determined, e.g. with electromagnetic procedures. These depend, e.g. on salinity beside the water content, porosity and the saturation point. Cross-linkage and pore shape distribution are further relevant influence factors etc. Other possible methods, e.g. the velocities of seismic waves, generally depend on the different matrix velocities are, the water content and, in unsaturated rocks on the distribution of the water (Yaramanci et al. 2000).

Table 3.8. Geophysical methods for the determination of the geohydraulic characteristics

Method	Range of application	Procedure	Remarks	Reference
Fluid logging	$10^{-5} \geq k_f \geq 10^{-9}$ m/s	Measurement of the electrical conductivity and the temperature of the borehole flushing along a borehole	Detecting water access from the rock formation	Schwarz et al. 2000
X-ray tomography	No indication	Rock core scans: porosity, permeability, saturation, fluid movement	New procedure, rarely applied	Brinkmann 1999
Electrical resistance measurement	$10^{-4} \geq k_f \geq 10^{-9}$ m/s	Characterisation of different rock formations over the measurement of the rock resistance and conclusion on rock porosity	Rock differences locatable	GLA 1994
Temperature distribution measurement (temperature logs)	$10^{-2} \geq k_f \geq 10^{-9}$ m/s	Admission of a temperature distribution during/briefly after a water extraction from the borehole	Determination of depth and amount of water inflows	GLA 1994
Fracture identification log (FI), sonic waveform log (BHC-WF), natural gamma spectrometry (NGS)	No indication	Different geophysical procedures for the joint recognition (e.g. measurement of the velocity of sound in the rock)		DVWK 1983; Lindner and Pretzschner 1998
Ultrasound measurements	No indication	Determination of the fissure/fracture distribution		Petzold 1976

3.2.4.2
Description of Geophysical Methods

Fluid logging. Fluid logging methods are commonly using electrolyte solutions as indicators. A borehole fluid is exchanged against water with clearly higher (or, in saline waters, lower) conductivity (Rosenfeld 1998b; Schwarz et al. 2000). The measured effect is a change in electrical conductivity in the bore hole. The water with added electrolyte is termed contrast water. In some situations, the water temperature may be used as an indicator. Subsequently, a pressure gradient is established causing inflows into the borehole from the surrounding rock formation. By lowering the static water level in an unlined borehole inflow zones (e.g. veins, fissures, fractures) become activated and feed water into the borehole proportionally to the fissure/fracture transmissivity.

If the inflowing water differs according to the electrical conductivity or the temperature of the borehole fluid, the inflow zones can be recognised by logging the borehole using a sensor probe. The temporal change of the electrical conductivity/temperature due to inflowing rock formation fluids is registered over time by the probe(s). The inflow zones become apparent as conductivity peaks. The transmissivity of the individual inflow zones can be determined from the temporal development of the conductivity logs with the help of analytical and numerical procedures (Rosenfeld 1998b). Applications of fluid logging are:

- Detailed localisation of inflow and discharge zones in the borehole;
- Determination of fissure/fracture transmissivities.

Fluid logging is a procedure that supplies without drilling under the use of packers all inflow zones contributing to the total transmissivity (Rosenfeld 1998b). From the inflow rates results for individual transmissible fissure/fracture zones can be computed by a geohydraulic evaluation of the pumping phase. The applicability of the method for inflow separation is depending on the distance of the inflowing aquifers, the conductivity difference between the fluid in the borehole and the contrast fluid and their relative contribution of each fluid to the total transmissivity (Rosenfeld 1998b). Single inflows can be identified if they form a distinguishable contribution to the total transmissivity, which may be in the range of $1/10$–$1/120\,000\,m^2/s$ (GLA 1994). To allow a numerical evaluation of the measured log data the following conditions are assumed to hold (a usually overly optimistic assumption; Rosenfeld 1998b):

- The aquifer is homogeneous, isotropic, from continuous thickness and has an apparently unlimited expansion, Darcy Law is applicable;
- The confined and/or free water level is almost horizontal, and the groundwater flow is horizontal over the entire aquifer;
- The aquifer does not receive surface water inflows within the test range.

The procedure is applicable in a relatively large permeability range of approximately $5 \cdot 10^{-4}$ to $10^{-9}\,m\,s^{-1}$. The fluid logging procedure is applicable in open drillings as well as in perfect and imperfect wells starting from a nominal size of 50 mm ($\sim$ 2 inches) with confined and free groundwater conditions. While the pumping phase inflow zones are in places noticeably, from which conductivity peaks are developing. The inflowing joint fluid is transported with the flow upward to the pump advectively. Additionally, the flow rate increases by the summation of the single inflows upward, so that a conductivity peak always represents the sum of the underlying inflows. The peaks increase in the inflow zones and spread with their fronts in direction towards the pump. Dispersion procedures cause an expansion and a flattening of the conductivity peaks (Rosenfeld 1998b).

If the joint fluid has a higher mineralisation compared to the contrast fluid, it will be subject to gravitational dropping in the borehole because of its higher density. Particularly peaks in deeper zones of the borehole spread more than can be explained by dispersion and diffusion downwards the borehole bottom. As a consequence, inflow zones are located a bit above their actual transmissivity maximum, where the early logs shows a characteristic rise of the conductivity. With increasing test time the conductivity peaks interfere. Especially in flat drillings, the duration of the front migration of the conductivity peaks is strongly shortened (Rosenfeld 1998b).

Nuclear magnetic resonance (NMR). Since the 1940s, nuclear magnetic resonance (NMR) is a sensitive standard technique for the structural analysis and quantification of hydrogen atoms in various matrices (including human tissues). NMR meanwhile is applied to geophysical investigations in the laboratory and, since about a decade, in the borehole (Yaramanci et al. 2000). The procedure offers a direct and quantitative access to the water content and, consequently, permeability (Kenyon 1992). For practical application, commercial equipment is available with surface NMR (SNMR) (Legchenko et al. 1995; Beauce et al. 1996).

With the SNMR procedure, a short magnetic pulse is produced in a coil which is corresponding with the Larmor frequency of the local earth's magnetic field. Following the pulse, the hydrogen's nuclear spin relax into natural distribution under emission of radiation. The associated weak signal is measured. The amplitude of the relaxation signal and the longitudinal relaxation time is a measure of the water content. The decay rate is proportional to the permeability and the phase shift relative to the excitation field is proportional to the electrical conductivity of the investigated rock volume (Yaramanci et al. 2000).

3.2.4.3
Discussion and Deficit Analysis of Geophysical Methods

To avoid major bias in the evaluation of fluid logs some fundamental flow and transportation characteristics in the borehole need to be taken into account.

The interpretation and evaluation of the results requires considerable expertise. The selection of a suitable evaluation procedure depends primarily on the size and form of the conductivity and/or concentration peaks. Evaluation procedures for overlapping signals exist (NTB 1993-47). In some investigation procedures, the concentration of the inflowing fissure/fracture fluid must be known, thereby requiring a rather complex depth-orientated logging.

The NMR method was shown to provide rather reliable water content values both in the saturated and unsaturated zone. For the estimation of the permeability from the magnetic signal decay times a numeric relationship for the dependence from particle size distribution was set up (Yaramanci et al. 2000). It was shown that the water content determined by SNMR corresponds to the flowing fraction of water (i.e. that the connate waters do not influence the measurements).

At present, an interpretation of the SNMR signals is possible only if a horizontal layering is assumed (1D). The influence of lateral variations of the excitation field and the effect of lateral changes of water content (due to convection and diffusion) were examined by a 3D-modeling method. The SNMR procedure is found useful as a supporting method for the determination of geohydraulic data. SNMR is nevertheless a rather modern technique and still under development (Yaramanci et al. 2000).

3.2.5
In-Situ Measurement of Permeability and Hydraulic Conductivity

3.2.5.1
Overview on Methods for In-Situ Measurement of Permeability and Hydraulic Conductivity

In-situ determination of the permeability/hydraulic conductivity is still an unsolved problem of geosciences. Nevertheless, improvements were achieved in recent years (Debschuetz 1995). The difficulties lie in the extreme sensitivity of the permeability to smallest changes in the fissure/fracture structure of a rock. In the last decade, considerable effort has been applied to the detailed hydraulic characterisation of rock formations. The modern hydraulic well testing and analysis techniques, originally developed in petroleum engineering, are increasingly applied to the field of low-permeability hydrogeology (Lavanchy et al. 1998).

The pressures and flow rates measured during hydraulic testing are controlled by the flow geometry in the coupled system borehole/formation, as well as the hydraulic properties of the geological formation (e.g. hydraulic conductivity, storativity, initial pressure). In low-permeability formations factors such as wellbore storage, pre-test pressure disturbances, equipment compressibility and temperature variations can gain such an importance that their effects will dominate the interval pressures and flow rates from or toward the forma-

Table 3.9. Field methods for the determination of hydraulic permeability

Method	Range of application	Procedure	Remarks	Reference
Pump test and	$10^{-1} \geq k_f$ $\geq 10^{-9}\,\mathrm{m\,s^{-1}}$	Lowering of the water level in lined boreholes, afterwards evaluation of hydraulic conductivity by various methods	Widely applied	Schneider 1987
Open end test	$10^{-4} \geq k_f$ $\geq 10^{-9}\,\mathrm{m\,s^{-1}}$	Heightening of the water level in lined boreholes with open bottom		Schneider 1987
Bail or fill-up test	$10^{-4} \geq k_f$ $\geq 10^{-9}\,\mathrm{m\,s^{-1}}$	Measurement of the water level development after lowering or artificial increase of the borehole water level	Approximation method, in case of low permeabilities time-demanding	Heitfeld 1989
Pulse test	$10^{-7} \geq k_f$ $\geq 10^{-13}\,\mathrm{m\,s^{-1}}$	Measurement of the pressure after brief cyclic pressure impulse jumps in a packered test range of the filter of the borehole	Sensitive to well effects, optimal geohydraulic connection to the rock necessary	Hayashi et al. 1987; Schwarz et al. 2000
Slug/bail test	$10^{-2} \geq k_f$ $\geq 10^{-9}\,\mathrm{m\,s^{-1}}$	Measurement of the water level after brief induced positive or negative changes of water level in a test track defined by packers	Sensitive to well effects, optimal geohydraulic connection to the rock necessary	Kraemer 1990; Schwarz et al. 2000
Water pressure test	$10^{-5} \geq k_f$ $\geq 10^{-9}\,\mathrm{m\,s^{-1}}$	Measurement of the water level development after pushing in a defined quantity of water		Schneider 1987
Gas water displacement test	No indication	Water displaces gas at different pressure gradients and/or displacement velocities (usually multi-cyclic)		Czolbe and Klafki 1998
Radiohydro-metric methods (tracer-tests)	$10^{-3} \geq k_f$ $\geq 10^{-9}\,\mathrm{m\,s^{-1}}$	Measurement of the concentration of natural or artificial tracers and determination of the residence time	Consideration of matrix diffusion necessarily	Delakowitz 1996; Guimera and Carrera 2000

Table 3.9. continued

Method	Range of application	Procedure	Remarks	Reference
Noble gas dating	No indication	Measurement of the noble gas content of pore waters degassed under vacuum from the rock sample as depth profiles	Determination of residence times, in case of well-known porosity indication of the permeability	Osenbrück 1996; Lippmann 1998; Rübel 1999
In-situ-permeameter (e.g. Guelph-permeameter)	$10^{-5} \geq k_f \geq 10^{-13} \, \text{m s}^{-1}$	Measurement of the quantity of water infiltrating per time unit from a cylindrical unlined borehole into the rock measured under stationary conditions (Mariotte bottle principle)	Usually used in low permeable soils	Salverda and Dane 1993

tion. Therefore, these factors must be considered in all the following steps of a measurement, in test design, in field testing and, if necessary, also in the final measurement data evaluation.

The under/overpressure induced by drilling activities and pre-test stand-by periods produces a disturbance of hydraulic or pneumatic pressure within the rock formation. In dense formations, the time necessary for this (unwanted, but technically unavoidable) under/overpressure to dissipate is a function of the pre-test pressure history. The dissipation may last several days or weeks. The transient pressure recovery back to initial conditions will be superimposed on the pressure transients induced during the active test period(s) on which the determination of the hydraulic parameters are eventually based. In addition, the lest interval wellbore storage, which is determined by the interval volume as well as the fluid and equipment compressibility, dominates the initial part of tests such as the constant rate test or a pressure recovery period after shut-in and may therefore hide the formation properties for days or even weeks.

As a result, it is necessary to elaborate appropriate testing procedures which allow borehole history effects to be accounted for and minimize the importance of the wellbore storage. The extent of the disturbance being dependent on borehole conditions such as interval volume as well as formation properties, it is essential to design the test on the basis of appropriate equipment and specific test types adapted to an expected range of formation permeability (Lavanchy et al. 1998).

The borehole history effect can only be reasonably detected and accounted for if the test starts with a pressure recovery period. Furthermore, constant rate tests, even though performed at the lowest rate afforded by the pump, cannot overcome the wellbore storage effect within a reasonable amount of time if performed in a very low permeability formation. Instead constant head tests followed by a recovery period have to be conducted. For very dense formations operational constraints (testing duration, flow control capacity) often reduce the choice to a slug lest, or even a pulse lest, which is the simplest test that can be conducted.

Two principal groups are distinguished for the geohydraulic determination of in-situ permeabilities: tests with constant pressure and tests with constant flow. A survey on the currently available procedures is given in Table 3.9.

Examples of the practical application of these procedures are given in Peterson et al. (1981), Beauheim (1993), Delakowitz (1996), Croise et al. (1998), Lavanchy et al. (1998), Rothfuchs et al. (1998), Schittekat (1998), Wieczorek (1998), Zenner (1998) and Gautschi (2001).

3.2.5.2
Description of In-Situ Measurement Methods for Permeability

Pump or recovery tests and open end tests. During pump test the water table in a borehole is reduced and the recovery of the water table is recorded. Similarly, a fill-up test involves the addition of water into a borehole and the dissipation of the water into the subsurface environment of the borehole is monitored. Fill-up and pump tests can be accomplished in all drillable rocks (Heitfeld 1998). Under most circumstances a practical depth limit is given by the depth of the borehole itself.

The open end test represents a special form of the fill-up test whereby an increase of the water level takes place in lined boreholes which are in contact with the host formation only at the bottom of the borehole. If no water level is present in the borehole, the method is termed an "infiltration test". For evaluation of the measurement data several theoretical or empirical relationships have been proposed. Both the saturated and/or unsaturated soil zone are accessible depending on the water level(s) during the test. The position of saturated and unsaturated zone has to be appropriately considered for the selection of the evaluation procedures.

Pump tests are limited to the depth below the ground-water level. Fill-up tests and pump tests can be realised in a stationary regime (i.e. with constant fill and/or pump rate at constant borehole water level) or in the instationary regime with a dropping or rising water level (Heitfeld 1998). A fill-up test in the stationary regime involves a constant amount of water to be added into the borehole. The rising water level implies a rising hydraulic pressure. The stationary regime is established if the fill rate equals the dissipation rate, characterized by a constant water level in the borehole. In the instationary case water is added into the borehole and the reestablishment of the previous water level is followed. In a bail test the procedure runs in the opposite way, but is otherwise closely related to the fill-up test.

In many cases fill-up and pump tests need to be performed during the drilling campaign. Hence, the drilling activity needs to be adapted appropriately to the needs of the fill-up/pump tests. A detailed description of the goals and requirements is already necessary in the planning phase, e.g. to avoid the flushing of additives which may have an adverse effect on the permeability of borehole walls. By generation of finely ground rock a sealing effect can take place causing incorrect information on the water exchange between borehole and host rock formation (Heitfeld 1998).

Fill-up and pump tests are feasible in all kinds of soils and rocks; in unstable rocks the infiltration can be by open end tests, whereby borehole lining need not to be removed from the borehole. Fill-up/pump tests open direct access to the in-situ rock permeability in the direct vicinity of a borehole. Furthermore, the change of the permeability with depth can be determined. The ranges of application are given with $10^{-4} \geq k_f \geq 10^{-9}\,\text{m/s}^{-1}$ (Heitfeld 1998).

Pulse test. The pulse test is applicable to the determination of low permeabilities, i.e. $10^{-7} \geq k_f \geq 10^{-13}\,\text{m s}^{-1}$ (Hayashi et al. 1987; Poier 1998; Schwarz et al. 2000). A pulse test is applicable only in the saturated zone. In a borehole, the test section separated by packers and a brief pressure (pressure pulse) is induced. The test section is closed by a valve, so that the pressure can dissipate only via the host rock. Despite the very low permeability ranges in which the pulse test can be accomplished, a test does not take more than 15–30 min (Poier 1998). The test serves for the determination of transmissivity, hydraulic conductivity and skin effects.

For the numerical evaluation of pulse tests, several procedures have been proposed (Ramey et al. 1975, Bredehoeft 1980; Peres 1989). Closer explanations can be found in Poier (1998).

The evaluation of experimental data assumes the perfect well. Besides host rock homogeneity isotropy and water saturation of the rocks are considered to be uniform, i.e. a radial symmetrical flow and rigidity of the rocks or the test equipment are assumed. Furthermore, the spatial extension of the aquifer is considered to be horizontally unlimited (Poier 1998).

Slug/bail tests. The concept of slug tests (also called slug injection test) and bail tests is based on sudden, artificially produced changes in the geohydraulic downward gradient between a test well and the surrounding aquifer (Rosenfeld 1998). The slug is a body which causes a rapid increase in the water level when immersed into the bore hole. For illustration, the principle of a slug/bail test is shown in Fig. 3.6.

A slug test changes of the geohydraulic downward gradient in the borehole by a sudden increase of the water level. The reverse procedure, i.e. the sudden decrease of the water level in the well, is called a bail test. The measurement

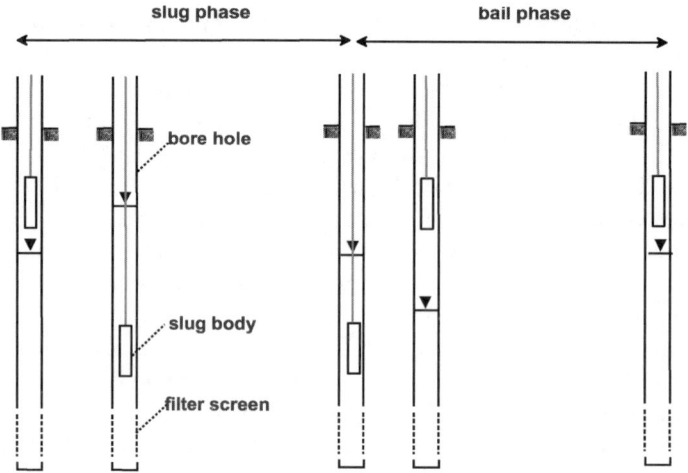

Figure 3.6. Schematic description of a slug/bail test. For permeability determinations the bore hole walls may have no lining in the depth range of interest. Filter screens are often necessary to ensure wall stability but influence the permeability determination

values of water level change in comparison to the static water level supplies the basis for the determination of the transitivity (Rosenfeld 1998). The test allows of the determination of transmissivity, hydraulic conductivity, storage coefficient and skin effect.

Slug/bail tests provide the permeability and/or transmissivity at the borehole location. For the evaluation of the experimental data, a set of procedures is available, the validity of which is limited by flow conditions, and aquifer and well characteristics. Evaluation methods are divided into rectilinear, type-curve and analytic procedures (Rosenfeld 1998).

Slug tests can be accomplished in lined groundwater wells only if the static water level lies above the filter. In the unlined borehole a dissipation of the water gauge into the saturated zone must be prevented. Bail tests can be applied both in lined and unlined boreholes. In principle the wells must be perfect, too. For the accurate mathematical calculation the following conditions are required:

• The aquifer is homogeneous, isotropic and has an apparently unlimited expansion. Darcy law (laminar flow) is valid.
• Inflow and outflow to and from the aquifer are completely horizontal over the whole aquifer.
• The aquifer does not receive surface water inflow within the test range.

Water pressure test. A water pressure test is performed in a test section of a borehole separated by two packers (Czolbe and Klafki 1998; Poier 1998). Under a constant pressure, a certain amount of water is injected into the rock. From the pressure difference, a value of the hydraulic conductivity k can be computed. In addition, informations about the deformation and erosion behaviour of the rock can be derived. The application is possible in hydraulic conductivity ranges $10^{-4} > k > 1 \cdot 10^{-8} \, \mathrm{m \, s^{-1}}$ (Heitfeld 1984; Schneider 1987). With appropriate equipment, it is possible to apply the test in rocks with higher and smaller permeable rocks. The measurement capabilities of water pressure tests can be summarised as follows:

- Determination of the pressure-dependent water absorption ability.
- Determination of the hydraulic conductivity.
- Determination of rock ranges with different permeabilities.
- Qualitative estimation of the permeability above the ground-water level.

For the evaluation of water pressure tests, several approaches are distinguished. Classical evaluation procedures are based on continuous processing (Poier 1998). Under stationary flow conditions the following requirements have to be met:

- The rock is water-saturated, homogeneous and isotropic.
- Radial symmetry, open system with the boundary condition h(R) = 0 with R = range, the injecting interval is constant, no change of the natural groundwater surface.
- The fissure/fracture body is rigid, the groundwater is incompressible, no erosion takes place during the test.

By applying water pressure tests, flow components are assumed to be directed only perpendicular to the borehole and Darcy law to be valid (laminar flow). The application of this test is limited to drillings with stable bore hole walls with small variations of the bore hole diameter. Therefore, careful preparation of the drilling campaign, and experienced personnel for its performance, is a necessary requirement to avoid misinterpretation of water pressure tests.

Design calculations of hydraulic tests. The purpose of design calculations of hydraulic tests is to set up appropriate test sequences which minimize the magnitude of disturbing effects (borehole pressure history, wellbore storage or temperature variations) and allow for the determination of the formation properties within a minimal time frame (Lavanchy et al. 1998). When the test objectives are clearly defined, several factors have to be considered when designing tests:

- Borehole conditions (diameter, test equipment, test fluid, history period), formation hydraulic parameters (Darcy flow in a poro-elastic medium).
- Physico-chemical processes that may have some impact on the pressure transients.

Factors such as borehole diameter and history period often result from operational constraints, and can therefore only be slightly influenced when designing the test. As a general rule, the testing fluid should be chosen according to the fluid expected in the formation investigated. Several options can be used to improve the test equipment. For instance, the test volume, controlling the well-bore storage, can be reduced by increasing the tubing volume and placing the opening/shut-in valves as well as the pressure transducers downhole instead of at the surface.

Once the initial borehole conditions are known, setting up a test sequence involving appropriate test types for an optimal time frame depends on the selected goal parameters and the expected range of formation properties (e.g. permeability, static pressure). Usually, a test sequence include at least two or three individual phases comprising:

- an initial recovery period (stabilization of the starting pressure distribution around the interval to reduce the impact of pre-test pressure history);
- an active flow phase involving a constant rate, constant head, slug or pulse test (inducing a pressure disturbance under controlled conditions);
- a final recovery phase allowing for observation of pressure stabilisation back to initial condition.

The initial and final recovery phases are passive tests during which, given well defined criteria, the only decision to be taken by the experimenter(s) is to stop or continue the test. Choosing a test type for the active flow phase depends mainly on the equipment performance (e.g. flow/pressure control and measurement detection limit) and formation hydraulic properties. Constant rate tests are usually preferred as long as the flow control equipment, the geomechanical state and the time allocated for testing allow them. By continuously monitoring the test using appropriate graphical techniques (e.g. log-log diagnostic plots), an objective determination of the flow model and of the decision criteria continuing or stopping a test is possible (Lavanchy et al. 1998). If the permeability is too low for a constant rate test, then constant head test (also known as constant pressure test) is performed. Like the constant rate test, this test also allows for an active flow period to be conducted with a constant inner boundary condition (pressure). Since the well pressure is constant, this test is not affected by wellbore storage disturbances. Furthermore, the pressure is easier to control than low flow rates, and very low flux variations can be measured. If constant head tests cannot be applied there remains the option using slug or pulse tests. For both tests, an instantaneous initial pressure change is applied to the interval, and the pressure recovery is then measured either with the shut-in valve open (slug) or closed (pulse). The wellbore storage is smaller in the latter case, improving the sensitivity of System and therefore allowing the determination of very low hydraulic conductivities (K-value about 10^{-13} m/s, or sometimes even lower).

Radiohydrometric methods (tracer tests). During radiohydrometric methods (tracer tests), short-living radioactive isotopes are added into a borehole. The radioactivity of natural or artificial tracers and their residence time in the borehole is followed. Single borehole and multi-borehole methods are distinguished. Further details are given in Drost (1989). In small-permeable rocks only the single borehole technique is suitable, since the residence times are very long in the groundwater (Guimera and Carrera 2000; NTB 1994-21).

The measurement equipment for the determination of the filter velocity and the flow direction of the groundwater forms a tracer probe (Delakowitz 1996). Into the central part of the borehole, the measuring or dilution volume (height about 0.5 m), a radioactive tracer is injected and homogeneous distributed due to spiral mixing. The detection of the tracer is accomplished by a scintillation detector. For measurements in small-calibre drillings, a probe with two independent detectors is used. The first scintillation counter, which is implemented in the measuring volume without screen, serves for the determination of the filter velocity. The second detector is implemented as direction detector to the lower end of the section. A typical tracer is the radioactive isotope ^{82}Br in NH_4Br-solution. ^{82}Br is an almost ideal hydrologic tracer because its flow behaviour is similar to that of the groundwater. Its gamma radiation can be registered also by outside of the filter pipe.

The application of the one-borehole method requires a vertical drilling, which represents the aquifer in its thickness. The borehole should be lined with a filter pipe and a gravel filter that it can be passed by the groundwater with small filter resistance. The groundwater flow through the filter pipe is assumed as stationary and horizontal. The filter velocity is determined from the tracer dilution process. For this reason, the distribution of the radioactive tracer in the water column of the investigated horizon of the filter pipe needs to be homogeneous. Due to horizontal ground-water flow, the tracer is carried away, which leads to an activity decrease in the measurement volume with time. From this data the filter velocity can be calculated. The lower detection limit for the determination of the filter velocity from the dilution log is given by the internal diffusion of the tracer on its way from the filter pipe into the surrounding gravel coat.

Tracer material and isotope investigations (noble gas dating). On the basis of tracer material and isotope investigations, hydraulic conductivity and/or permeability of rock formations can be determined. In this way, the variation of natural environmental tracers in the ground and pore water of rock formation is used. Of special interest are the noble gas isotopes ^{4}He as well as the stable water isotopes ^{2}H and ^{18}O. If the isotope investigations in the pore water are accomplished with samples collected from drill cores the groundwater profiles, and/or pore water residence times or at least a lower bound of the water age (residence time) can be determined. With a more detailed knowledge of the porosity of the rock formation as well as the upper and lower groundwater pressure level the hydraulic conductivity can be computed (Lippmann 1998; Osenbrück 1998) from the residence time (Rübel 1999).

Noble gas (helium, neon, argon, krypton and xenon) isotopes provide a key tracer in groundwater modeling studies. Gases from the crust, the mantle and the atmosphere (dissolved in groundwater at recharge) generally have a unique isotopic fingerprint that enables resolution and quantification of fluid contributions from these different sources (Ballentine 2002). Noble gases helium, neon, argon, krypton, and xenon are chemically sufficiently inert. These gases have a strong tendency to partition into gas or fluid phases and are used as tracers indicating origin and transport of fluids. In rocks these noble gases are present typically in very low concentrations of $\sim 10^{-9}$ to 10^{-6} cm^3 STP/g (He, Ar; 1 cm^{-3}) (STP is equivalent to $2.7 \cdot 10^{19}$ atoms) and $\sim 10^{-13}$ to 10^{-10} cm^3 STP/g (Ne, Kr, Xe). Therefore their concentrations and isotopic compositions may be modified to a measurable extent by nuclear processes such as radioactive decay or natural nuclear reactions. The relative abundance of the respective isotopes can thus be used as dating tools (e.g. U/Th/^{4}He, ^{40}K/^{40}Ar, surface exposure dating). Over the history of Earth, such processes have modified the noble gas isotopic compositions in distinct terrestrial reservoirs (mantle, crust, atmosphere). The isotopic signature of noble gases therefore yields important information about the origin and history of a rock or fluid sample.

To perform a noble gas dating, a core sample is kept in hermetic plastic bags to collect the pore gases (Lippmann 1998). However, standardised regulations for the extraction of representative samples from rock do not exist up to now. Only a few investigations on the influence of varying boundary conditions (sampling and sample preparation) have been performed. Determination of noble gas content is performed with a volume of gas extracted from freshly sampled drill cores. To minimise degassing effects, sampling must take place on site, immediately after withdrawal of the drill core from the borehole. A rock slide with a weight of 200–400 g is cut from the drill core. To remove potentially degased material the sample portions with contact to air will be removed. Then the sample is stored into a vacuum. Over a period of 4 weeks, the noble gases are allowed to diffuse from the pore waters into the vacuum chamber, where quantitative detection by mass spectrometry takes place. The amount of pore water, to which the measured noble gas content is referred, is gravimetrically determined by drying the sample in an oven at 105 °C. The isotopes ^{2}H and ^{18}O of the pore water are determined at separate samples of the drill core, e.g. by equilibration of the isotopic composition of a standard water sample with the unknown rock sample of pore water.

Measured profiles of noble gas content (^{4}He) and isotopy (^{2}H and ^{18}O) of pore waters are evaluated by computed diffusion/advection profiles. If the uranium and thorium contents (radioactive source of ^{4}He) as well as the porosity of the rock formation are known, a the diffusion constant of ^{4}He and H$_2$O results as well as the residence time of groundwater can be estimated. Vertical filter velocity and horizontal hydraulic conductivity of the rock formation are directly linked by porosity as well as upper and lower pressure level difference.

The method is applicable to rock formations of almost all hydrogeological characteristics (porous or fissure/fracture aquifers). If the transport of the isotopes takes place only by diffusion (for instance in case of missing groundwater circulation) only the upper limit of the hydraulic conductivity can be derived. This method is appropriate for rock formations with a hydraulic conductivity in the order of $k_f = 10^{-10}\,\mathrm{m\,s^{-1}}$ (Osenbrück 1996).

In-situ injection test

Gas injection. Gas injection tests a performed in a borehole section locked by a packer system. The borehole section is filled with a gas (e.g. nitrogen). During the injection phase the flow rate and pressure build-up in the test section are recorded. After the injection phase the test section is hermetically isolated and the relaxation of the pressure inside the test interval, caused by gas flow in the rocks, is followed (Wittke 1999). From the pressure curve and the flow rate, the effective rock permeability can be determined (Earlougher 1977).

Several test modifications are distinguished, depending on the test process (Miehe et al. 1993). A constant rate test injects the gas with a constant flow rate. If already during the injection process a significant flow from the borehole in the rocks can be observed, both the injection and relaxation phase can be used for the evaluation. Whether such a process occurs during the injection phase can be derived from the pressure build-up curve. If this curve is straight, it shows the storage capability of the borehole, if it deviates from straight line behaviour, then gas flows off from the test interval (Wittke 1999). In case of extremely small rock permeability, the pulse injection test is applied, where the flow phase is kept sufficiently short that no significant interaction between injected gas and the rock can take place. Here only the relaxation phase is used for geohydraulic interpretation. A special form of gas injection tests are water-gas displacement tests, where water displaces gas at different pressure gradients and/or displacement velocities (normally in multiple cycles) (Czolbe and Klafki 1998).

For the evaluation of all gas injection measurements the following requirements have to be met:

- The rock formation is homogeneous and unlimited,
- A partial saturation of the pore area with fluids is neglected,
- The borehole has a finite radius,
- The test section volume is representative for the storage capability of the complete borehole.

Stormont et al. (1991), Dale and Hurtado (1996) and Knowles et al. (1996) reported permeability tests with gas as test medium. In this case multiple packer systems of fixed sizes and equal test section lengths were used. The test sections varied between approximately 40 cm and approximately 1965 cm length. Test equipment with test section lengths of > 1.5 m have been reported by Miehe et al. (1994) and Wieczorek (1996). In most cases, injection tests in low permeable rocks are prepared with gas.

A major nuisance with gas injection test is gas loss (Stormont et al. 1991; Wieczorek 1996). As an alternative, fluid salt brines have been used in conjunction with double packer systems (e.g. Beauheim 1993; Dale and Hurtado 1996). Typical test section lengths were in the range 40 – 150 cm. BGR reported on permeability tests in the deepest borehole by drillings in the rock salt under application of a single packer systems. In this way extremely short test sections can be realised. This method is, however, rather complex, because after each test the packer systems have to be removed, the drilling equipment has to be reinstalled and the drilling process to be continued to the next level, where process must be repeated again.

Fluid injection test. Fluid injection tests are preferentially applied in salt with a salt brine as test fluid. The injection is made by a piston pump. For the measurement of the injection rate a flow meter (measuring range $0.2 – 1.8\,dm^3\,h^{-1}$) is used. The injected amount of fluid is determined by continuous weighing. Salt brines are highly corrosive. Therefore, clear water for rinsing the equipment is required.

Fluid tracer test in fissure/fracture and weak rock zones. The experiments described in the following were performed in the Grimsel Rock Laboratory of NAGRA/CH and the Rock Laboratory at Yucca Mountain Site by the Lawrence Berkeley Laboratory/USA. A gas or a tracer dissolved in water is introduced into an aquifer using a double packer system with a defined, constant pressure in a given borehole (primary borehole). Subsequently, the migration velocity in the weak rock zones is determined by the recovering the tracer material at distant sampling points in the same weak rock zone (secondary boreholes). Such tests give information on transportation velocities. The artificial pressure at the primary borehole provides a driving force for the migration of the tracer.

3.2.5.3
Discussion and Deficit Analysis of In-Situ Measurement Techniques

The fill-up and pump tests in boreholes represent technically simple permeability tests suitable only in coarse rocks. Fill-up tests have a practical advantage over pump test because pump test may cause the transfer of small-particle smudges into the borehole. Smudges already present in the hole tend to move with the pressure gradient towards the borehole walls and, if available, the borehole bottom. The disadvantage of fill-up tests is lateral flow in the line system often cannot be excluded, since a pulling off the lining system up to the level of the decreased water level is normally not possible (Heitfeld 1998). This leads to problems in the description of the boundary conditions during the evaluation. Fill-up and pump tests are evaluated using both theoretically founded and empirically established principles on basis of requirements, e.g. homogeneity, isotropy, radial symmetry etc. are rarely fulfilled. Thus numerical values are presented on basis of often rather complex and expensive test methods whose fitness-for-purpose is extremely difficult to judge.

Pulse tests are generally interpreted under the assumption of negligible deformations of both the bore hole walls and the equipment, despite the in part rather high pressures. This requirement is of special importance in the field of final waste disposal, because the generation artificial water paths by drilling exploration bore holes has to be avoided. This can however not completely be avoided by the use of lower pressure impulses (Poier 1998). Especially deformations of the test equipment induced by pressure pulses can never been excluded. A systematic study of these deformations and protocols for evaluation of such nuisance effects is still lacking.

Wall effects affect the optimal geohydraulic connection to the rock. Coverage of bore hole walls, e.g. by fine sludges may clog porous water/gas pathways. On the other hand, the release of sludge materials will contaminate the test section. During an evaluation by means of curves analysis (a common evaluation procedure in bore hole and well investigations), an experimental set of data has to be classified according to certain curve types. The classification is largely based on subjective judgement (experience and expertise). Misclassification therefore generates unaccounted errors (Poier 1998). The curve types represent idealised behaviour in the test sections and negligible test section/wall interaction. The assumption of homogeneous and isotropic aquifer is essential. Since these boundary conditions will be only approximate (often poorly approximate) in nature, deviations between idealised curve type and experimental data are common. Here the experience is a substantial condition for the minimisation of the errors resulting from the data interpretation (Poier 1998). Nevertheless, pulse tests within a hydraulic conductivity range $k_f < 10^{-7}\,\mathrm{m\,s^{-1}}$ are considered as the most robust and exact procedures (Poier 1998). Despite the low permeability where these tests are commonly applied, time demand is in the order of a few minutes.

Slug and bail tests (cf. Fig. 3.4) are highly sensitive to wall effects because an optimal geohydraulic connection to the rock is crucial (Kraemer 1990). This requirement results from the sudden pressure pulse. The movement of the slug body immersing in the bore hole water causes turbulence where damages to the walls cannot be excluded (Rosenfeld 1998). The most important influence factor in case of slug/bail tests is the evaluation process, especially by the empirical selection of an evaluation method. To illustrate the multiplicity of evaluation models, the following list gives the names associated with the major evaluation models:

- Hvorslev
- Bouwer-Rice

- Cooper
- Ramey
- Mönch
- Dougherty.

The selection of the optimum procedure depends furthermore on the level of knowledge on the underground rock conditions. Whatever evaluation methods is selected, non-compliance of the basic requirements of homogeneous and isotropic aquifers will cause bias. A typical cause of bias is the presence of fissures/fractures (Rosenfeld 1998). In addition, the validity of Darcy law for the study area must be examined. Aquifer, well and flow conditions have to be taken into account. Deviations from ideal behaviour in most cases can only be documented. There is very little discussion on how to account for such bias.

Special consideration require mixed rock formations, e.g. in clayey stones due to the complicated flow processes and/or the low permeability. By repeated tests in the same wells a satisfactory reproducibility of the results was found with the procedure of Bouwer (1989). Rosenfeld (1998) notes that the results using the evaluation procedures of Bouwer (1976) and Cooper (1967) forwarded "realistic" permeabilities, whereby differences of a half order of magnitude are not uncommon: "As long as no accurate procedure for the permeability test in low permeable fissured/fractured rocks is available, the results determined by slug and bail test have to quoted to be of sufficient accuracy for the most questions" (Rosenfeld 1998).

The water pressure test is applicable above and below the groundwater level. The application is possible in confined and free aquifers. Tests can be accomplished in stable unlined boreholes or in partly lined drillings on the bottom of the borehole, were the diameter of the drilling depends on the test equipment. The test is applicable into large depths, whereby the length of the test interval of the rock permeability can be adapted. If the test arrangement is once developed, then the water pressure test is in principle a test which can be accomplished by simple means. The evaluation is uncomplicated and possible without large data processing requirements. A comparison of the results with other test methods showed a satisfactory agreement (Poier 1998). A major disadvantage is the extensive equipment, which makes the test expensive. Because the test requires quasi-stationary flows, the results are sensitive to the factors skin effect, storage coefficient and porosity, which is usually the biggest problem with tests with not constant flow rates (Poier 1998). Beside the k value, the pressure-referred water absorption ability of the rock can be used also for the evaluation of a location.

The application of borehole methods requires a representative groundwater well. The well is installed with filter line and gravel filter, so that the groundwater flow can go sand-free through with small filter resistance. The filter lines cause a lowering and horizontal deformation of the flow net, so that the groundwater discharge in the filter pipe is larger than the discharge in the aquifer.

Test analysis. Flow model identification (i.e. identification of the dimensionality of the flow field around the borehole) is the first and most important step in well test analysis. If the wrong model is selected, the interpreted parameters from the analysis will be incorrect (Lavanchy et al. 1998). For classical constant rate and recovery tests conducted in relatively high-permeability formations, the flow model is usually obtained using graphical techniques. Direct flow model identification is more difficult in low-permeability formations because constant rate tests can no longer be systematically conducted, due to time constraints, and specific tests such as constant head tests and pulse tests are therefore performed. These tests do not have direct diagnostic procedures comparable to the constant rate test. However, the adequacy of an assumed flow model can be checked using specific graphical representations (Lavanchy et al. 1998). Recovery sequences following packer inflation, constant pressure tests or slug tests can also be studied with the log-log diagnostic technique, and therefore provide valuable information for the flow model recognition (Lavanchy et al. 1998). Beyond this, the flow model evaluation can be improved when a reasonable fit is obtained simultaneously on all test sequences during the subsequent matching of the pressure transients by numerical simulation.

Modern test design and analysis techniques rely on sophisticated, user friendly, borehole flow simulators. The flexibility of such codes allows for rapid flow model identification (see above) and test simulation. However, most of the commercial simulators available in the petroleum and ground-water fields handle standard tests, as usually applied in relatively high permeable formations (Lavanchy et al. 1998). Only a few of them are able to simulate a series of pulse and slug tests and to account for non-ideal conditions such as varying pre-test pressures and temperature effects.

The flexibility of a flow simulator allows for non-ideal conditions such as varying pre-test pressures and temperature effects. Moreover, a series of successive test events of any kind and in any order can be easily simulated. The data can be represented on Cartesian, semi-log or log-log plots. Usually all test phases are simulated simultaneously to yield a consistent set of Parameters. However individual test phases are also plotted individually at different scales, to allow checking of the quality of the fit on specific plots (log-log, semi-log).

The MULTISIM borehole simulator handles numerous flow models such as fractional dimension, dual porosity and composite flow (Lavanchy 1998). The code can also account for liquid and gas flow with constant pressure or no-flow outer boundary conditions (Tauzin and Johns 1997). The inverse modeling capability of MULTISIM allows for an automatic fitting of the measured data. Furthermore, using the simulator, sophisticated uncertainty analysis on hydraulic test data can be performed, bringing significant information on the reliability of the test parameters (Jaquet et al. 1998).

3.3
Cause-and-Effect Analysis for Geohydraulic Measurement Values

"Nature is full of infinite causes that have never occurred in experience."

(Leonardo da Vinci)

3.3.1
General Approach to Quality Assurance of Permeability Data

The International Standards Organisation (ISO) defines quality as "the ability of a service to satisfy stated needs of a client". Quality control (QC) involves monitoring and elimination of causes leading to unsatisfactory performance, while quality assurance (QA) regards systematic actions to provide confidence that a service will satisfy given quality re-quirements. Usually, both quality control and quality assurance are closely intertwined aspects in the communication of quality. In QA/QC the measurement of quality consists of quantifying the current level of performance according to expected standards. It is the systematic identification of the current level of quality the facility or system is achieving. The QA approach to "measuring quality" is inextricably linked with "defining quality", because the indicators for quality measurement are derived from the specific definition or standard under scrutiny. Quality cannot be measured without a clear definition or standard. Likewise, measuring quality leads directly to the identification of areas for improvement or enhancement–the first step in improving quality. Successful improvement ultimately contributes to attaining quality care, the goal of QA/QC. QA/QC activities include:

- Definition of quality
- Quality assessment
- Quality monitoring
- External evaluation of quality.

A quality assessment frequently combines various data collection methods to overcome the intrinsic biases of each method alone. Due to this fact, methods for the evaluation of the influence of possible uncertainty sources are needed. One of these methods is the cause-and-effects analysis.

3.3.2
Cause-and-Effects Analysis

A cause-and-effect analysis generates and sorts possible influence factors. This process may, for instance, be initiated by asking participants, experts, practitioners and theoreticians to list all of the possible causes and effects for a method of interest. This analysis organises a large amount of information by showing links between events and their potential or actual causes and provides a means of generating ideas about problem sources and possible effects (consequences). Cause-and-effect analysis is furthermore sensitizing problem-solvers to broaden their insight and to acquire a general survey of a procedure under scrutiny. Cause-and-effect diagrams can reflect either causes that block the way to the desired quality or helpful factors needed to reach the desired quality. Cause-and-effect analysis is, in fact, an element of constant process improvement (kaizen) and quality management systems.

A graphic presentation, with major branches reflecting categories of causes, a cause-and-effect analysis stimulates and broadens thinking about potential or real causes and facilitates further examination of individual causes. Because everyone's ideas can find a place on the diagram, a cause-and-effect analysis helps to generate consensus about causes. It can help to focus attention on the process where a problem is occurring and to allow for constructive use of facts revealed by reported events. However, it is important to remember that a cause-and-effect diagram is a structured way of expressing hypotheses about the causes of a problem or about why something is not happening as desired. It cannot replace empirical testing of these hypotheses: it does not tell which is the root cause, but rather possible causes.

3.3.2.1
Types of Cause-and-Effect Analyses

There are two ways to graphically organise ideas for a cause-and-effect analysis. They vary in how potential causes are organised: (a) by category: called a fishbone diagram (for its shape) or Ishikawa diagram (for the man who invented it), and (b) as a chain of causes: called a tree diagram. The choice of method depends on the team's need. If the team tends to think of causes only in terms of people, the fishbone diagram, organised around categories of cause, will help to broaden their thinking. A tree diagram, however, will encourage team members to explore the chain of events or causes.

Causes by categories (fishbone diagram). The fishbone diagram, see Fig. 3.7, helps teams to brainstorm about possible causes of a problem, accumulate

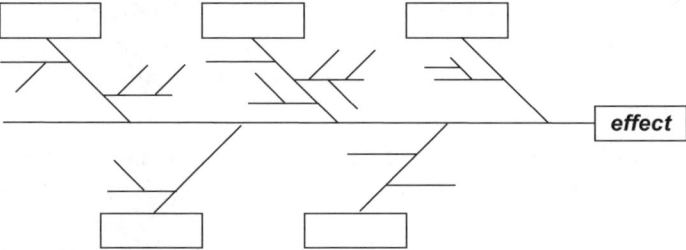

Figure 3.7. Example of a fishbone diagram

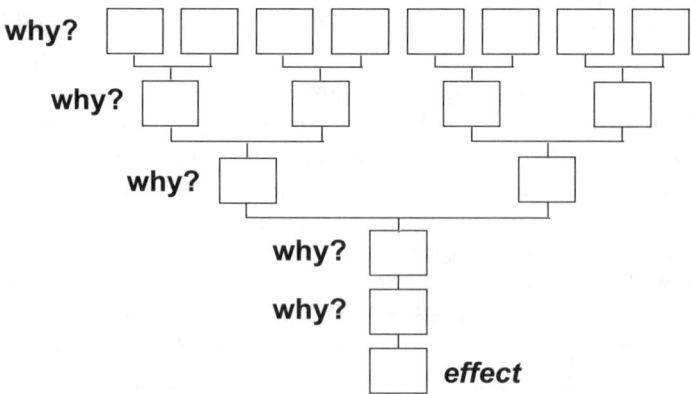

Figure 3.8. Example of a tree diagram

existing knowledge about the causal system surrounding that problem, and group causes into general categories. When using a fishbone diagram, several categories of cause can be applied. Some often-used categories in chemical and geohydraulic measurements are:

- analytical resources, methods, materials, measurements, and equipment;
- environment and procedures;

Causes by chains (tree diagram). A chain of causes (tree diagram; Fig. 3.8) and the "Five Why's" are a second type of cause-and-effect analysis diagram, which highlights the chain of causes. It starts with the effect and the major groups of causes and then asks for each branch, "Why is this happening? What is causing this?" The tree diagram is a graphic display of a simpler method known as the Five Why's. It displays the layers of causes, looking in-depth for the source of an influence quantity. This tool can be used as stand-alone approach or advantageously combined with fishbone diagram.

3.3.2.2
How to Use Cause-and-Effect Analysis

Although several ways to construct a cause-and-effect analysis exist, the steps of construction are essentially the same. Cause-and-effect analysis is almost always done by discussion. QA/QC activities on the professional level are usually done by discussion groups.

Step 1: Agree on the problem or the desired state and write it in the effect box. Try to be specific. Problems that are too large or too vague can bog the team down.

Step 2: If using a tree or fishbone diagram, define six to eight major categories of causes. Or the team can brainstorm first about likely causes and then sort them into major branches. The team should add or drop categories as needed when generating causes. Each category should be written into the box.

Step 3: Identify specific causes and fill them in on the correct branches or sub-branches. Use simple brainstorming to generate a list of ideas before classifying them on the diagram, or use the development of the branches of the diagram first to help stimulate ideas. Be sure that the causes as phrased have a direct, logical relationship to the problem or effect stated at the head of the fishbone.

Step 4: Keep asking "Why?" and "Why else?" for each cause until a potential root cause has been identified. A root cause is one that: (a) can explain the "effect", either directly or through a series of events, and (b) if removed, would eliminate or reduce the problem. Try to ensure that the answers to the "Why" questions are plausible explanations and that, if possible, they are amenable to action. Check the logic of the chain of causes: read the diagram from the root cause to the effect to see if the flow is logical. Make needed changes.

Step 5: The team chooses several areas they feel are most likely causes. These choices can be made by voting to capture the team's best collective judgement. Use the reduced list of likely causes to develop simple data collection tools to prove the group's theory. If the data confirm none of the likely causes, go back to the cause-and-effect diagram and choose other causes for testing.

Remember that cause-and-effect diagrams represent hypotheses about causes, not facts. Failure to test these hypotheses, e.g. by treating them as if they were facts, often leads to implementing the wrong solutions and to wasting time. To determine the root cause(s), the team must collect data to test these hypotheses. The "effect" or problem should be clearly articulated to produce the most relevant hypotheses about cause. QA/QC procedures always require detailed insight into a process or a procedure. Proven expertise is an essential requirement to achieve quality.

3.3.3
Quality Criteria of Permeability Data

3.3.3.1
Quality Assurance for Permeability Data

The most fundamental criteria for ensuring comparability of data are as follows:

- Traceability: characterises a result of measurement of a standard to be related to this standard by a continuous chain of uncertain measurements indicated by comparative measurements.

- Fitness for purpose: characteristic of data, which were generated by a measurement. The fitness for purpose permits the user of the data to make correct conclusions in connection with a given purpose.
- Indication of the measuring uncertainty: This is a parameter which characterises the measurement. It is expressed by the dispersion of the values, which could be assigned to the measured variable (GUM 1993, §2.2.3).

So the requirements for quality criteria of permeability data due to GUM can be summarised as shown in Fig. 3.9.

Permeability is an input quantity of major importance to geochemical modeling. From permeability conclusions on hydraulic conductivity are drawn. Hydraulic conductivity, in turn is a proportionality factor in Darcy law. Darcy law, in turn is the fundamental equation of geohydraulic modeling (cf. Eq. (2.1).).

Some effort has been made to illustrate the broad range of experimental methods in use for deriving permeabilities, hydraulic conductivities or porosities of rocks and soils. This discussion necessarily cannot be either exhaustive, or go into details. To limit the methods, discussed focus was given to methods suitable for investigations in rock salt. Rock salt attracts some attention because it is, next to clays and granite, a preferred medium for nuclear waste disposal in deep geological formations. Safety assessment of nuclear waste disposal caused at least a limited number of studies to be directed to the uncertainties and bias carried by the values obtained from these methods. Nevertheless, the situation is by no means satisfactory. Empirism and heuristics prevail.

This statement is also valid for the methods for data interpretation. Different experimental methods provide different types of numerical measurement results. These results can be evaluated by different procedures. And from each evaluation procedure different numerical values for the measurand (k, K_f, P)

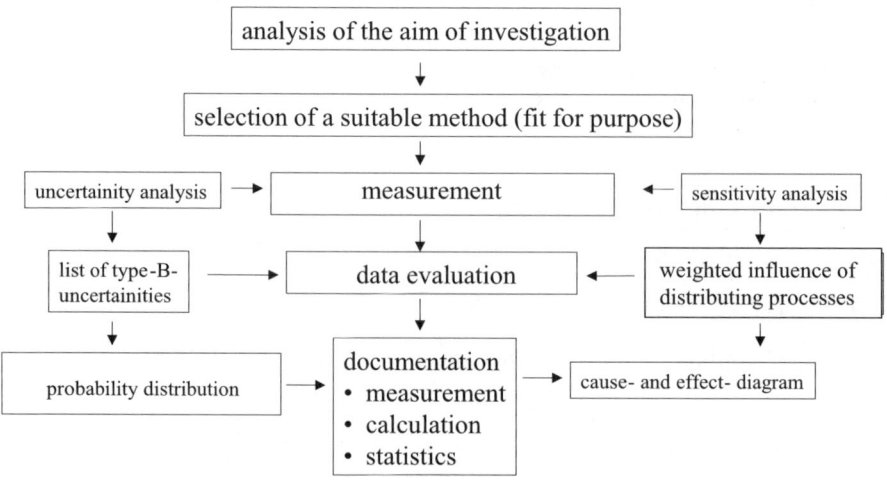

Figure 3.9. Scheme of quality assessment of permeability data

may be evaluated. With the exception of in-situ methods, which nevertheless are not necessarily representative for the complete rock body, the values obtained in laboratory have to be transferred to the model of a complete rock body. This transfer is a fundamental problem in the field of permeability data.

Conventional laboratory measuring procedures for the determination of the permeability assume stationary flow conditions (temporally constant flow rate with continuous pressure). Stationary conditions in small-permeable materials, e.g. clay/tone, salt rocks, porous rocks, concrete and cement, test periods up to several months. Investigations in materials intended as host rock for nuclear waste disposal sites, are especially affected by this time demand because for a geological barrier to be efficient a very low permeability is asked for.

For very small flow rates it is possible that the relationship between permeability, gradient and flow velocity is no longer linear. Since in this case, however, the velocities computed after Darcy are larger than in reality, the neglect of this still unproven nonlinearity effect may be considered to be conservative. Nevertheless, in the range of extreme low permeabilities the applicability of Darcy law is almost impossible to prove. Uncertainties from such fundamental aspects of geohydraulic modeling are only one type of doubt to be associated with subsurface flow simulation. A large number of influence factors affect geohydraulic data in small-permeable rocks. Considering the influence factors in data acquisition in the field, and relating these to the influence factors inferred onto a measurand value due to transport, sample handling, mounting and data collection, plus the influence factors to be considered as a consequence of data transfer from laboratory to the rock body, it seems reasonable to avoid laboratory measurements as a source of geohydraulic data for geohydraulic measurement. For this reason the criterion for the determination of geohydraulic data must be: rock representative data can only be determined with in-situ tests.

This statement does not intend to render laboratory tests useless in the determination of geohydraulic databases, since they represent the substantial instrument for the determination of porosity and the analysis of anisotropies in the rock. Fundamental investigations on the influence of the geopressures, pore pressures and flow pressures in laboratory permeation tests were given in Sachs (1982). There are also documented detailed recommendations for experimental set-up of laboratory permeation tests.

In principle both laboratory tests and on-site determination geohydraulic and geophysical data are necessary for representative field tests. Performance and evaluation of in-situ permeability tests require high expertise and sensitivity in handling and reporting experimental results, especially if expressed as numerical values of a measurand of interest.

Standardisation of field methods is rather difficult, because rock bodies, soils and sediments are impredictable, complex three-dimensional structures influenced by up to 4.5 billion years of geologic activity and alteration. A careful planning of a sampling campaign, comparable to chemical laboratory experiments with well-defined components and procedures, is impossible. Therefore, a careful documentation is essential for the comprehensibility and comparability of all investigations and test results. Nevertheless, in the following sections quality criteria for *porosity* and *permeability* measurements will be discussed. These criteria do not intend to overcome the difficulties but provide a means for comparing techniques, evaluation methods, and measurement values. The criteria form an essential basis for communicating measurement results and, as a side effect, may serve itself as a tool to identify influence factors which have been ignored, underestimated or overinterpreted.

Determination of geohydraulic permeability is usually coupled with porosity determination often using an empirical porosity-permeability-relationship. The porosity-permeability-relationship must be determined material-related. This relationship is rarely linear and depends on tortuosity, constrictivity and pore radius distribution. For the determination of a realistic porosity-permeability-relationship usually statistical procedures are used, like linear regression or a bivariate density distribution for normal or log-normal distributed variables. So the determination of permeability data requires not only an uncertainty analyses but also extensive statistic examinations. Suitable methods including the estimation of, at least, approximate uncertainty bounds, are given in Part I of this treatise.

Uncertainty analyses and/or extensive statistic examinations of permeability procedures were accomplished so far only in exceptional cases. Uncertainty analysis is further complicated by the use of a combination of different calculation methods (NTB 1994-02, NTB 1993-47). An allocation of identified uncertainty contributions to the respective numerical and/or statistical procedure is difficult without simulation.

The criteria given in the following paragraphs will later be applied to data reported in literature using results of uncertainty analyses with geohydraulic data in small-permeable rocks given in (Jaquet et al. 1998).

3.3.3.2
Quality Criteria for Pore Volume Measurement

One of the causes for the difficulties in the determination of the permeability lies in the extreme sensitivity of the permeability to smallest changes in the pore and/or fissure/fracture structure of a rock. Using measurement values of the geohydraulic data of a rock from laboratory studies to obtain more information on the fissure/fracture structure. While porosity indicates only the percentage by the pore volume in the rock, the internal structure of the pore volume is described by tortuosity, constrictivity and pore radius distribution.

Laboratory tests for the determination of porosity should implement the following considerations:

- Avoid circulation effects (short cut, e.g. along walls);
- Considering optimal pressure (the pressure must not generate additional flow paths in the sample);
- In case of the determination of gas permeabilities the associated water contents of the sample should be always indicated and documented;
- In porosity determination a measurement value should be confirmed by at least one additional, independent technique to confirm the porosity-permeability-relationship.

A clear definition of the purpose of a study, i.e. about the type of porosity (e.g. total porosity, effective porosity) to be determined, in order to select the suitable test (fitness for purpose). Characterisation and documentation of saturation processes and determination of phase distributions in a sample may become essential. Cause-and-effect analysis shows that water content measurement is also affected by uncertainty. To assess the uncertainty of ancillary information (e.g. water content) on a measurand value, repeated determinations may give a clue, e.g. by test series with different fluid concentrations.

Criteria for the determination of values for the measurand "porosity":

- Use of measurements, which are documented along (internationally) quality assurance criteria;
- Evaluation of a cause-and-effect-diagram for the entire measuring and evaluation process;
- Tabled list of type b evaluation uncertainties;
- Indication of the numerical and statistical procedures used in the evaluation of porosities from experimental data;
- Documentation of the standards and references to which the traceability refers;
- If possible a presentation of the empirical probability distribution of porosity. This step requires simulation.

3.3.3.3
Quality Criteria for Permeability Measurements

The procedures for permeability determination and the migration processes of water and dissolved species do not differ in coarse and solid rocks.

A common trend can be observed for solid and coarse rocks: the smaller the rock permeability, the larger the likely difference between rock sample permeability and permeability of the rock body. This is caused by an increasing probability that the rocks react to stress by inelastic processes, especially fissure/fracture formation.

The goal of permeability investigations is the determination of the water in the aquifer including all its dissolved substances (solution species, colloids) or other fluids (hydrocarbons, gases) under the theoretical assumption that interactions with solid and gaseous phases are negligible. Chemical reaction, dissolution and precipitation phenomena, sorption and desorption processes and – often forgotten – microbiological activity modify the surfaces, clog water passages or reopen clogged pathways, gas formation may expel water from capillaries and fractures and so on.

The methods for the determination of geohydraulic data and/or information can be divided in four groups:

1. Laboratory tests;
2. Field (in-situ) tests;
3. Indicator methods (noble gas dating);
4. Non-invasive procedures (e.g. Snmr).

Criteria for the determination of rock permeability are given as follows:

- Representative data for rocks can only be determined with in-situ (field) tests;
- Use of at least two different methods to cross-check the field results;
- Measurement documentation in accordance with international quality assurance criteria;
- Preparation of a cause-and-effect-diagram for the entire measuring and evaluation process;
- Listing the type b evaluation uncertainties and their magnitudes;
- Statement of the numeric and statistic procedures, on which the permeabilities were evaluated from the experimental data;
- Documentation of the standards and references to which the traceability refers (if available);
- If possible: presentation of the empirical probability distribution of permeability (requires simulation).

It is fundamental, that in-situ-investigations are necessary for validation and calibration of the laboratory data. Expertise and experience in the planning, performance and evaluation of in-situ permeability tests is essential, and should be a matter of course. However, even the most experienced experimenter cannot control all influence factors. Hence, it should be a matter of course, too, to inquire into the reliability (or its reciprocal: the uncertainty) of a reported value. A critical attitude is the more important as standardisation of field methods is (currently) not possible. International standards and reference materials for permeability determination are not in sight preventing traceable measurements. Standards in geology almost exclusively relate to mineral content standards, e.g. for prospection purposes.

Even so uncertainty analyses of permeability measurements currently are the exception a limited number of information resources exist (Autio et al. 1998; Behr 1998; Müller-Lyda et al. 1998; Bruck 1999). A systematic uncertainty analysis of geohydraulic data in small-permeable rocks were presented by Jaquet et al. (1998a,b). These analyses were accomplished by Monte Carlo simulations with consideration of a χ^2 distribution of the data and a confidence interval of 95%. Such extensive statistic analyses are the exception in the permeability measurements documented in the literature. The documentation both of lab and in-situ data in the literature is currently of a highly variable quality with considerable potential to develop.

3.4
Practical Approach to the Application of Metrological Concepts in Geohydrology

"Let us permit nature to have her way. She understands her business better than we do."

(M. de Montaigne)

3.4.1
Parametric Uncertainty Analysis of Permeability Data

The application of metrological methods to the determination of geohydraulic data will be demonstrated using geohydraulic measurement data documented in the literature. These studies have been performed in the context of nuclear waste disposal. The data include in-situ-injection tests and pulse tests for the determination of the permeability in small-permeable rocks, which were accomplished in the radioactive waste disposal site Morsleben (ERAM) in Northern Germany. A second topic in the practical application of metrological methods will be an uncertainty analysis of geohydraulic data. The analyses are reported in Jaquet et al. (1998). Both investigations will be used as a basis for the practical application of metrological criteria, thereby illustrating the criteria given in the previous chapter. The discussion will neither be directed to an assessment of the measured values nor to the evaluated permeabilities.

In the analysis of hydraulic borehole tests in Jaquet et al. (1998a), a theoretical model, numerically calculated, is fitted to the measured data (pressure and/or flowrates). As a result, best-fit parameters describing the tested formation (e.g. rock permeability) are obtained. An important issue in this fitting procedure is the estimation of the uncertainty affecting the best fit parameters. The sources of this uncertainty can be measurement errors and model uncertainty (e.g. related to the specific hypotheses made and to the types of numerical methods chosen). However, such an uncertainty analysis is often not performed, or is confined to a semi-quantitative analysis based on a limited set of additional simulations and expert opinion.

The primary goal of Jaquet et al.'s work was to develop a systematic quantitative uncertainty analysis technique, using Monte Carlo methods, and to evaluate the information yielded by such an approach in the analysis of field data issuing from the Morsleben mine.

The measured pressure data were fitted with a model using a parameter set a. The "goodness of fit" of the model is assessed on basis of the chi-square distribution, defined by:

$$\chi^2 = \sum \left[p_{\text{data}}\left(t_i\right) - p_{\text{sim}}\left(t_i, a\right) \right]^2 / \sigma^2 \tag{3.16}$$

where

χ^2:   objective function
p_{data}: pressure measurements
p_{sim}: simulated pressures
a_0:   vector of fitted parameters
σ: standard deviation (accounting for measurement and model errors)
t_i: time

The best guess parameter set, a_o, is obtained when, χ^2_{min}, the minimum χ^2 value, is reached:

$$\chi^2 = \sum \left[p_{\text{data}}\left(t_i\right) - p_{\text{sim}}\left(t_i, a_0\right) \right]^2 / \sigma^2 \tag{3.17}$$

with a_0 = vector of optimum parameters.

On the basis of the χ^2 distribution, confidence intervals of parameters can be obtained, once the distribution has been computed, using the χ^2 critical values for a confidence level of 0.25 and 97.5% (corresponding to a 95% confidence level) as a function of the degree of freedom as estimates for the uncertainty limits. The uncertainty analysis was performed via Monte Carlo process.

3.4.2
A Practical Example

3.4.2.1
Results

The method described is illustrated by a quantitative uncertainty analysis performed on l test T522210-11 issued from the Morsleben salt mine (BGR 1996). This gas test conducted in borehole RB522 is located in the interval 11.4–12.9 m away from the tunnel wall. This test was selected because a good fit of the middle/late time pressure data could be obtained with a simple flow model (radial homogeneous) in the prior analysis (Tauzin 1997). The main results obtained in the analysis are summarized in Table 3.10.

Several parameter combinations were considered for the uncertainty analysis. The value of the confidence interval obtained for the parameters using the Monte Carlo method are given in Table 3.10. A total of 5000 simulations were necessary to obtain 26 values of χ^2 in the joint-confidence region. The input distributions used in the Monte Carlo (MC) simulations and the corresponding χ^2 distribution are shown in Figs. 3.10 and 3.11.

Table 3.10. Test 522210-11. Results of the analysis performed (Tauzin 1997)

Parameter	Recommended value	Confidence interval
Flow model	Radially homogeneous	Flow model is uncertain
Permeability k	$2.2 \cdot 10^{-19}\,\mathrm{m}^2$	$8 \cdot 10^{-20}$ to $8 \cdot 10^{-19}\,\mathrm{m}^2$
Porosity p	0.2%	Not investigated
Flow dimension: n	2 (radial flow)	Not investigated
Formation pressure	100 kpa	Not investigated

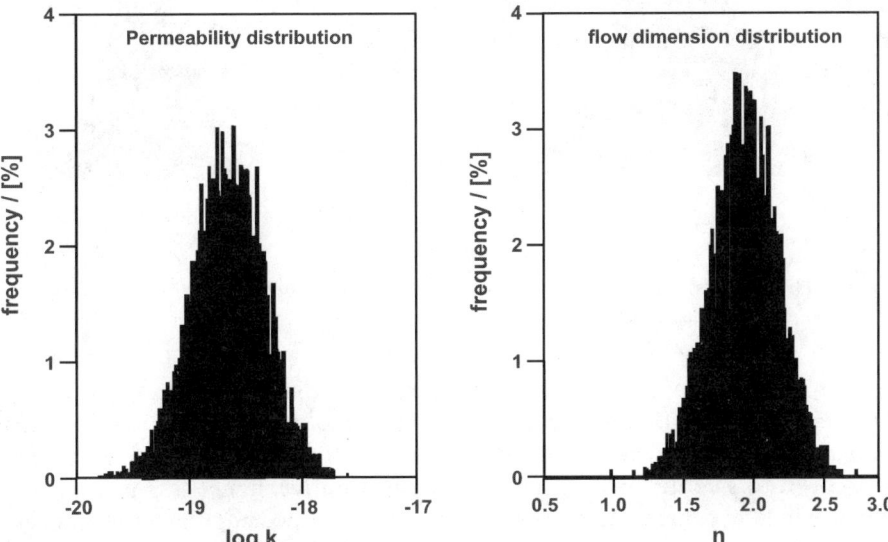

Figure 3.10. Binned distribution of permeability (*left*) and flow distribution (after Jaquet 1998a)

Table 3.11. Results of Monte Carlo simulations (Jaquet et al. 1998b)

Total number of computed points	5000
Number of points such as $X^2 < \Delta_{\min}^2 + \Delta X^2$	26
Permeability confidence interval	$1.1 \cdot 10^{-19} - 4.9 \cdot 10^{-19}\,\mathrm{m}^2$
Porosity confidence interval	$0.09 - 0.50\%$
Flow dimension confidence interval	$1.6 - 2.4$

The prior permeability confidence interval, with a range of about 1 order of magnitude (see Table 3.10) appears overestimated compared to the computed interval which covers approximately half an order of magnitude. However, it should be noted that these two confidence intervals do not take into account the same uncertainties. The prior confidence interval at-

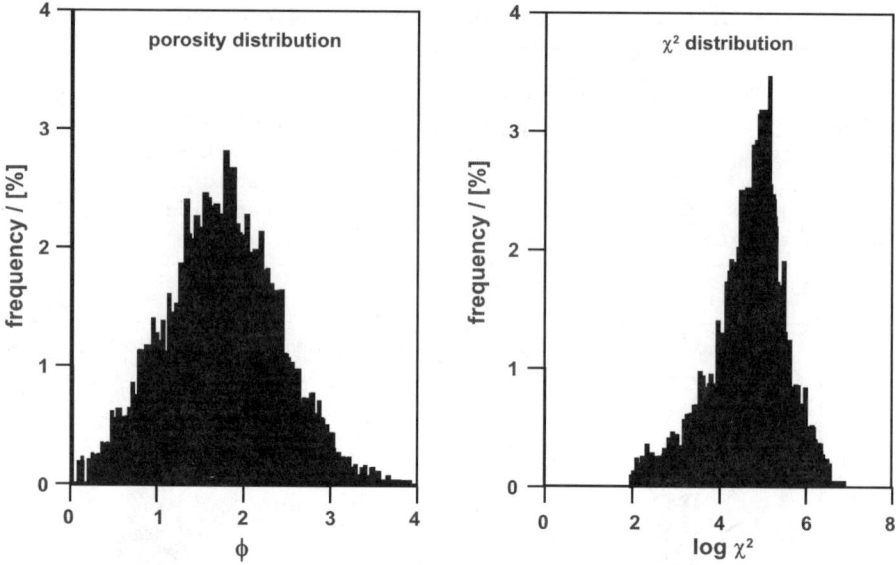

Figure 3.11. Binned distribution of porosity (*left*) with the associated optimization function χ^2 (after Jaquet 1998a)

tempts to take into account all the uncertainties that may have an influence on the derived formation parameters. These uncertainties include, in addition to the factors listed above, the borehole history and the flow model. Moreover, this confidence interval is only qualitative, hence less precise (i.e. wider, since a conservative approach must be respected) than the interval obtained with the MC simulations. In the MC case, the only uncertainties accounted for are the porosity, the flow dimension, and the poor fit quality at early time. The prior confidence interval given (qualitative and interpreted) includes the one (quantitative and not interpreted) computed by the MC method, since it attempts to take more uncertainties into account.

As a result, the uncertainty analysis performed on the test T522210-11 is helpful to define a new global permeability confidence interval for this test (see Table 3.12). This interval is global in the sense that it attempts to include all the information available from the test and from its analysis (test execution, qualitative determination of uncertainty resulting from borehole history and

Table 3.12. Final permeability best guess and confidence interval for test T522210-11 (Jaquet et al. 1998b)

Parameter	Expectation value	Confidence interval
Permeability	$2.2 \cdot 10^{-19}\,\mathrm{m^2}$	$9 \cdot 10^{-20} - 6 \cdot 10^{-19}\,\mathrm{m^2}$

flow model, quantitative determination of uncertainty resulting from porosity, flow dimension and poor fit quality at early time).

The Monte Carlo technique is a powerful way to determine confidence intervals and joint-confidence regions for fitted parameters, provided the number of parameters is limited. Correlations and dependencies usually cannot be taken into account. Furthermore, the number of repetitions necessary to achieve a satisfactory consideration of the tails of the (n-dimensional) distribution increases drastically with n. However, this uncertainty analysis clearly shows that hydraulic borehole tests are not different form other complex measurement processes. The methodology can be further improved and adapted to take additional effects into account.

The disadvantages of the MC method are well known: it is time-consuming and has a poor coverage if more than three parameters have to be considered. In order to overcome these limitations, a direct approach based on the analysis of the Hessian matrix could be applied for deriving confidence intervals. This method offers the advantage, if combined with an inverse modeling algorithm, of allowing the determination of the best guess set of Parameters together with their uncertainty. The MC approach is a straightforward way for a preliminary assessment of measurement results.

At the operational level, due to cost and time limitations, it does not seem adequate to perform systematically a full uncertainty analysis (including the non-fitted parameters), but rather to apply it whenever the studied hydraulic test requires a detailed interpretation.

3.4.2.2
Discussion

The criteria for permeability measurements presented in the previous chapter are separated into eight points:

1. Representative data for rocks can only be determined with in-situ (field) tests.
2. Use of at least two different methods to cross-check the field results.
3. Measurement documentation in accordance with international quality assurance criteria.
4. Preparation of a cause-and-effect-diagram for the entire measuring and evaluation process.
5. Listing the type B evaluation uncertainties and their magnitudes.
6. Statement of the numeric and statistic procedures, on which the permeabilities were evaluated from the experimental data.
7. Documentation of the standards and references to which the traceability refers (if available).
8. If possible: presentation of the empirical probability distribution of permeability (requires simulation).

The report (Jaquet et al. 1998b) will be scrutinized in the following according to these eight criteria.

Item 1:

Gas-injection test is in-situ test. As discussed in the previous chapters this kind of geohydraulic tests is one of the test methods beside pulse test and isotope and noble gas methods expected to produce representative results for permeability in small-permeable rocks.

Item 2:

The studies comply with this criterion. Additional studies, e.g. pulse tests, isotope analyses and noble gas determinations, have been performed in the region of interest. Relevant informations on the preparation and performance of the test are documented in Tauzin (1997). The experimental test equipment was described and outlined schematically. The measured variables of the tests were listed.

Item 3:

The documentation in Jaquet (1998) is extensive. There are, however, no references to documentation criteria given. Quality assurance in geohydraulics is a rather recent field. Specific documentation recommendations will hopefully become available in the future.

Item 4:

The cause-and-effect-diagram allows a concise presentation of influence factors and their interrelationship. In Jaquet et al. (1998b), no cause-and-effect-diagram of the tests was given. However, on basis of the informations a cause-and-effect diagram for a in-situ gas injection test (Fig. 3.12) and a pulse test (Fig. 3.13) can be derived. These diagrams may serve a basis for further discussion. The diagrams show the feasibility of cause-and-effect analysis for geohydraulic investigations.

Figure 3.12 shows the cause-and-effect-diagram for an in In-situ gas injection test. The left-side fishbone part holds the ISO Type B evaluation uncertainties. These uncertainties must be determined by separate experimentation, e.g. to assess the influence of temperature on the accuracy of gas volume measurement. The right-side box holds those uncertainty contributions which must be derived from the experimental data. These uncertainties are summarized as ISO Type A evaluation uncertainties. MC procedures (e.g. Krause et al. 2005), repetitious measurements, computer-intensive methods of statistics, Markov Chain Monte Carlo methods etc. are possible approaches for geohydraulic data. Such an analysis has been reported.

The major type B influence factors are time, pressure, viscosity, sample surface and gas volume determination in study section between the packers. In comparison to a laboratory test the branch referring to sample geometry is affected by fewer influences (the "sample" size is much larger. Skin effects have less influence). The temperature can be measured in the field with similar accuracy as in a laboratory environment (but is more difficult to control).

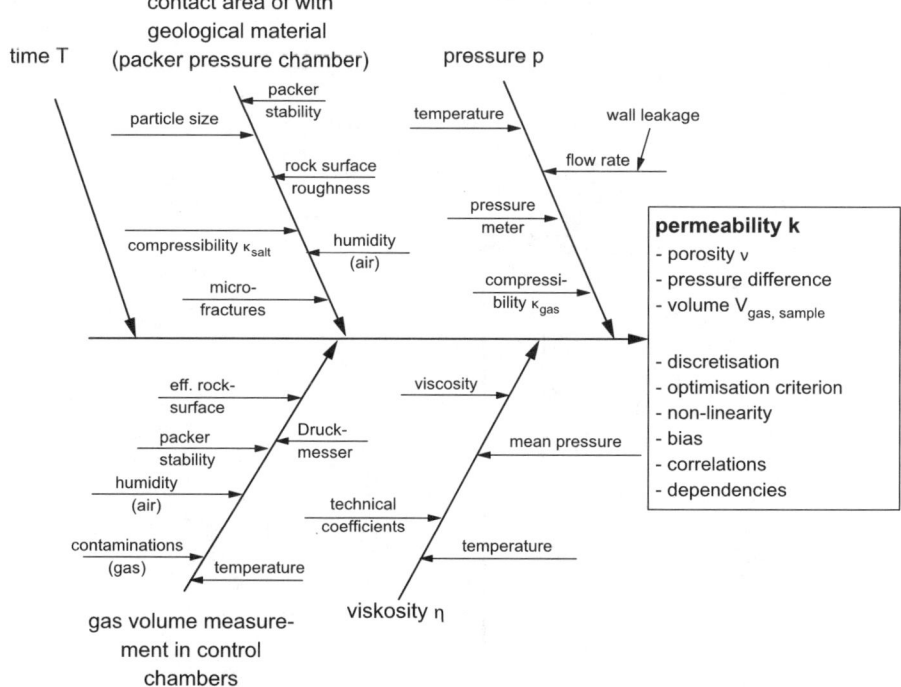

Figure 3.12. Cause-and-effect diagram for in-situ gas injection test

The humidity content of a sample and the associated effective porosity, are certainly subject to higher uncertainties, because they are less accessible in the field, while in the laboratory the respective conditions can be controlled. On the other hand, the easier accessibility also may give rise to further influence factors, e.g. by drainage and additional evaporation/condensation processes. Thus, a cause-and-effect diagram may direct to supplementary investigations which quantify the relative magnitudes of uncertainty contributions.

Figure 3.13 shows the cause-and-effect-diagram of a pulse test. The fishbone diagram gives the ISO Type B evaluation uncertainties, the right-side box the ISO type A evaluation uncertainties. Again, the relevant type B influence factors are time, pressure, viscosity and the sample surface. In case of a pulse test pressure is varied (alternatively with liquids or with gases) within sections closed by packers. The observed time-dependent process of the pressure decrease is registered. The measurands therefore are pressure and time. From this data by means of modeling the permeability is derived. A whole set of parameters contribute to the model as ancillary data: saturation points, specific weight of the fluid, capillary pressure, dynamic viscosities etc., but also characteristics of the technical equipment.

For the evaluation of a pulse test complex finite elements models are usually applied. The task is to interpret the coupled nonlinear differential equations of

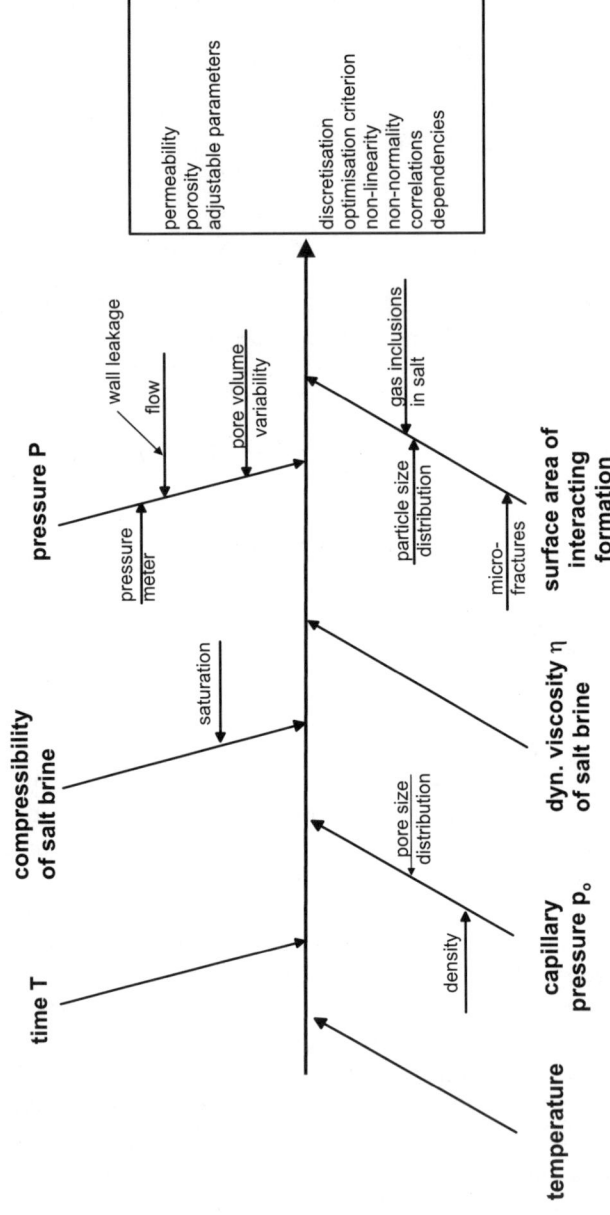

Figure 3.13. Cause-and-effect-diagram for a pulse test (in rock salt using brine)

fluid and gas flows in the underground rock. The result depends on multiple boundary conditions and assumptions as discussed in Chap. 3. The dominant influence factors for permeabilities derived from pulse test are physical properties of the fluids and/or the rock, e.g. dynamic viscosity and temperature.

The evaluation in Wittke (1999) was done by a three-dimensional finite element model, which fitted the flow equations and their parameters to the observed pressure time processes. From this complex fitting process values for the parameters porosity and permeability were obtained. Hence, the detailed data interpretation is, more or less, performed by a "black box" with multiple correlations and dependencies among the influence quantities. The compressibility of the salt solution, for instance, affects the dynamic viscosity, the measured pressure, as well as the capillary pressure.

Figure 3.13 does not show branch "equipment". The equipment is a source of uncertainty (e.g. variability of sealed volume under pressure, accuracy of the total volume within the packers, leakage), whose contributions can be regarded as stochastic errors (uncertainty) or as bias. However, the data source does not give any clue about such effects and, for sake of clarity, the respective branch has been omitted. The branch "surface are interacting rock" might play a role in case of rock salt if the components the salt brine exchange with salts of slightly different composition. The branch "time" might probably represent the most unproblematic influence. Time can be determined simply and very exactly, even though delays between pressure built-up, pressure release etc. may play a role in deep bore holes. Time as an influence factor does not play an important role for itself, but for the limited accuracy by which certain events in the bore hole (e.g. slug/bail rate) can be correlated with the time scale. The branch "temperature" needs a careful and situation-specific treatment. Gases have rather temperature-dependent properties. In summary, the classification of uncertainty contributions eases documentation and discussion of measurement uncertainty contributions.

Item 5:
    The methods selected for the investigation of rock salt with very low permeabilities are "fit for purpose". However, clear specifications what "purpose" was to be achieved is not given.

Item 6:
    In Jaquet et al. (1998a,b), no explicit listing the type B-uncertainties is given. A list of main influence factors of gas-injection test and pulse tests is available from Wittke (1999) with similar studies performed inside the ERAM nuclear waste repository (see Table 3.13). However, comparison of Table 3.13 with the respective cause-and-effect diagrams show that only a part of the influence factors are mentioned.

    Table 3.13, however, gives ranges for the uncertainty contributions of important physical parameters. Therefore, Table 3.13 may serve as an (incomplete) model to criterion 6. Together with other informations in the respective ref-

Table 3.13. Survey of influences factors for rock permeability (after Wittke 1999)

Aim: measurement of test medium flow into the geological formation for assessment of formation permeability

Applied techniques	In-situ gas injection	Pulse test	
Influence factors	Volume measurement	Pressure measurement	Assessment and evaluation of influence
Test medium Compressibility	No effect; p = const.	Has effect; $p \neq$ const.	Numerical
Gas inclusions in salt brine	No effect; p = const.	Has effect; $p \neq$ const.	Field tests + numerical
Temperature variation	No effect; T = const.	No effect; T = const.	–
Borehole effects Displacements in bore hole walls			
- Elastic	No effect; p = const.	Has effect; $p \neq$ const.	Numerical
- Creep	Has effect	Has effect	Numerical
Equipment Axial displacements due to unilateral pressures	No effect; no unilateral pressures applied	No effect; no unilateral pressures applied	–
Friction losses in tubing	No effect; pressure measurements flow-free in separate tubing	No effect; no flow	–
Distortions in packer tube and tubing due to experimental pressure	No effect; p = const.	Has effect; $p \neq$ const.	Field tests + numerical
Distortions inside the packer due to experimental pressure	No effect; p = const.	Has effect; $p \neq$ const.	Field tests + numerical

erence, the type B uncertainties in Table 3.11 would allow a reinterpretation of experimental results at any time.

If measured variables should have been ignored and/or under or overestimated, then a) this can be corrected, b) overestimation be proven, in that the measured variables actually has a significant influence on the result, and c) it can be compared with other measurements. The latter argument also applies to comparison with future measurements. Thus the evaluation process has become transparent. The numerical procedures are documented in detail in Jaquet (1998). The detailed flow charts of the models of Jaquet's analysis are documented in Tauzin (1997) and (Tauzin and Johns 1997), also the model runs, the sensitivity analysis and the calibration of the models.

Item 7:

There are currently no internationally accepted standards and references available for geohydraulic parameters like porosity and permeability. The relevant international organisation is ISO REMCO which has a geological branch. However, it is doubtful whether "certified permeability reference materials" will become available. A simple reason is the sensitivity of rock samples to environmental conditions and their limited stability over time.

Item 8:

The empirical distributions have been obtained by MC simulation and are given in Figs. 3.10 and 3.11.

In summary, the measurements reported and documented by Tauzin (1997) and Jaquet et al. (1998) are in close agreement with the general quality criteria derived for a wide variety of measurements. The MC method may easily be replaced by more advanced statistical techniques and methodologies, e.g. Latin Hypercube Sampling or Markov Chain Monte Carlo methods. The strongest aspect of the work reported in Tauzin (1997), Jaquet (1998) and Wittke (1999) is the proof of feasibility for quality assurance concepts in hydrogeological measurements.

3.4.2.3
Conclusions

The Guide to the Expression of Uncertainty in Measurement (GUM) (ISO 1993) applies to all kind of measurements. Its generality is ensured by the fact that all measurements are comparisons. All measurements do have certain elements in common. There is, however, a large variability in the accessibility of the object of interest. Pores in a hard rock buried deeply in a host formation definitively pose some limits in accessibility. Geology has made these objects more accessible but much of the inference about the properties of the objects of interest is associated with large uncertainties, that is doubt.

Therefore, assessment of measurement uncertainty for important quantities of in hydrogeology is the more relevant. The discussions given in Jaquet et al. (1998a,b) and Wittke (1999) show that the principles of the GUM are applicable to hydrogeological parameters, too. This finding is the more important as Jaquet et al. and Wittke seem to have been completely ignorant of the GUM. There is no indication in the extensive reports indicating a familiarity with the GUM.

Criteria 1–8 have been derived in a similar form for thermodynamic data of chemical reactions, for surface interaction parameters of solution species with surfaces, and in Part III of this treatise, for hydrogeological measurements. The consistency of the criteria is no surprise and is a result of the common property of all measurements to be comparisons. A major difficulty with hydrogeological parameters is the lack of a traceability chain. In this respect, hydrogeological measurement values share a similarity with values obtained from chemical analysis. There, the "sample" is varied, inaccessible and often not really representative. Here, the sample matrices are extremely variable, often unstable and of questionable representativity. The goal of metrology in chemistry is an assessment of reproducibility. Globalisation of trade, commerce, services and customers is the driving force to establish metrological rules. In hydrogeology, a strong driving force is, at present, missing. The reports of Tauzin (1997) and Jaquet et al. (1998) have been performed in the framework of nuclear waste disposal. Here, the licensing requirements of a waste repository for nuclear wastes provided a driving force, and funding. Due to the sensitivity of permeability data on performance assessment in nuclear waste disposal, it has to be discussed whether during the evaluation of the suitability of geological formations as waste deposit a decision regarding to primary and secondary flow paths should be made. This would simplify the selection of suitable geohydraulic research methods, which are fit for purpose. Such a catalogue could be provided, which geohydraulic methods under which conditions and with which goal are applicable. This represents a first step toward a standardisation of the tests and contributes for the fact, that permeability determinations should always be realised as goal and problem oriented. For the qualification of the results of permeability data the following aspects must be considered:

- The goal of the investigation must be clearly defined;
- Sensitivity analyses (also of the methods);
- Uncertainty analysis of the overall evaluation and its elements.

Since there is neither normative standard nor material references to ensure traceability of geohydraulic data, the establishment of a suitable reference laboratory has to be discussed.

A second important aim of Part III is to generate an understanding for the uncertainties contributed by the most basic measurement values, e.g. permeability, to a geohydraulic model. The most important experimental methods to determine values for these parameters have been discussed. Most of these methods are heuristic and empirical. They are affected by a wide variety of influence factors which are rather difficult to control. A clear and systematic understanding of these influence factors is, at present, missing by large. Here, cause-and-effect analysis demonstrated by Figs. 3.12 and 3.13 is an approach with a considerable potential.

The geohydraulic model is an essential part of a geochemical model. Both elements, the transport model and the chemical speciation code, contribute to the computer model output. There is little reward in developing one component to high precision and accuracy while the other is a rough-and-dirty approximation method. The decision, where to invest resources to achieve an over-all improvement of performance, cannot be reasonably made without a clear understanding of the contributing uncertainties and their respective magnitudes. Therefore, this book and its discussion may contribute to a more profound understanding of uncertainties, the need to assess uncertainties. And it hopes to spark a glimpse of insight that assessment of measurement uncertainty offers considerable benefits: e.g. optimised allocation of personnel, economic and time resources, efficient ranking of priorities, increased gain in efficiency with the associated revenues etc.

A Appendix

A.1
Overview of CD Content

This book is accompanied by a CD providing some additional features. The content of the CD is provided "as is" and no further warranty whatsoever is given. The intention of providing the material on CD is to ease the access to the information. At the time of composing the CD all items were also available from the WWW free of charge.
Please note:

1. There is additional information in the text resources directory on CD, especially in the manuals section.
2. The ReadMe Text and the Licensing Agreement on the CD-ROM and printed as Chap. A.1 of this Appendix contain important information on system requirements, installation and on the license.
3. Using the programs implies acceptance of the licensing agreements.
4. The CD is an addition to the book where it is enclosed. But the content in the book is completely independent from the content of the CD. Great care has been given in the preparation of the programs and related informations. However, these codes have been written non-professional programmers for scientific demonstration purposes. There is no liability what-so-ever that these codes will work on a specific computer equipment.

The CD holds three directories:

a) LJUNGSKILE_S
b) TBCAT_S
c) Text Resources

and the ReadMe Text and the License Agreement (also printed as Chap. A.1 of this Appendix).

The directories a) and b) provide installation routines for computer programs together with additional material, e.g. data collections, input files and measurement data. The reader may use the material to reproduce or even further inquire into topics discussed in the respective sections of the book. Both

LJUNGSKILE_S and TBCAT_S come with a detailed manual. The manual is installed together with the codes. The codes are written in Visual Basic and C++.

Directory c) collates reports and text documents in PDF format including the manuals for the programs TBCAT_S and LJUNGSKILE_S. This directory also includes the documents concerning the dispute within OECD/NEA reviewers about thermodynamic data on Neptunium. These text resources are given as downloaded from the WWW and provided without further manipulation. Directory c) does not require further discussion.

A.2
A Brief Introduction into the Computer Codes

Please note that using the programs implies acceptance of the licensing agreements (see Chap. A.1 and A.3).

a) The LJUNGSKILE program (Ödegaard-Jensen et al. 2004)

The program is installed by a set-up routine (setup.exe). Several executables are installed: Ljungskile_S.exe, Simulation.exe, phreeqc.exe and LDP20.exe. The names of the executables must not be changed because the codes call each other during execution.

Chemical equilibria in aqueous solutions are commonly modeled on the assumption of thermodynamic equilibrium. Hence, kinetic aspects do not play a role. Some more details are given in 2.1.2.1. Numerical modeling of chemical systems is generally based on the Law of Mass Action. Chemical thermodynamics provides a formal way to describe even complex equilibria mathematically. Modern computer algorithms allow to solve the resulting numerical problems, at least in principle.

The practical approach requires considerable experience in combination with computing skills and insight into the related numerical algorithms. Therefore, few people will endeavour the development of their own numerical code but prefer to use available computer programs instead. A widely used computer code for numerical solution of chemical equations is PHREEQC (Parkhurst 1995) from US Geological Survey. It is available on-line (USGS 2006) together with detailed manuals and further software (e.g. the PHREEQCI graphical user interface). Next to being publicly available, PHREEQC is well maintained and very powerful. An excellent guide to PHREEQC is available (Merkel and Planer-Friedrich 2005).

Like all speciation codes, PHREEQC is a mean-value based program. It has no provision to work with uncertainty-affected data. Until recently there has been no interest in considering measurement uncertainty in numerical speciation. In contrary, relevance of uncertainty in formation constants on chemical speciation calculations was even denied (cf. §2.4). Nevertheless, the relevance of accounting for uncertainty in chemical speciation is more and more acknowledged.

PHREEQC provides numerical features that address a wide range of problems. Most features are invoked by creating batch input files using key words. The calculation results are subsequently written to an output file. If errors occur during execution of the code, a log file is generated automatically by PHREEQC. This batch processing feature is very useful when including PHREEQC into a shell code where PHREEQC has the task to solve a chemical system.

Here the term "solving a chemical system" means to find a numerical solution satisfying the boundary conditions (e.g. chemical equilibrium constants for all species involved, solubility products for all possible solid phases, Henry constants for gaseous phase equilibria, consideration of ionic strength and temperature effects and total concentrations of chemical components in the system etc.).

The LJUNGSKILE code is such a shell program. It provides a graphical user interface where a database may be selected and formation constants of chemical species may be specified. The specification of chemical species and their formation constants also includes the statement of measurement uncertainties. LJUNGSKILE allows one to specify measurement uncertainty as a standard deviation of a normal distribution or as a range of a uniform distribution.

The installation routine also installs a manual, two databases and several LJUNGSKILE projects (with the extension *.prj). Figure A.1 shows the main form of the LJUNGSKILE code. Five projects and two databases are available. After selecting the project JESS_Fe.prj and the database example.dat, the respective species, formations constants and the specified uncertainties can be

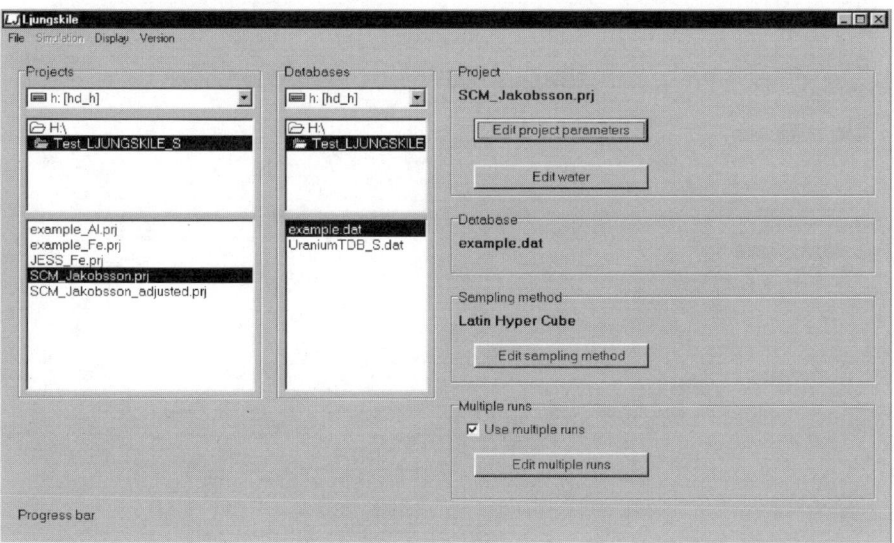

Figure A.1. The main screen of the LJUNGSKILE code

found by selecting the "Edit project parameters" button (Fig. A.2). It is important that a species specified in a project file is also available in the database. This requirement is also true for solid and gaseous phases.

The species of the JESS_Fe project are available in both databases. The SCM_Jakobsson project for example includes uranium species which are only available in the database UraniumTDB_S.dat. There, a message is displayed upon closing the Project parameter window if the wrong database is selected. Of course, the user can add species to the database by manipulating the respective database files following the PHREEQC conventions. A field in the Project parameter window is manipulated by clicking into it. By left-clicking a window pops up allowing to add and to remove a species and to select a sampling distribution.

Before the LJUNGSKILE code can start its task, it requires some details about sampling. Upon clicking to the "Edit sampling method" button (Fig. A.3), information can be specified on the general sampling method (Monte Carlo or Latin Hypercube Sampling), a seed for the random number generator, the details of the cumulative distribution from which the data are sampled and the number of runs. Details about these parameters are found in the manual.

The JESS_Fe project calculates the solution composition at a pH value specified in the Water description screen. This screen is available via the "Edit water" button. It is also possible to specify a range of pH values where the solution composition is calculated. For this basic introduction, a single pH run will do.

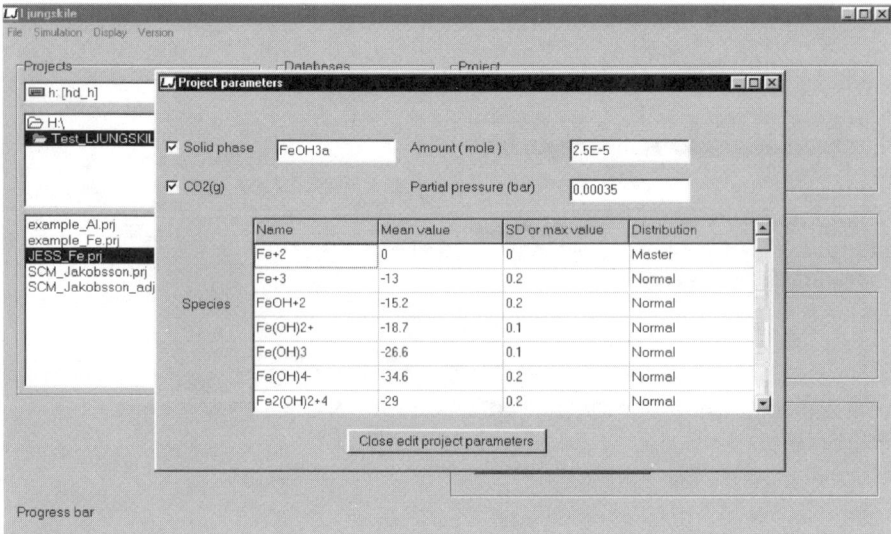

Figure A.2. The "Edit project parameters" window

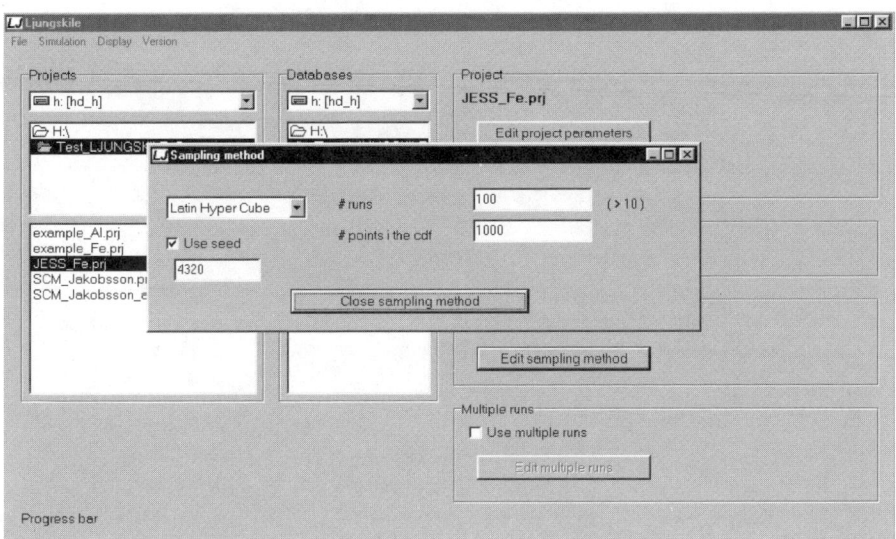

Figure A.3. The "Sampling method" window

The LJUNGSKILE code is started by the Simulation menu item. PHREEQC processes in a DOS window. It will take some minutes. After finishing, a click on the menu item "Display" should forward automatically a picture similar to Fig. A.4.

Figure A.4 is generated by the Ljungskile Display Program (LDP). LDP gets information on the data for display from the file info.lju in the LJUNGSKILE installation directory. If LDP fails to show the diagram, deleting info.lju and searching via the File: Open menu item can be helpful. The JESS_Fe.ldp file should be found in the Results subdirectory of the LJUNGSKILE installation directory.

LDP uses the LJUNGSKILE output to provide a series of options to the user. In case of a single run output, the concentrations are presented as modified Box plots. The center square gives the mean value, the box represents the 68% confidence region and the whiskers enclose the total range. The user may choose between several representations. Figure A.4 gives the linear graph. Upon selecting the logarithmic presentation from LDP's Diagram menu, the display should be similar to Fig. A.5.

Figure A.5 holds the information given in Fig. 1.37. The installation directory includes a manual with further details on the LJUNGSKILE code and a guide how to generate multiple run diagrams.

b) The TBCAT_S program

The code TBCAT_S analyses multivariate UV-Vis spectra. It is a common task for a spectroscopist to extract from spectra information on the species giving rise to a set of spectral observations, the single component of the species

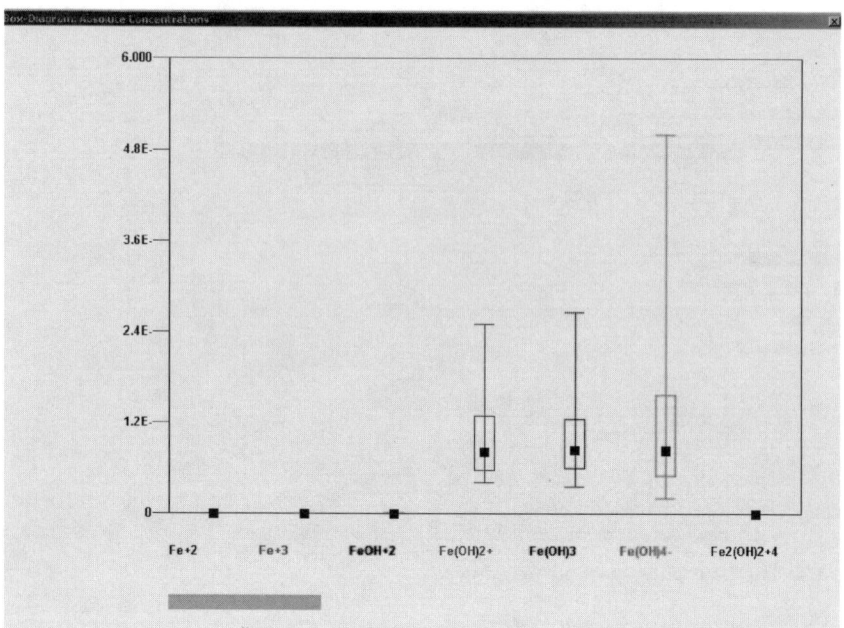

Figure A.4. Iron species concentrations at pH 7.95 with uncertainties calculated by the LJUNGSKILE code

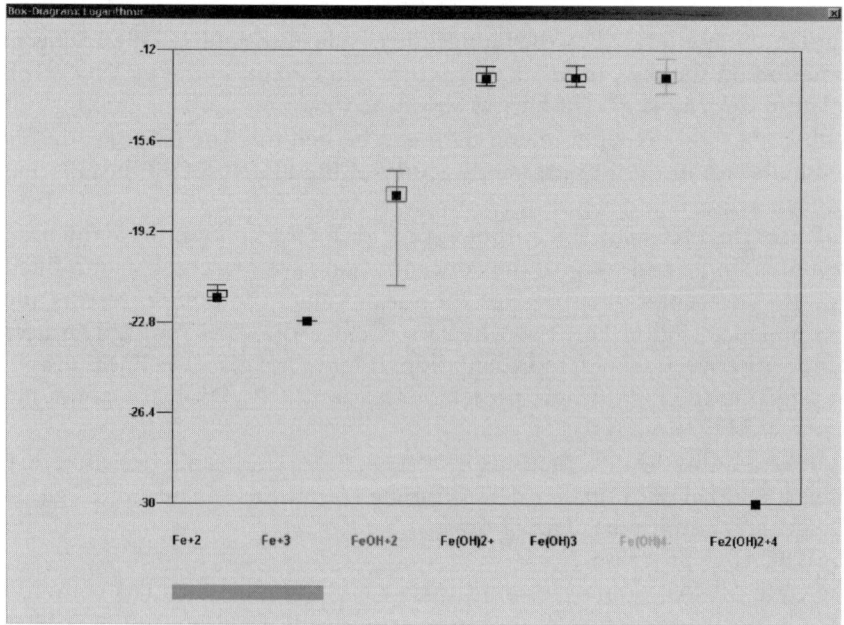

Figure A.5. Logarithmic concentrations of iron species at pH 7.95 with uncertainties

and, eventually, the formation constants of the absorbing species in the system. Factor analysis is a suitable numerical method. Eigen analysis takes advantage from the data structure of most UV-Vis spectra: They can be analysed on the assumption of linear equations. TBCAT_S uses the Singular Value Decomposition (SVD) for that task.

All experimental data is a mixture between a signal (the "true value"), unwanted random noise and bias. Thus there is no univocal method to assess how many species give rise to the individual spectra in a spectroscopic data set. The crucial point in factor analysis is the determination of the number of species or, more mathematically spoken, the true rank of the data matrix. As a consequence, a wide range of methods and procedures has been proposed in literature. The large number (the topic is still prolific) of methods is a good indicator that the available procedures are not fully satisfactory. TBCAT does not propose a new method but is based on the assumption that all suitable possibilities should be tested. In fact it is not uncommon to find that a system may have several equally suitable interpretations (Meinrath et al. 2004, 2006). Factor analysis is a developed field of chemometrics and the respective literature should be consulted.

The CD holds a set of 34 sample spectra that allow to get a direct practical grip to the procedure. Subdirectory \Text_Resources\Manuals_Resources \Spectra_Evaluation holds these spectra (background corrected spectra in TBCAT_S have the extention *.bkg). The spectra must be copied to a directory on hard disk. A possible write-protection should be removed. It should be possible to repeat the analysis of these spectra with the following instructions and the "Example_Guide.pdf". This guide is also found in the \Manuals_Resources subdirectory.

TBCAT_S analysis is a rather complex procedure that may take considerable time. The program may be seen as a tool box allowing to perform a series of tasks which otherwise would require several other programs. It is not foolproof and, despite its GUI, should not be compared to a commercial computer program. To abort the code and to restart is a common procedure.

The spectra have to be obtained from the same chemical system. The user has to provide a series of UV-Vis spectra collected under different conditions. The ASCII files of these have to include a header providing relevant information on the chemical conditions. Details can be found in the manual. The manual (file TBCAT_S_Manual.pdf) is found in the \Manuals_Resources directory on the CD.

The first task is to install TBCAT_S to a directory on the hard disk. Second, the subdirectory directory "Spectral Evaluation" should be copied from the CD to the hard disk. Any write protection of the directories and files must be removed (files copied from a CD are commonly write-protected by the operating system).

After starting up TBCAT_S, the spectra (with extension *.bkg) in the "Spectral Evaluation" subdirectory on the hard disk should be loaded into TBCAT_S.

The procedure takes some seconds while a table with the concentration properties of the spectra is displayed. After the data have been read and analysed by the program the spectral information is graphically displayed. Note that it is of crucial importance that the spectrum "UO2Std.bkg" appears at the top of the list.

The first step of the analysis procedure is to get an idea about the number of species giving rise to the recorded spectra. The "DO" menu item provides a number of actions which can be performed with the data. The extension *.bkg of the spectral data files indicates that the spectra have already been background-corrected. The second option "Analyze" should be selected.

The Analyze Input Form will open. Three groups of information are required: "Search Interval", "SIMPLEX Input Parameters" and "Result Filename". Search interval accepts the user's guess for the maximum and minimum number of species. For the minimum the number 2 is appropriate while for the maximum number of species the number 4 is sufficient in all but the most unusual circumstances. TBCAT_S starts to search for suitable solutions of the linear equation systems with two to four factors. Finally, each factor stands for a single component species. At the Analyze level, however, only numerically acceptable solutions are searched. The search is performed by SIMPLEX analysis using random starting values. A system is repeatedly analysed using different starting values at each repetition. The number of repetitions is specified by the user in the respective field. The number entered here should between 10 and 25. The meaning of the SIMPLEX input parameters may be unclear to those with less experience in optimization. The SIMPLEX is a powerful optimization algorithm. While Newton-Raphson or Marquard algorithms require numerical or analytical derivative information, SIMPLEX does without. In the field "Iterations" a figure between 1000 and 2000 should be entered. The convergence criterion field should get a value between $5 \cdot 10^{-3}$ to 10^{-4}. TBCAT_S will generate a considerable amount of data in the Analyze step. These data are stored in a user-specified file for subsequent analysis. The Analyze files have extension "*.lyz". A suitable, characteristic name should be provided in the field where "enter filename" appears and the field should be left with a Return key. The Start button becomes available and the procedure can be started.

Two graphic windows appear showing the optimization work of the TBCAT_S Analyze routine. The green left-side window shows the agreement between the known spectrum (here the absorption spectrum of the UO_2^{2+} species) and the TBCAT_S estimate, while the right-side window shows the estimated single components. These single components give rise to a sum of squared residuals (SOR) which should be minimized. As usual the lowest SOR indicates the optimum combination of parameters. Each combination of species and starting values is stored in the *.lyz file. The relevant information is displayed in a Table on screen. In the top left corner, the total number of runs performed during the Analyze procedure is shown. Upon termination of the procedure, this table will be sorted with ascending SOR. The Analyze procedure may take several hours to complete. The amount of numerical operations performed during this step is enormous.

The sorted list of results in the Table "Analyze: Key vs. SOR" is the source of information for the next step. The concept of the key may seem a bit obscure but results purely from the necessities of a special method of factor analysis: target factor analysis. While factor analysis decomposes a given matrix of information into its singular vectors and singular values (if the matrix is square, the singular values are the roots of the eigen values and the singular vectors are the eigen vectors) this purely information is of limited value to an experimenter. However, the singular vectors may be transformed into physically meaningful information. A short tutorial has been given by Hopke (1989) and as a more complete account Malinowski's (1991) treatise can be recommended.

The essential point is to find a matrix which rotates the singular vectors into the physically meaningful vectors. While the input matrix consisted of 34×2051 individual data, the fact that only two to four components are relevant reduces that amount of data into $(2-4) \times 2051$ data points. The present example should indicate that three factors are important with the key 000 being at the top of the ordered list in the Table "Analyze: Key vs. SOR". Hence, the 69 734 spectral data have been reduced to 6153 data – everything else is noise. Because there are only three singular vectors, the target transformation matrix is just 3×3. The key just gives the diagonal elements of that transformation matrix. The diagonal elements should be either $+1$ or -1. The symbol "-1" is a bit clumsy and therefore replaced by "0". Hence, a key "100" just stands for the diagonal elements 1, -1, -1. The lengths of the key there gives the number of single components.

Clicking into a field of Table "Analyze: Key vs. SOR" opens a window asking whether a "default file" should be created. It is appropriate to store default files for the first three or four *different* keys. A default file holds the best fit parameters of all values in the target transformation matrix giving rise to the indicated SOR. Thus, these values can be used as starting values in subsequent calculations. A 4×4 matrix has 16 entries; setting these entries manually (even repeatedly) takes some efforts. A default file helps to avoid these efforts. Nevertheless, the window does provide a button allowing to generate arbitrary random values.

Having conclude the first analysis step, TBCAT_S should be closed and restarted. The spectral files should be reloaded. Now, from the "Do" menu "CAT" should be selected. The CAT Simplex Optimization window (cf. Fig. A.6) appears. Selecting a default file provides appropriate input for all fields. The number of iterations may be varied if necessary. The default value is usually appropriate. Clicking on the "run CAT" button starts the procedure. The action on screen is not much different from the previous step, but performed only for the data specified in the CAT input window. After convergence the spectral shapes of the single components are shown together with the respective data tables.

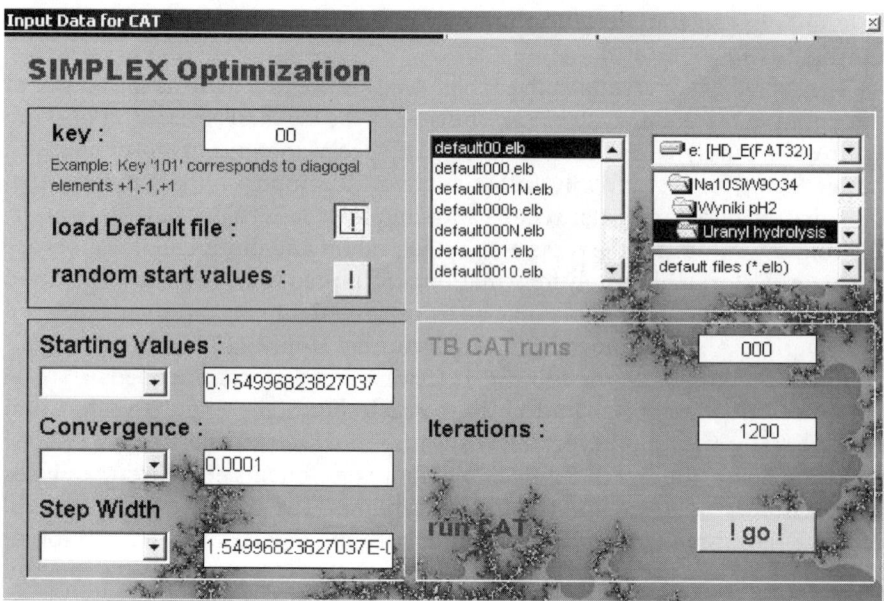

Figure A.6. CAT: SIMPLEX Optimisation window

The numerical data now requires chemical interpretation. The spectral curves are calculated but the computer cannot know which species these spectra represent. This information the user has to specify by selecting "Molar absorption" from the "Do" menu item. Figure A.7 gives a representation of the respective windows.

In the present case three components (UO_2^{2+}, SO_4^{2-} and pH) have been given. The chemical species in solution must be made up from these components. Note that a component "pH" is internally transformed into OH^- concentrations on basis of the Debye–Hückel law. Higher ionic strengths are not appropriate. If the H^+ concentrations are desired, the component "H+" should be specified and these concentrations entered manually in the headers of the respective spectra.

The "Thermodynamic Input" window requires input only in the top "Chemical Informations – Input" section. Under b), names for the species can be specified. TBCAT_S assumes that a metal ion is giving rise to the absorption while the ligands just modify the metal ion's absorption. Therefore, species 2 has the name "Metal ion" as default name.

Note that the second species MUST always be the species whose absorption spectrum is used as the known spectrum (the spectrum loaded first and appearing on top of the input file list). Otherwise, TBCAT_S cannot handle the information appropriately. If the absorbing compound is a ligand (e.g. in case of complexation of Arsenazo III with a metal ion), the second species must be the absorption spectrum of uncoordinated Arsenazo III.

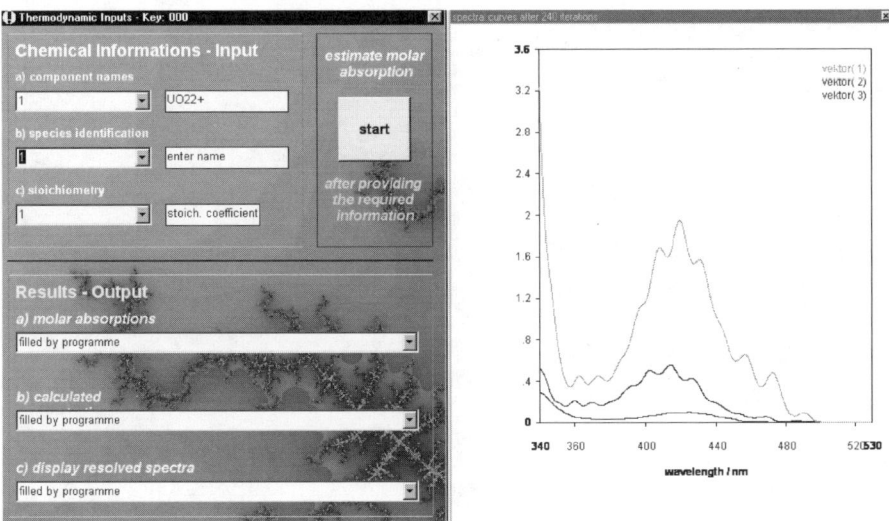

Figure A.7. Molar absorption window allows specification of chemical information

Hence, under b) the following input should be made:

1. UO2SO4
2. UO22+
3. U22

The most crucial input is specified under c). Here, the stoichiometric coefficients for the species must be given. Under c) the following input should be made:

1. 110
2. 100
3. 202

With this information, the procedure may be started. After a few seconds the screen should look like in Fig. A.8.

The "Results – Output" section of the CAT window now holds a) the molar absorptions for each species, b) the difference between calculated and specified U(VI) concentration (opened) and c) the formation constants calculated for each species on basis of the mean value analysis.

Clicking on a line in c) "display resolved spectra" lets TBCAT_S give a graphical representation of the results in the graphics window including the contributions of single components and the calculated sum spectrum. The graphics window does show a grey button "LSR". This button provides a least-squares analysis using the difference between calculated and measured data as a basis for a statistical analysis. The QR decomposition is used. Note that this analysis may take some minutes. Figure A.9 shows an example. The LSR analysis also forwards the correlation matrix.

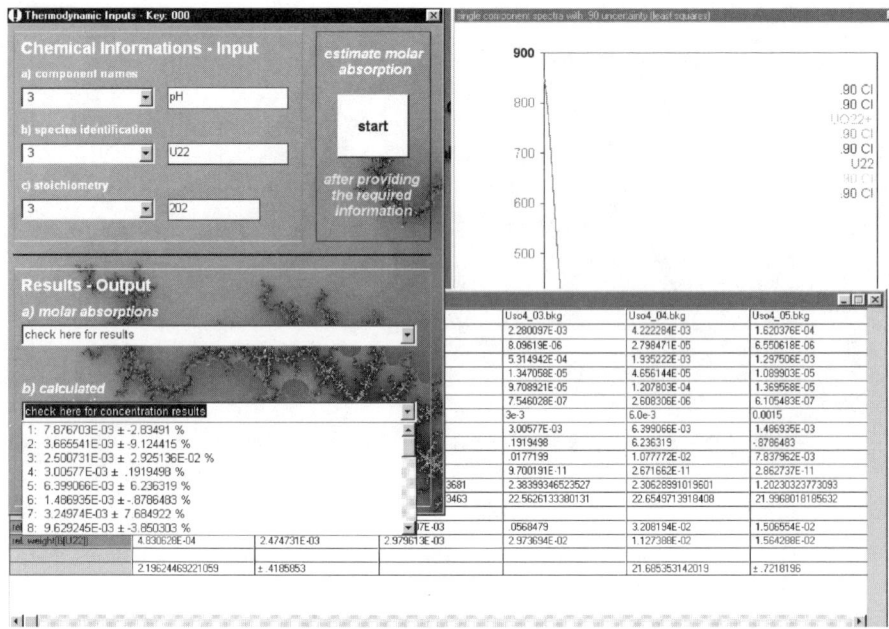

Figure A.8. The result screen of the CAT procedure

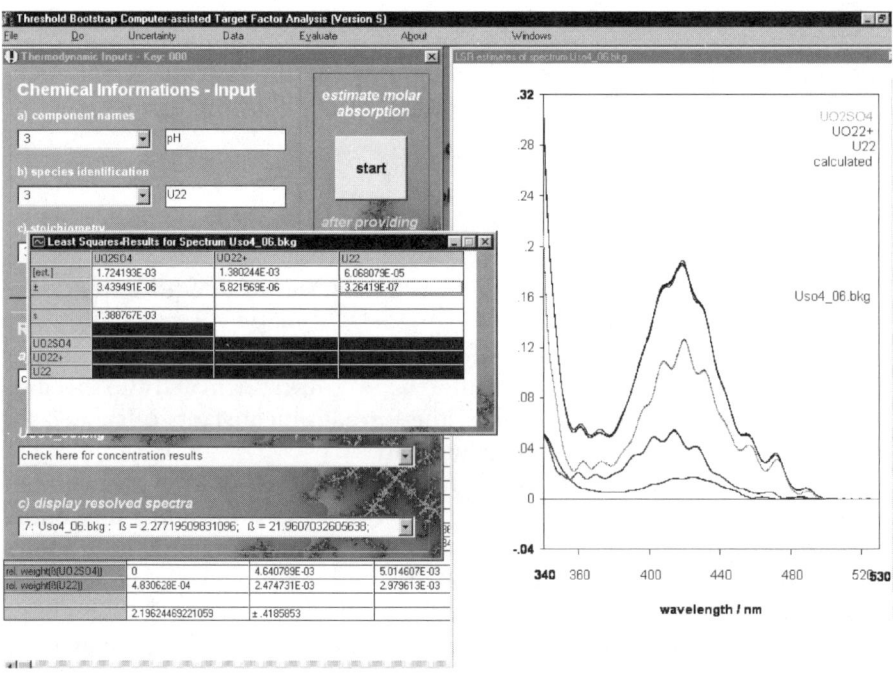

Figure A.9. An example of CAT analysis output

All graphical results can be exported as x,y ASCII data by setting the focus to the graphics window and using the "Save" item in the "File" menu.

The Table "Summary of calculated data" gives the detailed data for each spectrum. Table entries with red background indicate physically meaningless values (e.g. negative concentrations). The total amount of information can be saved as ASCII file by setting focus to the Table and selecting "Save" from the "File" menu item.

CAT analysis may indicate whether the interpretation is satisfactory. It may also indicate extraneous data or misinterpretations. Note that there is no use to expect highly consistent data if, say, the relative species concentrations in a sample are very different or the free metal concentration is almost zero. The bottom row of the Table also summarizes the formation constants over all spectra. The first values are for the first species, the second values are valid for the second species.

If the respective interpretation has been identified as the most reasonable, probably etc. one, the final lag of the complete analysis may be taken: the TB CAT analysis.

TB CAT analysis stands for threshold bootstrap computer-assisted target factor analysis. The threshold bootstrap is a modification of the moving block bootstrap (MBB) discussed in 1.7.5. TB CAT analysis repeats the CAT analysis a large number of times, say 1000 times. Each time, the input data are slightly varied according to the results of the metrological analysis of an analytical method.

For the TB CAT analysis, the TBCAT_S should be terminated and restarted. The spectra should be reloaded. To start the TB CAT analysis, the "TB CAT" menu item should be chosen from the "Do" menu item. The "SIMPLEX Optimization" window appears. The same information as in the CAT step should be entered. Now, the field "TB CVAT runs" is enabled. A value between 1000 and 2000 should be entered and the "Start" button should be clicked.

The "Thermodynamic Input" window appears. Again, the same information as before should be entered. Then, the "Run" button should be pressed. Now, the TB CAT analysis takes place. There are two formation constants to be calculated: first for species $UO_2SO_4^\circ$, second for $(UO_2)_2(OH)_2^{2+}$. Each run of the TB CAT cycles forwards a value. The values will differ slightly because the input parameters vary slightly. These values are visible in the two tables at the main window's right side.

TB CAT analysis may take a long time. For larger data sets and higher number of TB CAT cycles, a duration of 24 h is not uncommon. A total of 1000–2000 repetitions is not a small number. However, when compared with the time necessary to collect the information, a few days may be invested into the data evaluation.

The menu item "Evaluate" has two relevant entries: "Spectral uncertainty" and "Differentiate". These menu items provide algorithms to summarize the enormous amount of information calculated during a TBCAT analysis. The

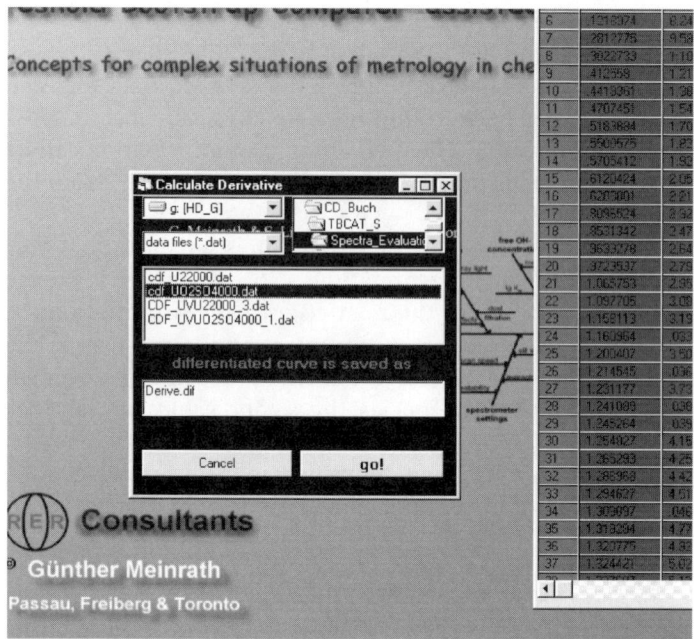

Figure A.10. "Derivative" window

Differentiate item (cf. Fig. A.10) allows the user to select a formation constant file (name convention cdf_*key.dat; where * = species name; key = a sequence of 0 and 1). The formation constants are sorted, the empirical cumulative probability distribution (CDF) is determined and the CDF is numerically differentiated. The result is graphically displayed and automatically stored as a dif_*key.dat file in ASCII format.

The spectral information is evaluated using the "Spectral uncertainty" item. From, say, 1000 repetitions 1000 different single component spectra are obtained. A system with three species thus generates 3000 files with estimates of the single component spectra. TBCAT_S collects the information for each wavelength, sorts the data and obtains the mean value as well as the respective values several, user-specified, confidence limits (cf. Fig. A.11). This information is written to a CDF file. Hence, from the 3000 spectra three files are created holding the mean values for each single component as well as some selected upper and lower confidence ranges.

The Uncertainty menu item provides several input masks to communicate the magnitude of an influence quantity's contribution to measurement uncertainty (Fig. A.12). Once a set of uncertainties has been entered the information may be saved in an ASCII file. Thus, several independent sets of uncertainties may be handled conveniently, e.g. to compare the influence of different choices on the output quantities, for instance formation constants (Fig. A.13).

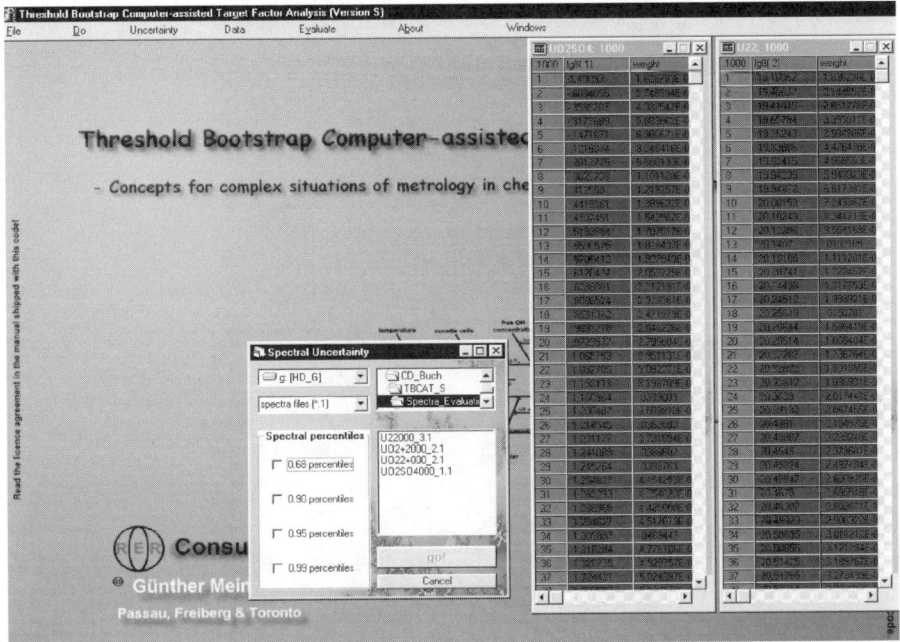

Figure A.11. "Spectral uncertainty" window. The user may select several confidence levels

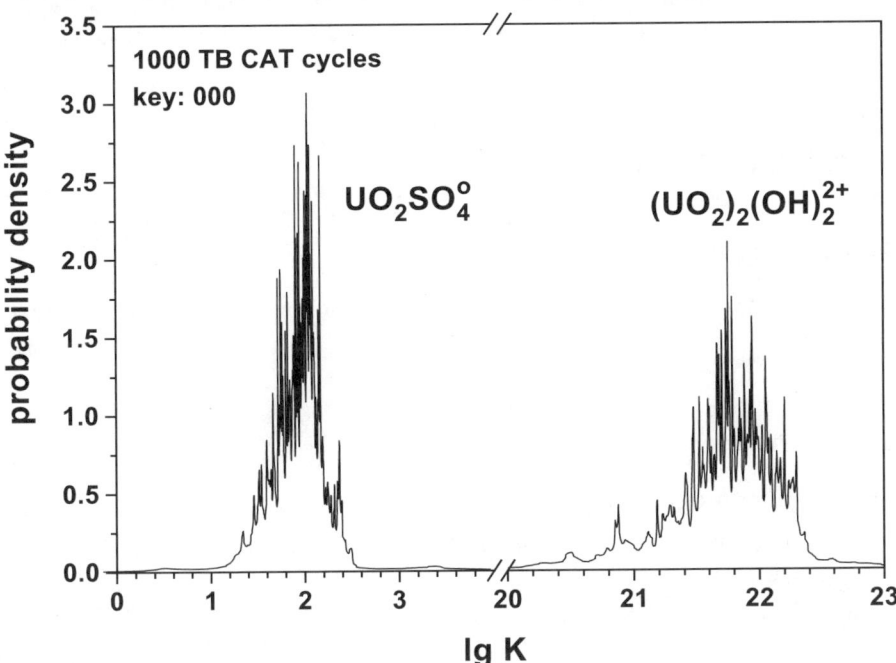

Figure A.12. Probability distribution of formation constants $\lg K_{110}$ (*left*) and $\lg K_{202}$ (*right*) obtained from 1000 TBCAT cycles

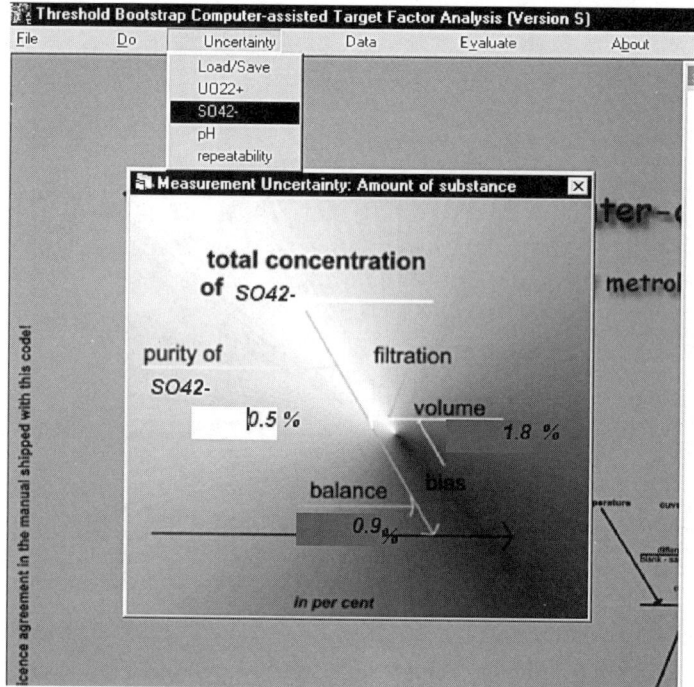

Figure A.13. Example of an input window for influence quantities (here total sulfate concentration)

A.3
ReadMe Text and License Agreement

A.3.1
Description of the Contents of the CD-ROM

The CD holds two groups of additional data: computer programs and text resources.

A.3.1.1
Computer Programs

Two computer programs are available:

- the LJUNGSKILE_S code for probabilistic speciation in subdirectory \Ljungskile_S.
- the TBCAT_S code for measurement uncertainty evaluation of UV-Vis spectroscopic measurement data in subdirectory \TBCAT_S.

The use of both programs is described in detail in Chap. A.2 of this Appendix and in the respective manuals on the CD-ROM.

Both programs are installed on a computer by set-up routines (setup.exe). Manuals and additional information are available in the \Manuals subdirectory directory of the \Text Resources subdirectory. In addition, example files are included. These files refer to the respective discussions in the book. Therefore, the interested user can reproduce and modify the calculations. Both codes are unrestricted and, therefore, can be applied to other problems.

Both codes have been tested on various computers before. Great care has been given to avoid problems. Due to the existence of a large variety among computer operating systems, it is impossible to foresee and prevent all problems.

A.3.1.2
Text Resources

In the subdirectory \Manuals_Resources, the manuals for the computer programs included with this CD are provided as platform-independent PDF files.

In the subdirectory \EURACHEM Guides, three EURACHEM Guides are available as PDF documents:

- Quality Assurance for Research and Development and Non-routine Analysis
- Quantifying Uncertainty in Analytical Measurement, 2nd ed.
- Traceability in Chemical Measurement

Please also note the Copyright Acknowledgement (copyright_acknowledgement.txt).

In the subdirectory \Neptunium_Dispute an account of the dispute within the OECD/NEA Thermodynamic Database Project on the review process on Neptunium thermodynamic data is available. These files are of minor scientific interest but document the difficulty to achieve consensus in the presence of conflict of interests if no appropriate protocols and references are available. The files are also available to the public on http://www.vitorge.name/pierre/ insultes (last accessed November 2006).

A.3.2
Hard- and Software Requirements

Please make sure that the following minimum hard- and software requirements are in compliance with your computer:

Operating system Windows 98SSE and higher versions
CPU clock speed 500 MHz
50 MB of free hard disk space
CD-ROM drive
Display with a resolution of 1024×768

To read the PDF files an appropriate reader, e.g. the Acrobat 4.0 reader from Adobe must be installed.

A.3.3
Help Desk

In case of problems with the programs, please send an e-mail to the Springer Help Desk: _Springer.com/helpdesk-form_.

A.3.4
Licence Agreement, Including Warranty and Liabilities

A.3.4.1
Licence Agreement

The programs are distributed free of charge with the book but they are not freeware or public domain. The manuals include additional licensing agreements. The User is obliged to read these licensing agreements. The installation and use of the codes implies acceptance of the respective licensing agreement(s). Only if a user agrees to these licence and warranty requirements and, in addition, to all agreements listed in chap. 3.4 of this ReadMe-Text, the codes and programs may be used.

The codes and the respective manuals are one entity. Therefore the manuals must be read and accepted by the user. With the exceptions mentioned in the licence agreements, the codes must not be distributed to third parties.

All copyrights remain with the author(s) of the codes and with Springer-Verlag.

A.3.4.2
Warranty

A.3.4.2.1 Springer-Verlag is not the originator of the data and programs but only makes them available. The User should be aware of the fact that it is impossible to create faultless software; therefore, Users must take appropriate steps to verify the correctness of the results of their calculations. Therefore, there is no warranty whatsoever that the codes and the results generated by the codes are capable or meaningful for the intended purpose.

A.3.4.2.2 In the case of faulty material, manufacturing defects, absence of warranted characteristics, or damage in transit, Springer-Verlag shall exchange the CD-ROM. Further claims shall only be admitted if the User has purchased the book including the CD-ROM from Springer-Verlag directly. The warranty requires the User to supply a detailed written description of any fault immediately.

A.3.4.3
Liabilities of Springer-Verlag

A.3.4.3.1 Springer-Verlag will only be liable for damages, whatever the legal ground, in the case of intent or gross negligence and with respect to warranted characteristics. A warranty of specific characteristics is given only in individual cases to a specific User and requires explicit written representation. Liability under the product liability act is not affected hereby. Springer-Verlag may always claim a contributory fault on the part of the User.

A.3.4.3.2 The originator or manufacturer named on the CD-ROM will only be liable to the User, whatever the legal ground, in the case of intent or gross negligence.

A.3.4.3.3 Additional conditions for Users outside the European Community: Springer-Verlag and authors will not be liable for any damages, including any lost profits, lost savings, or other incidental or consequential damages arising from the use of, or inability to use, this software and its accompanying documentation, even if Springer-Verlag and/or the authors have been advised of the possibility of such damages.

A.3.4.4
Liabilities of the User

A.3.4.4.1 The User agrees to comply with the rules outlined herein. Violations of these rules may be criminal offences and may also give rise to claims for damages against the User from the licensers of Springer-Verlag.

A.3.4.4.2 In the case of serious violations committed by the User, Springer-Verlag may revoke the license.

A.3.4.4.3 Upon installation of the codes the User(s) implicitly accept the licensing agreements provided with the manuals of the respective code(s). Therefore, the manuals are to be read immediately after installation of the respective program(s).

A.3.5
Final Provisions

A.3.5.1) If any provision of the entire Agreement in chap. 4) is or becomes invalid or if this Agreement is incomplete, the remainder of the Agreement is not affected. The invalid provision shall then be replaced by a legally valid provision, which comes as close as possible to the invalid provision as far as its economic effect is concerned. The same applies to possible gaps in the Agreement.

A.3.5.2) This Agreement falls under the jurisdiction of the courts at Heidelberg, if the User is a merchant who has been entered in the commercial register, a legal entity under public law, or a public special fund, or if the User has no residence or place of business in Germany.

A.3.5.3) This Agreement is subject to the laws of Germany to the exclusion of the Uncitral Trading Rules.

Additional conditions for Users outside the European Community see Chap. A.3.4.3.3.

References

Academic Software Sourby Old Farm, Timble, Otley, Yorks, LS21 2PW, UK

Ades AE, Lu G (2003) Correlation between parameters in risk models: estimation and propagation of uncertainty by Markov Chain Monte Carlo. Risk Anal 23:1165–1172

AG BODENKUNDE (2005) Bodenkundliche Kartieranleitung. 5. Aufl 438 p. Hannover/FRG

Agmon S (1954) The relaxation method for linear inequalities. Can J Math 6:382–392

Albrecht H, Langer M (1974) The rheological behavior of rock salt and related stability problems of storage caverns, ADE606360 DOE London/UK

Allison JD, Brown DS, Novo-Gradac KJ (1991) MINTEQ2A, a geochemical assessment database and test cases for environmental systems. REPORT EPA 600-3-91 US Environmental Protection Agency Athens/USA

Alpers JS, Gelb RJ (1990) Standard errors and confidence intervals in nonlinear regression: Comparison of Monte Carlo and parametric statistics. J Phys Chem 94:4747–4751

Alpers JS, Gelb RI (1991) Monte Carlo methods for the determination of confidence intervals: Analysis of nonnormally distributed errors in sequential experiments. J Phys Chem 95:104–108

Alpers JS, Gelb RJ (1993) Application of nonparametric statistics to the estimation of the accuracy of Monte Carlo confidence intervals in regression analysis. Talanta 40:355–361

Analytical Measurement Committee (1995) Uncertainty of measurement: implications of its use in analytical scieResults and discussion

Anderson RK, Thompson M, Culbard E (1986) Selective reduction of arsenic species by continuous hydride generation – I. Reaction media Analyst 111:1153

Antoniadi EM (1910) Sur la nature des "canaux" de Mars. Astronomische Nachrichten 183:221

Aregbe Y, Harper C, Norgaard J, de Smet M, Smeyers P, van Nevel L, Taylor PDP (2004) The interlaboratory comparison "IMEP-19 trace elements in rice" – a new approach for measurement performance evaluation. Accred Qual Assur 9:323–332

Atkinson AC, Fedorov VV (1975) The design of experiments for discriminating between two rival models. Biometrika 62:57–70

Autio J, Siitari-Kauppi M, Timonen J, Hartikainen K, Hartikainen J (1998) Determination of the porosity, permeability and diffusivity of rock in the excavation-disturbed zone around full-scale deposition holes using the 14C-PMMA and the He-gas methods, J Contam Hydrol 35:19–29

Avilia R, Broed R, Pereira A (2003) Ecolego – a toolbox for radioecological risk assessment. Int Conf Protection of the Environment from the Effect of Ionizing Radiation. 6–10 Oct 2003 Stockolm/S

Aziz K, Settari A (1979) Petroleum reservoir simulation. Applied Science, London/UK

Baber R (1987) The spine of software; designing provable correct software: theory and practice. Wiley, Chichester, UK

Baes CF, Mesmer RE (1976) The hydrolysis of the cations. Wiley and Sons, New York, USA

Bakr AA, Gelhar LW, Gutjahr AL, Mac-Millan JR (1978) Stochastic analysis of variability in subsurface flows: I. Comparison of one- and three-dimensional flows. Water Resour Res 14:263–271

Ball JW, Nordstrom DK (1991) User's manual for WATEQ4F. US Geological Survey Open-File Report 91–183

BAM (2003) 30. Ringversuch Wasseranalytik. Lösung 3N1. Bundesanstalt für Material-forschung und -prüfung Berlin/FRG

Bamberg HF, Häfner F (1981) Laborative Bestimmung von Porosität (Porenanteil) und Durchlässigkeit an Locker- und Festgesteinsproben, Zeitschrift für angewandte Geologie 27:218–226

Banks J (1989) Testing, understanding and validating complex simulation models. In: Proceedings of the 1989 Winter Simulation Conference

Barnett V (1978) The study of outliers: purpose and model. Appl Stat 27:242–250

Bates DM, Watts DG (1988) Nonlinear regression analysis and its application. Wiley Series in probability and mathematical statistics. Wiley, Chichester, UK

Baucke FGK (1994) The modern understanding of the glass electrode response. Fresenius J Anal Chem 349:582–596

Baucke FGK (2002) New IUPAC recommendations on the measurement of pH – background and essentials. Anal Bioanal Chem 374:772–777

Baucke FGK, Naumann R, Alexander-Weber Ch (1993) Multiple-point calibration with linear regression as a proposed standardization procedure for high-precision pH measurements. Anal Chem 65:3244–3251

Bays C, Durham SD (1976) Improving a poor random number generator. ACM Trans Math Software 2:59–64

Bear J, Verruijt A (1987) Modeling groundwater flow and pollution. Reidel Dordrecht/NL 414 pp

Beauce A, Bernard J, Legchenko A, Valla P (1996) Une nouvelle méthode géophysique pour les études hydrogéologiques: l'application de la résonance magnétique nucléaire. Hydrogéologie 1:71–77

Beauheim RL, Roberts RM, Dale TF, Fort MD, Stensrud WA (1993) Hydraulic testing of Salado Formation evaporites at the Waste Isolation Pilot Plant site: Second interpretive report, SANDIA National Laboratories, SAND92-0533 Albuquerque, USA

Beckman RJ, Cook RD (1983) Outlier..........s. Technometrics 25:119

Behr A, Förster S, Häfner F, Pohl A (1998) In-situ Messungen kleinster Permeabilitäten im Festgestein, Freiberger Forschungshefte A 849:274–285

Bennet DG, Liew SK, Mawbey CS (1994) CHEMTARD manual. WS Atkins Engineering Sciences Epsom Surrey, UK

Beran J (1992) Statistical method for data with long-range dependence. Stat Sci 7:404–427

BGR (1996) Hydraulische Untersuchungen im Grubengebäude Morsleben. Zwischenbericht

BIPM (1980) Recommendation INC-1. BIPM Sèvres/F

BIPM (2005) http://www.bipm.fr

Bjerrum J, Schwarzenbach G, Sillèn LG (1957) Stability constants of metal ion complexes. Special publication 6 Royal Chemical Society London, UK

Blesa MA, Maroto AJG, Regazzoni AE (1984) Boric acid adsorption on magnetite and zirconium dioxide. J Colloid Interface Sci 99:32–40

Boebe M (1994) Permeabilitätsbestimmungen mit Hilfe eines Minipermeameters. Erdöl Erdgas Kohle 110:58–61

Bolt GH, Groeneveld PH (1969) Coupling phenomena as a possible cause for non-Darcian behaviour of water in soil. Bull Int Assoc Sci Hydrol 14:17–26

Bolt GH, van Riemdijk WH (1982) Ion adsorption on organic variable charge mineral surfaces. In: Soil Chemistry B. Physico-chemical models (Bolt ed) 2nd edn. Elsevier, Amsterdam, pp 459–503

Bond AM, Hefter GT (1980) Critical survey of stability constants and related thermodynamic data of fluoride complexes in aqueous solution. IUPAC Chemical data series 27. Pergamon Press Oxford, UK

Borgmeier M, Weber JR (1992) Gaspermeabilitätsmessungen an homogenen Modellsalz-kernen, Erdöl Erdgas Kohle, 108:412–414

Bousaid IS, Ramey Jr HJ (1968) Oxidation of crude oil in. porousmedia. Soc Petrol Eng J (June), 137–148

Bouwer H (1989) The Bouwer and Rice slug test – an update. Ground Water 27:304–309

Box GEP (1960) Fitting empirical data. Ann NY Acad Sci 86:792–816

Box MJ (1971) Bias in nonlinear estimation. J R Stat Soc B33:171–201

Box GEP, Draper NR (1987) Empirical model building and response surfaces. Wiley series in probability and mathematical statistics. Wiley, Chichester

Box GEP, Muller ME (1958) A note on the generation of random normal deviates. Ann Math Stat 29:610–611

Boyle R (1661) The Sceptical Chymist. http://oldsite.library.upenn.edu/text/collections/science/boyle/chymist, (last accessed November 2005)

Bredehoeft JD, Papadopulos SS (1980) A method for determining the hydraulic properties of tight formations. Water Resources Res 16:233–238

Brinkmann J (1999) 225 kV Röntgentomograph ermöglicht, Gesteine auf Porosität und Permeabilität hin zu "durchleuchten". Internet publication, Technische Universität Clausthal, Clausthal-Zellerfeld, FRG

Bronstein IN, Semendjajew KA (1979) Taschenbuch der Mathematik 19th ed. (Grosche, Ziegler eds). Verlag Harri Deutsch Frankfurt 860 pp

Bruck J von der (1999) Zum Permeabilitätsverhalten von kompaktiertem Salzgrus, Diss. TU Bergakademie Freiberg, FRG

Brumby S (1989) Exchange of comments on the Simplex algorithm culminating in quadratic convergence and error estimation. Anal Chem 61:1783–1786

Burden SL, Euler DE (1975) Titration errors inherent in using Gran plots. Anal Chem 47:793–797

Burmester DE, Anderson PD (1994) Principles of good practice for the use of Monte Carlo techniques in human health and ecological risk assessments. Risk Anal 14:477–481

Burneau A, Tazi M, Bouzat G (1992) Raman spectroscopic determination of equilibrium constants of uranyl sulphate complexation in aqueous solution. Talanta 39:743

Caceci MS (1989) Estimating error limits in parametric curve fitting. Anal Chem 61:2324

Caceci MS, Cacheris WP (1984) Fitting curves to data: The Simplex is the answer. Byte 340

CAEAL (2005) Accred Qual Assur 10:197–183

Cammann K, Kleiböhmer W (1998) Need for quality management in research and development. Accred Qual Assur 3:403–405

Cardano G (1526) Liber de ludo alea. This 15 page note appeared in the 1st edition of Cardano's complete works in 1663

Carlstein E (1986) The use of subseries values for estimating the variance of a general statistic from a stationary sequence. Ann Stat 14:1171–1179

Carman PC (1937) Fluid flow through a granular bed. Trans Inst Chem Eng 15:150–156

CCQM-9 (1999) http://www.bipm.fr, KCDB, (last accessed November 11 2005)

Cerny V (1985) A thermodynamic approach to the travelling salesman problem: An efficient simulation. J Optim Theory Appl 45:41–51

Chalmers RA (1993) Space age analysis. Talanta 40:121–126

Chandratillake MR, Falck WE, Read D (1992) CHEMVAL. project – guide to the CHEMVAL thermodynamic database. DOE Report DOE, HMIP, RR, 92.094, London, UK

Chatterjee S, Hadi AS (1986) Influential observations, high leverage points, and outliers in linear regression (with discussion). Stat Sci 1:379–393

Choppin GR, Mathur JN (1991) Hydrolysis of actinyl(VI) ions. Radiochim Acta 52, 53:25–28

Christensen JM, Holst E, Olsen E, Wilrich PT (2002) Rules for stating when a limiting value is exceeded. Accred Qual Assur 7:28–34

Christiansen RL, Howarth SM (1995) Literature review and recommendation of methods for measuring relative permeability of anhydrite from the Salado formation at the Waste Isolation Pilot Plant, REPORT SAND93-7074, UC-721

Cicero MT (80 BC) De inventione. Reprinted by E Stroebel Leipzig, FRG (1981)

Clapham PB (1992) Measurement for what is it worth. Eng Sci Educ J 1:173–179

Clark RW (2003) The physics teacher: To a physics teacher a mass is a mass. To a chemistry teacher a mass is a mess. J Chem Educ 80:14–15

Comas Solá J (1910) Observaciones de Marte. Astronomische Nachrichten 183:219–221

Cooper HH, Bredehoeft JD, Papadopulos IS (1967) Response of a finite-diameter well to an instanteous charge of water. Water Resources Res 3:263–269

Courant L, Isaacson E, Rees M (1952) On the solution of nonlinear hyperbolic differential equations by finte differences. Comm Pure Appl Math 5:243–255

Covington AK, Bates RG, Durst RA (1983) Definition of pH scales standard reference values, measurement of pH and related terminology. Pure Appl Chem 55:1467

Cox JD, Wagman DD, Medvedev VA (eds) (1987) CODATA key values for thermodynamics, Taylor and Francis Basingstoke, UK (1987) p 271

Crank J, Nicolson P (1947) A practical method for numerical evaluation of solutions of partial differential equations of the heat conduction type. Proc Cambridge Phil Soc 43:50–64

Criscenti LJ, Laniak GF, Erikson RL (1996) Propagation of uncertainty through geochemical calculations. Geochim Cosmochim Acta 60:3551–3568

Croise J, Marschall P, Senger R (1998) Gas-water flow in a shear zone of a granitic formation: interpretation of field experiments. In: International Conference On Radioactive Waste Disposal, Hamburg, FRG, pp 88–93

Cross JE, Ewart FT (1991) HATCHES – A thermodynamic database and management system. Radiochim Acta 52, 53:421–422

Czolbe P, Klafki M (1989) Laboruntersuchungen zu Zweiphasenvorgängen und Bestimmung von Mobilitätsänderungen in Aquifergasspeichern und Gaslagerstätten. In: Eigenschaften von Gesteinen und Böden, Freiberger Forschungsheft A 849, TU BA Freiberg, FRG

Dale T, Hurtado LD (1996) WIPP air-intake shaft disturbed-rock zone study. Paper presented at the 4th Conference on the Mechanical Behaviour of Salt (SALT IV), Montreal

Dathe M, Otto O (1996) Confidence intervals for calibration with neural networks. Fresenius J Anal Chem 356:17–20

Davison AC, Hinkley DV (1997) Bootstrap methods and their application. Cambridge series in statistical and probabilistic mathematics. Cambridge University Press Cambridge

Dean RB, Dixon WJ (1951) Simplified statistics for small numbers of observations. Anal Chem 23:636–638

de Angelis D, Hall P, Young GA (1993) Analytical and bootstrap approximations to estimator distributions in L1 regression. J Am Stat Soc 88:1310–1316

de Bièvre P (1999) Reproducibility is never proof of accuracy... Accred Qual Assur 4:387

de Bièvre P (2000) When do we stop talking about "true values"? Accred Qual Assur 5:265

de Bièvre P (2000a) Is not traceability about the link from value to value (anymore)? Accred Qual Assur 5:171–172

de Bièvre P (2000b) Traceability of (values carried by) reference materials. Accred Qual Assur 5:224–230

de Bièvre P (2000c) The key elements of traceability in chemical measurement: agreed or still under debate? Accred Qual Assur 5:423–428

de Bièvre P (2001) Metrology is becoming the language of the (inter) national marketplace. Accred Qual Assur 6:451

de Bièvre P (2002) Too small uncertainties: the fear of looking "bad" versus the desire to look "good". Accred Qual Assur 8:45

de Bièvre P (2004a) Reply form the editor to the letter to the editor. Accred Qual Assur 10:62–63

de Bièvre P (2004b) Target measurement uncertainties I: are they coming Accred Qual Assur 10:589

de Bièvre P (2006) Editor's reply to a letter to the editor by Xavier Fuentes-Arderiu. Acrred Qual Assur 11:260

de Bièvre P, Williams A (2004) The basic features of a chemical measurement. Accred Qual Assur 10:64

Debschütz WG (1998) Hydraulische Untersuchungen zur Charakterisierung der Porenstruktur natürlicher Gesteine Int Conf Geo-Berlin 98 (Ann Meeting German Geological Societies) Oct 6–9 1998, Berlin, FRG

Delakowitz B (1996) Bestimmung von Grundwasserströmungsgeschwindigkeit und -richtung mittels radiohydrometrischer Einbohrlochmethode. Unveröffentlichter Bericht an das Bundesamt für Strahlenschutz zum Projekt BfS 8232-6, 9M, PSP 212.250.50 Salzgitter, FRG

Delakowitz B, Meinrath G, Spiegel W (1996) A literature survey of mineral-specific sorption data on radionuclides with relevance to the disposal of radioactive waste. J Radioanal Nucl Chem 213:109–125

de Morgan A (1838) An essay on probability. Longman, London

Denison FH, Garnier-Laplace J (2004) The effects of database parameter uncertainty on uranium(VI) equilibrium calculations. Geochim Cosmochim Acta 69:2183–2191

De Vahl Davis G, Mallinson GD (1972) False diffusion in numerical fluid mechanics. School of Mechanical and Industrial Engineering, University of New South Wales Sidney, Au REPORT FMT, p 122

Delambre JBJ (1810) Base du système métrique décimal, ou mesure de l'arc du méridien compris entre les parallèles de Dunkerque et Barcelone [Texte imprimé], exécutée en 1792 et années suivantes, par MM. Méchain et Delambre. Baudouin Paris, F (1806, 1807, 1810)

Denison FH, Garnier-Laplace J (2005) The effects of database uncertainty on uranium(VI) speciation calculations. Geochim Cosmochim Acta 69:2183–2192

Dettinger MD, Wilson JL (1981) First order analysis of uncertainty in numerical models of groundwater flow Part 1: mathematical development. Water Resources Res 17:149–161

Dixon WJ (1950) Analysis of extreme values. Ann Math Stat 21:488–506

Drake S (1980) Galileo. Oxford University Press, Oxford

Draper N, Smith H (1981) Applied regression analysis, 2nd edn. Wiley series in probability and mathematical statistics. Wiley, Chichester

Drglin T (2003) An estimation of the measurement uncertainty for an optical emission spectrometric method. Accred Qual Assur 8:130–133

Drost W (1989) Single-well and multi-well nuclear tracer techniques – a critical review. Technical Documents in Hydrology, SC-89, WS 54, UNESCO Paris, F

DVWK (1983) Beiträge zu tiefen Grundwässern und zum Grundwasser-Wärmehaushalt, Schr. DVWK, 61, Hamburg–Berlin

Dzombak DA, Morel FMM (1990) Surface complexation modeling. Hydrous ferric oxide. Wiley, New York

EA (2002) Expression of uncertainty in calibration. Report EA-4, 02 European Accreditation. (http://www.european-accreditation.org)

Earlougher RC (1977) Advances in well test analysis. Henry L. Doherty Memoriai Fund of the American Institute of Mining, Metallurgical, and Petroleum Engineers (AIME) and Society of Petroleum Engineers of AIME, Dallas, USA

Efron B (1979) Computers and the theory of statistics: thinking the unthinkable. SIAM Review 21:460–480

Efron B (1981) Nonparametric estimates of standard error: the jackknife, the bootstrap and other resampling methods. Biometrika 68:589–599

Efron B (1981b) Censored data and the bootstrap. J Am Statist Assoc 76:312–319

Efron B (1986) Why isn't everyone a Bayesian? (with discussion) Am Stat 40:1–11

Efron B (1987) Better Bootstrap confidence intervals. J Am Statist Assoc 82:171–200

Efron B, Tibshirani R (1986) Bootstrap methods for standard errors, confidence intervals and other measures of accuracy. Stat Sci 1:54–77

Efron B, Tibshirani RJ (1993) An introduction to the bootstrap. monographs on statistics and applied probability, vol 57. Chapman and Hall, London, 436 pp

Eisenberg NA, Rickertsen LD, Voss C (1987) Performance assessment, site characterization, and sensitivity and uncertainty methods: Their necessary association for licensing. In: Buxton (ed) Proceedings of Conference on Geostistical, Sensitivity, and Uncertainty Methods for Ground-water Flow and Radionuclide Transport Modeling. DOE, AECL Batelle Press San Francisco, USA

Ekberg C (1996) Uncertainty and sensitivity analyses of chemical modelling approaches applied to a repository for spent nuclear fuel. Thesis, Chalmers University of Technology Göteborg, 61 pp

Ekberg C (2006) Uncertainties in actinide solubility calculations illustrated using the Th-OH-PO_4 system. Thesis 2nd ed. Chalmers University of Technology Göteborg, 111 pp

Ekberg C, Emrèn AT (1996) SENVAR: a code for handling chemical uncertainties in solubility calculations. Comput Geosci 22:867

Ekberg C, Lunden-Burö I (1997) Uncertainty analysis for some actinides under groundwater conditions. J Statist Comput Simul 57:271–284

Ellison SLR (2005) Including correlation effects in an improved spreadsheet calculation of combined standard uncertainties. Accred Qual Assur 10:338–343

Ellison SLR, Barwick VJ (1998) Using validation data for ISO measurement uncertainty estimation. Part 1. Principles of an approach using cause and effect analysis. Analyst 123:1387–1392

Ellison SLR, Holcombe DG Burns M (2001) Response surface modelling and kinetic studies for the experimental estimation of mesurement uncertainty in derivatisation. Analyst 126:199–210

Ellison S, Wegscheider W, Williams A (1997) Measurement uncertainty. Anal Chem 69:607A–613A

Emrén A (2002) The CRACKER program and modelling of Äspö groundwater chemistry. Chalmers University of Technology, Göteborg, 170 pp

EPA (1999) Reliability-based uncertainty analysis of groundwater contaminant transport and remediation: REPORT EPA, 600, R-99, 028 EPA Washington, USA

EPTIS (2005) http://www.eptis.bam.de

EURACHEM, CITAC (1998) Quality assurance for research, development and non-routine analysis. EURACHEM, CITAC. (http://www.measurementuncertainty.org)

EURACHEM, CITAC (2002) Quantifying uncertainty in chemical measurement. EU-RACHEM, CITAC. (http://www.measurementuncertainty.org)

EURACHEM, CITAC (2004) Traceability in chemical analysis. EURACHEM, CITAC. (http://www.measurementuncertainty.org)

Evans JE, Maunder EW (1903) Experiments as to the actuality of the "Canals" observed on Mars. Monthly Notices of the Royal Astronomical Society 63:488–499

Ewing RC, Tierney MS, Konikow LF, Rechard RP (1999) Performance assessment of nuclear waste repositories: A dialogue on their value and limitations. Risk Anal 19:933–958

Faust CR, Mercer JW (1984) Evaluating of slug tests in wells containing a finite-thickness skin. Water Resources Research 20 (4):504–506

Fein E, Müller-Lyda, I, Storck R (1996) Ableitung einer Permeabilitäts-Porositätsbeziehung für Salzgrus und Dammbaumaterialien, GRS-Bericht Nr. 132 Braunschweig, FRG

Ferson S (1995) Quality assurance for Monte Carlo risk assessment. Proc 3rd Int Symp Uncertainty Modelling and Analys ISUMA-NAFIPS '95 Sept 17–20 University of Maryland Maryland, USA 14–19

Filella M, May PM (2003) Computer simulation of the low-molecular-weight inorganic species distribution of antimony(III) and antimony(V) in natural waters. Geochim Cosmochim Acta 67:4013–4031

Filella M, May PM (2005) Reflections on the calculation and publication of potentiometrically-determined formation constants. Talanta 65:1221–1225

Finifter BM (1972) The generation of confidence. Evaluating research findings by random subsample replication. In: Costineau (ed) Sociological methodology. Jossey-Bass, San Francisco, pp 112–175

Finkel AM (1994) Stepping out of your own shadow: a didactic example of how facing uncertainty can improve decision-making. Risk Anal 14:751–761

Finsterle S, Persoff P (1997) Determining permeability on rock sample using inverse modelling. Water Resources Res 33:1803–1811

Fisher RA (1937) The design of experiments, 2nd edn. Oliver and Boyd, Edinburgh, 260 pp

Fisher RA (1951) Statistical methods for research workers, 11th edn. Oliver and Boyd Edinburgh

Förster S (1972) Durchlässigkeitsuntersuchungen an Salzgesteinen. Dissertation Bergakademie Freiberg, FRG

Fortin V, Bernier J, Bobée B (1997) Simulation, Bayes, and Bootstrap in statistical hydrology. Water Resources Res 33:439–448

Francis JGF (1961) The QR transformation Part I. Comput J 4:265–271

Francis JGF (1962) The QR transformation Part II. Comput J 4:332–345

Frank IE, Friedman JH (1993) A statistical view of some chemometric regression tools. Technometrics 35:109–135

Frank IE, Friedman JH (1993) Response. Technometrics 35:143–148

Fredlund DG, Dahlmann AE (1972) Statistical geotechnical proper-ties of glacial lake Edmonton sediments. Statistics and probability in civil engineering, Hong Kong International Conference, Hong Kong University Press, distributed by Oxford University Press, London

Freeze RA, Cherry JA (1979) Groundwater. Prentice Hall, New York

Friedrichs KO (1954) Symmetric hyperbolic linear differential equations. Comm Pure Appl Math 7:345–392

Fröhlich H, Hohentanner CR, Förster F (1995) Bestimmung des Diffusions- und Perme-abilitätsverhaltens von Wasserstoff in kompaktiertem Salzgrus. Abschlussbericht der BMBF-Forschungsvorhaben 02 E 849, 3 und 02 E 846, 9 Karlsruhe, FRG

Fuentes-Arderiu X (2006) True values may be known in certain cases. Accred Qual Assur 11:29

GAEA (2006) www.gaea.ca (last accessed August 2006)

Garthwaite PH, Jolliffe IT, Jones B (2002) Statistical inference. Oxford University Press, Oxford

Garvin D, Parker VB, White HJ (eds) (1987) CODATA thermodynamic tables: Selections for some compounds of calcium and related mixtures – a prototype set of tables. Springer Berlin, 356 pp

Gauss CF (1809) Theoria motus corporum coelestium in sectionibus conicis solem ambi-entium. Perthet, Hamburg

Gautschi A (2001) Hydrogeology of a fractured shale (Opalinus clay): Implications for deep geological disposal of radioactive wastes. Hydrogeol J 9:97–107

Gilbert PA (1992) Effects of Sampling disturbance on laboratory measured soil properties; final report. Miscellaneous Paper GL-92–35 US Army Engineer Waterways Experiment Station Geotechnical Laboratory. Vickburg MS, USA

GLA (1994) Ergiebigkeitsuntersuchungen in Festgesteinsaquiferen. Geologisches Lan-desamt Baden-Württemberg, Karlsruhe, FRG

Gleason JR (1988) Algorithms for balanced bootstrap simulations. Am Stat 42:263–266

Gloth H (1980) Entwicklung eines Triaxialgerätes für hohe Drücke und Durchführung von Festigkeitsuntersuchungen an Gesteinen unter besonderer Berücksichtigung des Ein-flusses von Porenraumdrücken, Freiberger Forschungsheft FFH A 607, TU BA, Freiberg, FRG

Gluschke M, Woitke P, Wellmitz J, Lepom P (2004) Sieving of sediments for subsequent analysis of metal pollution: results of a German interlaboratory study. Accred Qual Assur 9:624–628

Goldberg S (1991) Sensitivity of surface complexation modeling to the surface site density parameter. J Colloid Interface Sci 145:1–9

Golub GH, Reinsch C (1970) Singular value decomposition and least squares solutions. Numer Math 14:403–420

Golze M (2003) Why do we need traceability and uncertainty evaluation of measurement and test results? Accred Qual Assur 8:539–540

Gommlich G, Yamaranci U (1993) Untersuchung zur Schallgeschwindigkeit des Steinsalzes der Asse in-situ und im Labor, GSF-Bericht 22, 93 Braunschweig, FRG

Grader AS, Ramey HJ (1988) Slug test analysis in double-porosity reservoirs. SPE Formation Evaluation, June 1988, pp 329–339

Grauer R (1997) Solubility limitations: An "old timer's view". In: Modelling in aquatic chemistry. NEA, OECD Gif-sur-Yvette, France, 131–152

Guedj D (2000) Le mètre du monde. Editions du Seuil, Paris, France

Guimera J, Carrera J (2000) A comparision of hydraulic and transport parameters measured in low-permeability fractured media. J Cont Hydrol 41:261–281

Gutenberg Projekt (2006) http://www.gutenberg.org/etext/7781 (last visited September 2006)

Häfner F (1985): Geohydrodynamische Erkundung von Erdöl-, Erdgas-, und Grundwasser-lagerstätten. WTI des ZGI, 26, 1, Berlin, FRG

Häfner F, Boy S, Wagner S, Behr A, Piskarev V, Palatnik B (1997) The "front-limitation" algorithm. A new and fast finite-difference method for groundwater pollution problems. J Contam Hydrol 27:43–61

Hahn O (1962) Die falschen Transurane. Zur Geschichte eines wissenschaftlichen Irrtums. Naturwiss Rundschau 15:43

Hamed MM, Bedient PB (1999) Reliability-based uncertainty analysis of groundwater contaminant transport and remediation. REPORT EPA, 600, R-99, US Environmental Protection Agency Cincinnati, 71 pp

Hamer WJ, Wu YC (1972) Osmotic coefficients and mean activity coefficients of uni-univalent electrolytes in water at 25 °C. J Phys Chem Ref Data 1:1047–1099

Hampel FR, Ronchetti, Rousseeuw P, Stahel WA (1986) Robust statistics. Wiley, New York

Hammitt GM (1966) Statistical Analysis of Data from a Comparative Laboratory Test Program Sponsored by ACIL. Miscellaneous Paper 4–785. US Army Engineering Waterways Experiment Station, Corps of Engineers

Hardyanto W, Merkel B (2006) Introducing probability and uncertainty in groundwater modeling with FEMWATER-LHS. J Hydrol doi:10.1016, j.jhydrol.2006.06.035

Harel D (2000) Computers Ltd: what they really can't do. Oxford University Press, Oxford, 207 pp

Harned HS, Scholes Jr SR (1941) The ionization constant of HCO_3^- from 0 to 50 °C. J Am Chem Soc 63:1706

Harned HS, Davies Jr R (1943) The ionization constont of carbonic acid in water and the solubility of carbon dioxide in water and aqueous salt solutions from 0 to 50 °C. J Am Chem Soc 65:2030

Harned HS, Bonner FT (1945) The first ionization constant of carbonic acid in aqueous solution of sodium chloride. J Am Chem Soc 67:1027

Hartley FR, Burgess C, Alcock RM (1980) Solution equilibria. Ellis Horwood, Chichester, 358 pp

Harr ME (1987) Reliability based design in civil engineering. McGraw Hill, New York

Hartley J (1998) NAPSAC (release 4.1), Technical summary document, AEA Technology Harwell

Hartley FR, Burgess C, Alcock RM (1980) Solution equilibria. Ellis Horwood, Chichester, 361 pp

Hässelbarth W (1998) Uncertainty – the key concept of metrology in chemistry. Accred Qual Assur 3:115–116

Hässelbarth W, Bremser W (2004) Correlation between repeated measurements: bane and boon of the GUM approach to the uncertainty of measurement. Accred Qual Assur 9:597–600

Hastie T, Mallows C (1993) Discussion of a paper by Frank and Friedman. Technometrics 35:140–143

Hayashi K, Ito T, Abe H (1987) A new method for the determination of in-situ hydraulic properties by pressure pulse test and application to the Higashi Hachimantai geothermal field J Geophys Res 92:9168–9174

Heitfeld KH (1984) Ingenieurgeologie im Talsperrenbau. In: Bender F (Hrsg.) Angewandte Geowissenschaften Vol. 3 Enke-Verlag Stuttgart, FRG

Heitfeld M (1998) Auffüll- und Absenkversuche in Bohrungen, In: Handbuch zur Erkundung des Untergrundes von Deponien und Altlasten, Vol. 4–Geotechnik, Hydrogeologie; BGR (Bundesanstalt für Geowissenschaften und Rohstoffe), Springer, Heidelberg

Hellmuth KH, Klobes P, Meyer K, Röhl-Kuhn B, Sitari-Kauppi M, Hartikainen J, Hartikainen K, Timonen J (1995) Matrix retardation studies: size and structure of the accessable pore space in fresh and altered crystalline rock. Z Geol Wiss 23:691–706

Helton JC (1994) Treatment of uncertainty in performance assessments for complex systems. Risk Anal 14:483–511

Hermann M (2004) Numerik gewöhnlicher Differentialgleichungen, Anfangs- und Rand-
    wertprobleme. Oldenbourg Verlag München, FRG
Hibbert DB (2001) Compliance of analytical results with regulatory or specification limits:
    a probabilistic approach. Accred Qual Assur 6:346–251
Hibbert DB (2005) Further comments on the (miss-)use of r for testing the linearity of
    calibration functions. Accred Qual Assur 10:300–301
Hibbert DB (2006) Metrological traceability: I make 42; you make it 42; but is it the same
    42? Accred Qual Assur 11:543–549
Hiemstra T, van Riemdijk WH, Bolt GH (1989) Multisite proton adsorption modeling at the
    solid, solution interface of (hydr)oxides: A new approach. J Colloid Interface Sci 1:133
Hirano Y, Imai K, Yasuda K (2005) Uncertainty of atomic absorption spectrometer. Accred
    Qual Assur 10:190–196
Hoffman FO, Hammonds JS (1994) Propagation of uncertainty in risk assessments: the
    need to distinguish betwen uncertainty due to lack of knowledge and uncertainty due to
    variability. Risk Anal 14:707–712
Holland TJB, Powell R (1990) An enlarged and updated internally consistent thermodynamic
    dataset with uncertainties and correlations: the system K_2O-Na_2O-CaO-MgO-MnO-FeO-
    Fe_2O_3-Al_2O_3-TiO_2-SiO_2-C-H_2-O_2. J Metamorph Geol 8:89–124
Holmgren M, Svensson T, Johnson E, Johansson K (2005) Reflections regarding uncertainty
    of measurement, on result s of a Nordic fatigue test interlaboratory comparison. Accred
    Qual Assur 10:208–213
Hölting B (1995) Hydrogeologie. Einführung in die allgemeine und angewandte Hyd-
    rogeologie, Enke-Verlag, Stuttgart
Holzbecher E (1996) Modellierung dynamischer Prozesse in der Hydrologie. Springer Berlin,
    210 pp
Hopke PK (1989) Target transformation factor analysis. Chemom Intell Lab Syst 6:7–19
Hubbard S, Rubin Y (2000) Hydrogeological parameter estimation using geophysical data:
    a review of selected techniques. J Contam Hydrol 45:3–34
Hubbert MK (1956) Darcy law and the field equations for the flow of underground fluids.
    Trans Am Inst Metal Ing 207:222–239
Huber W (2004) On the use of the correlation coefficient r for testing the linearity of
    calibration functions. Accred Qual Assur 9:726
Hummel W (2000) "Comments On the influence of carbonate in mineral dissolution:1.
    The thermodynamics and kinetics of hematite dissolution in bicarbonate solutions at
    T=25 °C" by J. Bruno, W. Stumm, P. Wersin, and F. Brandberg. Geochim Cosmochim
    Acta 64:2167–2171
Hummel W, Berner U, Curti E, Pearson FJ, Thoenen T (2006) Chemical thermodynamic
    database 01, 01. Universal Publishers, New York, 589 pp
Huston R, Butler JN (1969) Activity measurements in concentrated sodium chloride-
    potassium chloride electrolytes using cation-sensitive galss electrodes. Anal Chem
    41:1695–1698
Imam RL, Conover WJ (1980) Small sample sensitivity analysis techniques for computer
    models with application to risk assessment. Commun Stat A9:1749–1842
Imam RL, Helton JC (1988) An investigation of uncertainty and sensitivity analysis tech-
    niques for computer models. Risk Analysis 8:71–90
IRMM (2000) Trade relations 1999 between the EU and the USA: News from "The New
    [1995] Transatlantic Agenda". Accred Qual Assur 5:74
ISO (1993) Guide to the expression of uncertainty in measurement. ISO Geneva, CH

ISO (1999) General requirements for the competence of testing and calibration laboratories. ISO 17025. ISO Geneva, CH

IT Corporation (1996a) Underground test area subproject Phase I: Data analysis taskbreak (vol. I)–Regional geologic model data documentation package: REPORT ITLV, 10972-181 Las Vegas, USA

IT Corporation (1996b) Underground test area subproject Phase I: Data analysis taskbreak (vol. IV)–Hydrologic parameter data documentation package. Report ITLV, 10972-181 Las Vegas, USA

IUPAC (1983) (Working Party on pH: Covington AK, Bates RG, Durst RA) Definition of scales, standard reference values, measurement of pH and related terminology. Pure Appl Chem 55:1467

IUPAC (1985) (Working Party on pH: Covington AK, Bates RG, Durst RA) Pure Appl Chem 57:531–542

IUPAC (2002) (Working Party on pH: Buck RP, Rondinini S, Baucke FGK, Brett CMA, Camões MF, Covington AK, Milton MJT, Mussini T, Naumann R, Pratt KW, Spitzer P, Wilson GS) The measurement of pH definitions, standards and procedures. Pure Appl Chem 74:2169–2200

IUPAC WP (2006) Meinrath G, Camoes MF, Spitzer P, Bückler H, Mariassy M, Pratt K, Rivier C Traceability of pH in a metrological context. In: Combining and Reporting Analytical Results (Fajgelj, Belli, Sansone eds) RSC Publishing London/UK

Jacobi CGJ (1846) Über ein leichtes Verfahren, die in der Theorie der Säkularstörungen vorkommenden Gleichungen leicht zu lösen. Crelles J 30:297–306

Jakobsson MA (1999) Measurement and modelling using surface complexation of cation (II to VI) sorption onto mineral oxides. Department of Nuclear Chemistry Chalmers University of Technology Göteborg

Jakobsson AM, Rundberg RS (1997) Uranyl sorption onto alumina. Mat Res Soc Symp Proc 465:797–804

Jaquet O, Tauzin E, Resele G, Preuss J, Eilers G (1998a) Uncertainity analysis of permeability measurements in salt. In: International conference on radioactive waste disposal Hamburg, FRG, pp 149–155

Jaquet O, Schindler M, Voborny O, Vinard P (1998b) Uncertainty assessment using geostatistical simulation and groundwater flow modeling at the Wellenberg site. In: International conference on radioactive waste disposal. Hamburg, FRG 507

Jaynes ET (2003) Probability theory – the logic of science. Cambridge University Press Cambridge, USA

Jenks PJ (2003) Accreditation but to which standard? Accred Qual Assur 8:428

Jenks PJ (2004) RMs and regulation putting the cart before the horse? Spect Europe p:31 JESS (2006) Joint Expert Speciation System. http://jess.murdoch.edu.au (last accessed 12 August 2006)

Johnson JW, Oelkers EH, Helgeson HC (1992) SUPCRT92: a software package for calculating the standard molar thermodynamic properties of minerals, gases, aqueous species, and reactions from 1 to 5000 bar and 0 to 1000 °C. Comput Geosci 18:899–947

Kalin M, Fyson A, Meinrath G (2005) Observations on the groundwater chemistry in and below a pyritic tailings pile and microbiology. Proc IMWA Conference, Oviedo, July 5–7, 2005 Kammerer WJ, Nashed MZ (1972) On the convergence of the Conjugate Gradient Method for singular linear operation equations. Numer Anal 9:165–181

Kaplan S, Garrick BJ (1981) On the quantitative definition of risk. Risk Anal 1:11–27

Kappei G, Schmidt MW (1983) Untersuchungen über das Tragverhalten von Salzschüttkörpern und deren Permeabilität, Vortrag anläßlich des Informationsseminars "Dammbau im Salzgebirge", DBE Peine, FRG

Karasaki K, Long JCS, Witherspoon PA (1988) Analytical models of slug tests. Water Resources Res 24:115–126

Kato Y, Meinrath G, Kimura T, Yoshida Z (1994) A study of U(VI) hydrolysis and carbonate complexation by time-resolved laser-induced fluorescence spectroscopy (TRLFS). Radiochim Acta 64:107–111

Kaus R (1998) Detection limits and quantitation limits in the view of international harmonization and the consequences for analytical laboratories. Accred Qual Assur 3:150–154

Källgren H, Lauwaars M, Magnusson B, Pendril L, Taylor Ph (2003) Role of measurement uncertainty in conformity assessment in legal metrology and trade. Accred Qual Assur 8:541–547

KCDB (2005) Key Comparison Data Base. (http://www.bipm.org/KCDB/)

Kellner R, Mermet JM, Otto M, Widmer HM (1998) Analytical chemistry. Wiley-VCH Weinheim

Keynes JM (1921) Treatise on probability. MacMillan New York, 466 pp

Kenyon WE (1992) Nuclear magnetic resonance as a petrophysical measurement. Int J Radiat Appl Inst E6:153–171

Kharaka YK (1988) SOLMINEQ.88, a computer program for geochemical modeling of water-rock interactions. US Geological Survey, Menlo Park, Denver, USA

Kim H-T, Frederick WJ Jr (1988) Evaluation of Pitzer interaction parameters of aqueous mixed electrolyte solutions at 25 °C. 2. Ternary mixing parameters. J Chem Eng Data 33:278–283

Kinzelbach W (1986) Groundwater modelling. Elsevier, Amsterdam

Kinzelbach W (1989) Numerical groundwater quality modelling: Theoretical basis and practical application. In: Proceedings of International Workshop on Appropriate Methologies for Development and Management of Groundwater Resources in Developing Countries, Volume III, Oxford and IBH Publishing Co. Pvt. Ltd., New Delhi, pp 223–241

Kleijnen JPC (1995) Verification and validation of simulation models. Eur J Op Res 82:145–162

Klemes V (1986) Dilletantism in hydrology: transition or destiny? Water Resources Res 22:177S–188

Klingebiel G, Sobott R (1992) Permeabilitätsmessungen an zylindrischen Proben aus gepresstem Salzgrus, Zentraltechnikum Preussag, Berghöpen, FRG

Knowles MK, Borns D, Fredrich J, Holcomb D, Oprice R, Zeuch D, Dale T, Van Pelt RS (1998) Testing the disturbed zone around a rigid inclusion in salt. Proc 4th Conf Mechanical Behaviour of Salt (SALT IV), Montreal

Koh KK (1969) Frequency response characteristics of flow of gas through porous media, Chem Eng Sci 24:1191

Königsberger E (1991) Improvement of excess parameters from thermodynamic and phase diagram data by employing Bayesian excess parameter estimation. CALPHAD 15:69–78

Kolassa JE (1991) Confidence intervals for thermodynamic constants. Geochim Cosmochim Acta 55:3543–3552

Kolb M, Hippich S (2005) Uncertainty in chemical analysis for the example of determination of caffeine in coffee. Accred Qual Assur 10:214–218

Kosmulski M (1997) Adsorption of trivalent cations on silica. J Coll Interface Sci 195:395–403

Kozeny J (1927) Über kapillare Leitung des Wassers im Boden. Sitzungsber Akad Wiss Wien, A 136:271–306

Kraemer CA, Hankins JB, Mohrbacher CJ (1990) Selection of single well hydraulic test methods for monitorings wells. Groundwater and vadose zone monitoring ASTM STP 1053. (Nielsen, Johnson eds) pp 125–137. ASTM, Philadelphian

Kragten J (1994) Calculating standard deviations and confidence intervals with a universally applicable spreadsheet technique. Analyst 119:2161–2165

Krause P, Boyle DP, Bäse F (2005) Comparison of different efficiency criteria for hydrological model assessment. Adv Geosci 5:89–97

Kufelnicki A, Lis S, Meinrath G (2005) Application of cause-and-effect analysis to potentiometric titration. Anal Bioanal Chem 382:1652–1661

Kunzmann H, Pfeifer T, Schmitt R, Schwenke H, Weckenmann A (2005) Productive metrology – adding value to manufacture. Annals CIRP 54:691–705

Kuselman I (2004) CITAC mission, objectives and strategies. Accred Qual Assur 9:172

Künsch HR (1989) The jackknife and the bootstrap for general stationary observations. Ann Stat 17:1217–1241

Lavanchy JM, Croise J, Tauzin E, Eilers G (1998) Hydraulic testing in low permeability formations – test design, analysis procedure and tools application from site characterization programmes. In: International Conference On Radioactive Waste Disposal, Hamburg, pp 139–148

Lawson CL, Hanson RJ (1974) Solving least squares problems. Prentice-Hall, Englewood Cliffs, 340 pp

Legchenko AV, Shushakov OA, Perrin JA, Portselan AA (1995) Noninvasive NMR study of subsurface aquifers in France. Proc of 65th Ann Meet Soc Exploration Geophysicists, pp 365–367

Legendre AM (1805) Nouvelles méthodes pour la détermination des orbites des comètes. "Sur la méthode des moindres quarrés" appears as an appendix

Lerche I (2000) The worth of resolving uncertainty for environmental projects: does one add value or just waste money? Env Geosci 7:203–207

Lewis D (1969) Convention. A philosophical study. MIT Press, Cambridge, MA

Li W, Mac Hyman J (2004) Computer arithmetics for probability distribution variables. Rel Eng System Safety 85:191–209

Libes SM (1999) Learning quality assurance, quality control using US EPA techniques. J Chem Educ 76:1642–1648

Liedtke L (1984) Standsicherheitskriterien für das Endlagerbergwerk Gorleben (PSE II), BMFT-Forschungsvorhaben KWA 51062 Bundesanstalt für Geowissenschaften und Rohstoffe, Hannover

Lindner H, Pretschner C (1989) Bohrlochgeophysikalische Bestimmung von Porosität und elastischen Modulen im Festgestein. In: Eigenschaften von Gesteinen und Böden, Freiberger Forschungsheft A 849, TU BA Freiberg, FRG

Lindner K (2002) Carl Friedrich von Weizsäcker-Wanderungen ins Atomzeitalter. Mentis Paderborn, FRG

Lippmann J (1998) Isotopenhydrologische Untersuchungen zum Wasser- und Stofftransport im Gebiet des ERA Moralleben, Dissertation Universität Heidelberg, FRG

Lisy JM, Cholvadova A, Kutej J (1990) Multiple straight-line least-squares analysis with uncertainty in all variables. Computers Chem 14:189–192

Luckner L, Schestakow WM (1976): Modelirovanie geofiltracii. Izd. nedra, Moskau, 467 S

Ludewig M (1965) Die Gültigkeitsgrenzen des Darcyschen Gesetzes bei Sanden und Kiesen. Wasserwirtschaft-Wassertechnik 15:415–421

Magnusson B, Näykki T, Hovind H, Krysell M (2004) Handbook for calculation of measurement uncertainty in environmental laboratories. Nordtest Espoo, FIN (http://www.nordtest.org/register/techn/tlibrary/tec537.pdf)

Maier-Leibnitz H (1989) Akademische Ethik und Abwägen als Hilfsmittel der Entscheidungsfindung. In: Noelle-Neumann, Maier-Leibnitz: Zweifel am Verstand. Edition Interform Zürich, pp 65–100

Malinowski ER (1991) Factor analysis in chemistry. Wiley, New York, 351 pp

Mandel J, Linnig FJ (1957) Study of accuracy in chemical analysis using linear calibration curves. Anal Chem 29:743–749

Marquard DW (1963) An algorithm for least-squares estimation of non-linear parameters. J Soc Indust Appl Math 11:413

Marsaglia G, Tsang WT (1984) A fast, easily implemented method for sampling from decreasing or symmetric unimodal density functions. SIAM J Sci Stat Comp 5:349–359

Marschal A (2004) Measurement uncertainties and specified limits: what is logical or common sense in chemical measurement. Accred Qual Assur 9:642–643

Martell AE, Motekaitis RJ (1992) Determination and use of stability constants. 2nd edn. VCH Weinheim, FRG

Martell AE, Smith RM, Motekaitis RJ (2003) NIST critically selected stability constants of metal complexes Ver 7 NIST Gaithersburg

Marshall SL, May PM, Hefter GT (1985) Least-squares analysis of osmotic coefficient data at 25 °C according to Pitzer's equation 1. 1:1 electrolytes. J Chem Eng Data 40:1041–1052

Massey FJ Jr (1951) The Kolmogorov–Smirnov test for goodness of fit. J Am Stat Soc 46:68–78

Mateen K, Ramey HJ (1984) Slug test data analysis in reservoirs with double porosity behaviour. Paper PSE 12779:459–468

Mathews JH (1989) Symbolic computational algebra applied to Picard iteration. Math Comp Educ J 23:117–122

Mavko G, Nur A (1997) The effect of a percolation threshold in the Kozeny–Carman relation. Geophysics 62:1480–1482

May PM, Murray K (1991) JESS, A joint expert speciation system – I. Raison d'être. Talanta 38:1409–1417

May PM, Murray K (1991a) JESS, A joint expert speciation system – II: The thermodynamic database. Talanta 38:1419–1426

May PM, Murray K (1993) JESS, A joint expert speciation system – III: Surrogate functions. Talanta 40:819–825

May PM, Murray K (2001) Database of chemical reactions designed to achieve thermodynamic consistency automatically. J Chem Eng Data 46:1035–1040

Mazurek M, Lanyon GW, Vomvoris S, Gautschi A (1998) Derivation and application of a geologic dataset for flow modelling by discrete fracture networks in low-permeability argillaceous rocks. J Contam Hydrol 35:1–17

McDonald MG, Harbaugh Aw (1988) MODFLOW – a modular three-dimensional finite-difference groundwater flow model. Water-Resources Investigations Report. US Geological Survey, Menlo Park, Denver, USA

McKay MD, Conover WJ, Beckman RJ (1979) A comparison of three methods for selecting values of input variables in the analysis of output from computer code. Technometrics 21:239–245

Medawar PB (1990) Is the scientific paper fraudulent? Yes; it misrepresents scientific thought. Saturday Review, 1 August 1964, pp. 42–43. Reprinted in: The threat and the glory: reflections on science and scientists. Harper and Collins, New York

Medawar PB (1979) Advice to a young scientist. Basic Books New York, 109 pp

Meinrath G (1997) Chemometric and statistical analysis of uranium(VI) hydrolysis at elevated U(VI) concentrations. Radiochim Acta 77:221–234

Meinrath G (1997a) Uranium(VI) speciation by spectroscopy. J Radioanal Nucl Chem 224:119–126

Meinrath G (1998) Direct spectroscopic speciation of schoepite-aqueous phase equilibria. J Radioanal Nucl Chem 232:179–188

Meinrath G (2000) Robust spectral analysis by moving block bootstrap designs. Anal Chim Acta 415:105–115

Meinrath G (2000a) Comparability of thermodynamic data – a metrological point of view. Fresenius J Anal Chem 368:574–584

Meinrath G (2000b) Computer-intensive methods for uncertainty estimation in complex situations. Chemom Intell Lab Syst 51:175–187

Meinrath G (2001) Measurement uncertainty of thermodynamic data. Fresenius J Anal Chem 369:690–697

Meinrath G (2002) The merits of true values. Accred Qual Assur 7:169–170

Meinrath (2002b) Extended traceability of pH: an evaluation of the role of Pitzer's equations. Anal Bio Anal Chem 374:796–805

Meinrath G, Kalin M (2005) The role of metrology in making chemistry sustainable. Accred Qual Assur 10:327–337

Meinrath G, Kimura T (1993) Carbonate complexation of the uranyl(VI) ion. J Alloy Comp 202:89–93

Meinrath G, Lis S (2001) Quantitative resolution of spectroscopic systems using computer-assisted target factor analysis. Fresenius J Anal Chem 369:124–133

Meinrath G, Lis S (2002) Application of cause-and-effect diagrams to the interpretation of UV-Vis spectroscopic data. Anal Bioanal Chem 372:333–340

Meinrath G, May PM (2002) Thermodynamic prediction in the mine water environment. Mine Water Environ 21:24–35

Meinrath G, Nitzsche O (2000) Impact of measurement uncertainty in chemical quantities on environmental prognosis by geochemical transport modelling. Isotopes Environ Health 36:195–210

Meinrath G, Schweinberger M (1996) Hydrolysis of the uranyl(VI) ion – a chemometric approach. Radiochim Acta 75:205–210

Meinrath G, Spitzer P (2000) Uncertainties in determination of pH. Mikrochim Acta 135: 155–168

Meinrath G, Kato Y, Kimura T, Yoshida Z (1996) Solid-aqueous phase equilibria of Uranium(VI) under ambient conditions. Radiochim Acta 75:159–167

Meinrath G, Klenze R, Kim J (1996a) Direct spectroscopic speciation of uranium(VI) in carbonate solutions. Radiochim Acta 74:81–86

Meinrath G, Kato Y, Kimura T, Yoshida Z (1998) Stokes relationship in absorption and fluorescence spectra of U(VI) species. Radiochim Acta 82:115–120

Meinrath G, Helling C, Volke P, Dudel EG, Merkel B (1999) Determination and interpretation of environmental water samples contaminated by uranium mining activities. Fresenius J Anal Chem 364:191–202

Meinrath G, Kato Y, Kimura T, Yoshida Z (1999a) Comparative analysis of actinide(VI) carbonate complexation by Monte Carlo resampling methods. Radiochim Acta 84:21–29

Meinrath G, Ekberg C, Landgren A, Liljenzin JO (2000) Assessment of uncertainty in parameter evaluation and prediction. Talanta 51:231–246

Meinrath G, Hnatejko Z, Lis S (2004) Threshold bootstrap target factor analysis study of neodymium with pyridine 2,4 dicarboxylic acid N-oxide – an investigation of traceability. Talanta 63:287–296

Meinrath G, Hurst S, Gatzweiler G (2000a) Aggravation of licencing procedures by doubtful thermodynamic data. Fresenius J Anal Chem 368:561–566

Meinrath G, Merkel B, Ödegaard-Jensen A, Ekberg C (2004a) Sorption of iron on surfaces: modelling, data evaluation and measurement uncertainty. Acta Hydrochim Hydrobiol 32:154–160

Meinrath G, Lis S, Piskua Z, Glatty Z (2006) An application of the total measurement uncertainty budget concept to the thermodynamic data of uranyl(VI) complexation by sulfate. J Chem Thermodyn 38:1274–1284

Meinrath G, Lis S, Böhme U (2006a) Quantitative evaluation of Ln(III) pyridine N-oxide carboxylic acid spectra under chemometric and metrological aspects. J Alloy Comp 408–412:962–969

Meinrath G, Kufelnicki A, Ewitek M (2006b) Approach to accuracy assessment of the glass-electrode potentiometric determination of acid-base properties. Accred Qual Assur 10:494–500

Meis T, Marcowitz U (1978) Numerische Behandlung partieller Differential-Gleichungen. Springer Berlin, FRG, 452 pp

Melchers RE (1987) Structural liability analysis and prediction. Ellis Horwood Ltd, Chichester, UK

Merkel B, Planer-Friedrich B (2005) Groundwater geochemistry. Springer Verlag, Berlin, 200 pp

Metcalf RC (1987) Accuracy of ROSS pH combination electrodes in dilute sulphuric acid standards Analyst:1573–1577

Metropolis N, Rosenbluth AW, Rosenbluth MN, Teller AH, Teller E (1953) Equation of state calculations by fast computing machines. J Chem Phys 21:1087–1092

Miehe R, Harborth B, Klarr K, Ostrowski L (1994) Permeabilitätsbestimmungen im Staßfurt-Steinsalz in Abhängigkeit von einer Streckenauffahrung. Kali Steinsalz, vol. 11

Miller JC, Miller JN (1993) Statistics for analytical chemistry. Ellis Horwood Chichester, UK

Morshed J, Kaluarachichi JJ (1998) Application of artificial neural network and genetic algorithm in flow and transport simulations, Adv Water Resources Res 22:145–158

MRA (1999) Reconnaissance mutuelle. BIPM Sèvres, France

Müller-Lyda I, Birthler H, Fein E (1998) A permeability-porosity relation for crushed rock salt derived from laboratory data for application within probabilistic long-term safety analysis. In: International conference on radioactive waste disposal Hamburg, FRG, pp 553–558

Mualem Y (1976) A new model predicting the hydraulic conductivity of unsaturated porous media. Water Resources Res 12:513–522

Nagel E (1939) Principles of the theory of probability. In: International encyclopedia of unified science. Vol. 1 No. 6. University of Chicago Press, Chicago

Narasimhan TN, Witherspoon PA (1976) An integrated finite difference method for analyzing fluid flow in porous media.Water Resources Res 12:57–64

Narasimhan TN, Witherspoon PA (1977) Numerical model for saturated-unsaturated flow in deformable porous media. 1. Theory. Water Resources Res 13:657–664

Narasimhan TN, Witherspoon PA, Edwards AL (1978) Numerical model for saturated-unsaturated flow in deformable porous media. 2. The algorithm, Water Resources Res 14:255–260

Narasimhan TN, Witherspoon PA (1978) Numerical model for saturated-unsaturated flow in deformable porous media. 3. Applications. Water Resources Res 14:1017–1034

Nash JC (1981) Compact numerical methods for computers. Adam Hilger, Bristol, 314 pp

Naumann R, Alexander-Weber Ch, Baucke FGK (1994) The standardization of pH measurement. Fresenius J Anal Chem 349:603–606

Naumann R, Alexander-Weber Ch, Eberhardt R, Giera J, Spitzer P (2002) Traceability of pH measurements by glass electrode cells: performance characteristic of pH electrodes by multi-point calibration. Anal Bioanal Chem 374:778–786

Neidhardt B, Mummenhoff W, Schmolke A, Beaven P (1998) Analytical and legal aspects of the threshold limit value concept. Accred Qual Assur 3:44–50

Nelder JA, Mead R (1965) A simplex method for function minimization. Computer J 7: 308–313

Nelson PH (1994) Permeability-porosity relationships in sedimentary rocks. Log Analyst 3:38–62

Neuzil CE (1994) How permeable are clays and shales? Adv Water Resources Res 30:145–150

Nguyen V, Pinder GF (1984) Direct calculation of aquifer parameters in slug test analysis. In: Rosenheim J, Bennet GD (eds): Groundwater hydraulics. Water Resources Monograph 9:222–239

Nielsen DR, Biggar JW, Erh KT (1973) Spacial Variability of Field-Measured Soil-Water Properties, J Agr Sci 42:215–260

Nitzsche O (1997) TReAC-Modell zur Beschreibung des reaktiven Stofftransports im Grundwasser. Wissenschaftliche Mitteilungen Institut für Geologie, Technische Universität Bergakademie Freiberg Freiberg, vol. 1, 135 pp

Nitzsche O, Meinrath G, Merkel B (2000) Database uncertainty as a limiting factor in reactive transport prognosis. J Contam Hydrol 44:223–237

Novakowski KS (1989) Analysis of pulse interference tests. Water Resources Res 25:2377–2387

NTB 93–47 (1993) Marschall P, Vomvoris S (eds) Grimsel Test Site: Developments in hydrotesting, fluid logging and combinded salt, head tracer experiments in the BK site (Phase III), NAGRA Report Villigen, CH

NTB 94–02 (1994) Liedke L, Götschenberg A, Jobmann M, Siemering W – BGR Hannover (1994) Felslabor Grimsel – Bohrlochkranzversuch: Experimentelle und numerische Untersuchungen zum Stofftransport in geklüftetem Fels, NAGRA Report Villigen, CH

NTB 94–21 (1994) Jacob A, Hadermann J INTRAVAL Finsjön test: Modeling results for some tracer experiments, NAGRA Report Villigen, CH

NTB 01–03 (2001) Mont Terri Underground Rock Laboratory – RA Experiment: Rock mechanics analyses and synthesis: Conceptual model of the Opalinus Clay. Mont Terri Project, Technical Report 2001–03. 6, 37, NAGRA Report Villigen, CH

NUREG (2006) Modeling adsorption processes: Issues in uncertainty, scaling, and prediction (Criscenti, Eliassi, Cygan, Jové Cólón, eds) US Nuclear Regulatory Commission Washington DC, USA

Oberkampf WL, Helton JC, Joslyn CA, Wojtkiewicz SF, Ferson S (2004) Challenge problems: uncertainty in system responses given uncertain parameters. Reliability Eng Sys Safety 85:11–19

Ödegaard-Jensen A, Ekberg C, Meinrath G (2003) The effect of uncertainties in stability constants on speciation diagrams. Mat Res Soc Symp Proc. 757:509–514

Ödegaard-Jensen A, Ekberg C, Meinrath G (2004) LJUNGSKILE: a program for assessing uncertainties in speciation calculations. Talanta 63:907–916

Ödegaard-Jensen A (2006) Uncertainties in chemical speciation calculations. Licentiate Thesis. Chalmers University of Technology Göteborg, 55 pp

OECD (1999) Quality assurance and GLP REPORT ENV, JM, MONO(99)20. OECD Paris, F

OECD, NEA (1991) Chemical Thermodynamics of uranium. NEA Paris, France

OECD, NEA (2003) Chemical Thermodynamics Vol. 5. Elsevier, Amsterdam, 919 pp

OECD, NEA (2000a) TDB-1: guidelines for the review procedure and data selection. OECD, NEA Issy-les-Moulineaux, France

OECD, NEA (2000b) TDB-3: Guidelines for the assignment of uncertainties. OECD, NEA Issy-les-Moulineaux, France

OECD, NEA (2000c) TDB-6: guidelines for the independent peer review of tdb reports. OECD, NEA Issy-les-Moulineaux, France

Malow G, Offermann P, Haaker RF, Müller R, Schubert P (1983) Characterization and comparison of high activity waste products, WAS-322-83-53-D (B), Arbeitskreis HAW-Produkte REPORT FZKA 6651 Forschungszentrum Karlsruhe, Karlsruhe, FRG

Olbricht W, Chatterjee ND, Miller K (1994) Bayes estimation: A novel approach to Derivation of internally consistent thermodynamic data for minerals, their uncertainties, and correlations. Part I: Theory. Phys Chem Minerals 21:36–49

Onysko SJ, McNearny RL (1997) GIBBTEQ: A MINTEQA2 thermodynamic data error detection program. Groundwater 35:912–914

Oreskes N, Shrader-Frechette K, Belitz K (1994) Verification, validation, and confirmation of numerical models in the earth sciences. Science 263:641–646

Oreskes N, Shrader-Frechette K, Belitz K (1994a) Comments and response (with letters to the editor). Science 264, 329–331

Osborne Ch (1991) Statistical calibration. Int Stat Review 59:309–336

Osenbrück K (1996) Alter und Dynamik tiefer Grundwässer – eine neue Methode zur Analyse der Edelgase im Porenwasser von Gesteinen, Dissertation Universität Heidelberg, FRG

Osterc A, Stibilj V (2005) Measurement uncertainty of iodine determination in radiochemical neutron activation analysis. Accred Qual Assur 10:235–240

Pahl J (1995) Die Bestimmung extrem niedriger Permeabilitäten an Steinsalzkernen mittels Gaschromatographie. Diplomarbeit, Institut für Bohrtechnik und Fluidbergbau, TU BA Freiberg, FRG

Palmer W (1989) Über die Partikelgröße und Konzentration des Eisens(III) in Flußwässern. REPORT KfK 4289. Kernforschungszentrum Karlsruhe, Karlsruhe, FRG, 102 pp

Papadakis I, Taylor DPD (2001) Metrological value in participation in interlaboratory comparisons. Accred Qual Assur 6:466–468

Papdakis I, van Nevel L, Vendelbo E, Norgaard J, Taylor P (2004) International measurement evaluation programme (IMEP); IMEP-14: Picturing the performance of analytical laboratories measuring trace elements in sediment. Accred Qual Assur 9:615–623

Papadopulos IS, Bredehoeft JD, Cooper HH (1973) On the analysis of slug test data. Water Resources Res 9:1087–1089

Pape H, Clauser C, Iffland J (1999) Permeability prediction based on fractal pore-space geometry. Geophys 64:1447–1460

Parkhurst DL (1995) Uer's guide to PHREEQC, a computer program for speciation, reaction-path, advective transport, and invese geochemical calculations. Water-Resources Investigations Report 95-4227, US Geological Survey Lakewood, USA

Parkhurst DL, Thorstenson DC, Plummer LN (1980) PHREEQE, a computer program for geochemical calculations. US Geological Survey, Lakewood, USA

Parkhurst DL, Kipp KL, Engesgaard P, Charlton SR (2002) PHAST – a computer program for simulating ground-water flow, solute transport, and multicomponent geochemical reactions U.S. Geological Survey Techniques and Methods 6-A8. USGS Boulder, USA

Peres AMM, Onur M, Reynolds AC (1989) A new analysis procedure for determining aquifer properties from slug test data. Water Resources Res 25:1591–1602

Peres AMM (1989) Analysis of slug and drillstem test. Dissertation, University of Tulsa, Order Number 9004161

Peterson EW, Lagus PL, Broce RD, Lie K (1981) In-situ permeability testing of rock salt. SANDIA; National Laboratories, SAND81-7073 Albuquerque, USA

Petzold H (1976) Spektrale Untersuchungsmethodik zur Klüftigkeitsbestimmung mit Ultraschall, Freiberger Forschungsheft C 327b, TU BA, Freiberg

Peyret R, Taylor DT (1985) Computational Methods for Fluid Flow. Springer, Heidelberg, 358 pp

Pickens JF, Gillham RW, Cameron DR (1979) Finite-element analysis of the transport of water and solutes in tile-drained soils. J Hydrol 40:243–264

Pitzer KS (1981) The treatment of ionic solutions over the entire miscible range. Ber Bunsenges Phys Chem 85:952

Pitzer KS, Simonsen JM (1986) Thermodynamics of multicomponent miscible ionic systems: theory and equations. J Phys Chem 90:3005

Plzák Z (2000) Are there two decks on the analytical chemistry boat? Accred Qual Assur 5:35–36

Poier V (1998): Pulse-Test. In: Handbuch zur Erkundung des Untergrundes von Deponien und Altlasten, Vol. 4-Geotechnik, Hydrogeologie; BGR (Bundesanstalt für Geowissenschaften und Rohstoffe), Springer, Heidelberg

Poincaré H (1908) Chance. in: Science et méthode. Flammarion Paris, France

Powell DR, Macdonald JR (1972) A rapidly convergent iterative method for the solution of the generalized nonlinear least squares problem. Computer J 15:148–155

Price G (1996) On practical metrology and chemical analysis: etalons and traceability systems. Accred Qual Assur 1:57–66

Price G (2000) Pragmatic philosophies of measurement in chemistry? Chemistry in Australia October 38–40

Price G (2001) On the communication of measurement results. Measurement 29:293–305

Price G (2002) An arrogance of technicians. Accred Qual Assur 7:77–78

Prichard E (1999) Basic skills of analytical chemistry: do we take too much granted? Accred Qual Assur 4:37–39

Pusch G, Schweitzer P, Gaminger O (1986) Stationäre und instationäre Gespermeabilitätsmessungen an niedrigpermeablen Gesteinen, Erdöl Erdgas Kohle 102:235–239

Quinn T (2004) The Metre Convention and world-wide comparability of measurement results. Accred Qual Assur 9:533–538

Quintessa Ltd (2002) AMBER v4.4. Quintessa Ltd, Henley-on-Thames, UK

Ramey HJ Jr, Argarwal RG, Martin I (1975) Analysis of slug test or dst flow period data. J Canad Petrol Technol 14:37–47

Reaction Design (2006) http://www.reactiondesign.com (last accessed August 2006)

Rechberger WH (2000) Fact finding beyond all reasonable doubt–Legal aspects. Fresenius J Anal Chem 368:557–560

Richter W, Güttler B (2003) A national traceability system for chemical measurement. Accred Qual Assur 8:448–453

Rietzschel I (1996) Porositäts- und Durchlässigkeitsmessungen an geringdurchlässigen Gesteinen im Vergleich zwischen herkömmlichen Verfahren und dem Ein- bzw. Zweikammerverfahren mit den Fluiden Gas und Flüssigkeit, Diploma Thesis TU BA Freiberg, FRG

Ringbom A (1958) The analyst and the inconstant constants. J Chem Educ 35:282–288

Ripley BD, Thompson M (1987) Regression techniques for the detection of analytical bias. Analyst 112:377–383

Riu J, Rius FX (1996) Assessing the accurancy of analytical methods using linear regression with errors in both axis. Anal Chem 68:1851–1857

Römpp (1999) Römpp Chemie Lexikon. (Falbe, Regitz eds.) Georg Thieme Verlag Stuttgart, FRG

Rosenfeld M (1998a): Slug- und Bail-Tests. In: Handbuch zur Erkundung des Untergrundes von Deponien und Altlasten, vol. 4 – Geotechnik, Hydrogeologie; BGR (Bundesanstalt für Geowissenschaften und Rohstoffe), Springer, Heidelberg

Rosenfeld M (1998b): Fluid-Logging. In: Handbuch zur Erkundung des Untergrundes von Deponien und Altlasten, vol. 4 – Geotechnik, Hydrogeologie; BGR (Bundesanstalt für Geowissenschaften und Rohstoffe), Springer, Heidelberg

Rossotti H (1978) The study of ionic equilibria: an introduction. Longman, New York

Rothfuchs T, Droste J, Wieczorek K (1998) Sealing of HLW disposal drifts and boreholes with crushed salt backfill. In: International conference on radioactive waste disposal Hamburg, FRG, pp 222–227

Rousseeuw PJ (1993) A resampling design for computing high-breakdown regression. Statistics Prob Lett 18:125–128

Rousseeuw PJ, van Zomeren BC (1990) Unmasking multivariate outliers and leverage points. J Am Stat 85:633–639

Roy T (1994) Bootstrap accuracy for nonlinear regression models. J Chemometrics 8:37–44

Rorabacher DB (1991) Statistical treatment for rejection of deviant values: Critical values of Dixon's "Q" parameter and related subrange ratios at the 95% confidence level. Anal Chem 63:139–146

Rubinstein RY (1981) Simulation and the Monte Carlo method. Wiley, Chichester

Rübel A (1999) Stofftransport in undurchlässigen Gesteinsschichten. Dissertation Universität Heidelberg, FRG

Ruth T (2004) A model for the evaluation of uncertainty in routine multi-element analysis. Accred Qual Assur 9:349–354

Rydén L, Migula P, Anderson M (eds) (2003) Environmental science. Baltic University Press Uppsala, 820 pp

Salsburg DS (1985) The religion of statistics as practiced in medical journals. Am Stat 39:220–223

Sachs W (1982) Untersuchungen zur Abhängigkeit der Gaspermeabilität poröser Sedimentgesteine von Hangenddruck, Porendruck und Fliessdruck, Diplomarbeit TU Clausthal Clausthal-Zellerfeld, FRG

Salverda AP, Dane JH (1993) An examination of the Guelph permeameter method for measuring the soil's hydraulic properties. Geoderma 57:405–421

Samper J, Juncosa R, Delgado J, Montenegro L (2000) CORE2D A code for non-isothermal water flow and reactive solute transport. User manual 2.0 ETS Ingenieros de Vaminos Universidad de La Coruna, E

Satterwhite FE (1959) Random balanced experimentation. Technometrics 1:111–137

Schneebeli G (1955) Experiences sur la limite de validate de la loi de darcy et l'apparition de la turbulence dans un écoulement de filtration. La Houille Blanche 10:141–149

Scheidegger AE (1961) General theory of dispersion in porous media. J Geophys Res 66:3273

Scheidegger AE (1974) The physics of flow through porous media, 2nd edn. University of Toronto Press, Toronto

Schildknecht F, Schneider W (1987) Über die Gültigkeit des Darcy-Gesetzes in bindigen Sedimenten bei kleinen hydraulischen Gradienten – Stand der wissenschaftlichen Diskussion, Geol Jb C48:3–21

Schittekat J, Minon JP, Manfroy P, Van Echelpoel E, Bontemps T (1998) Low radioactivity waste disposal site selection in Belgium. In: International Conference On Radioactive Waste Disposal, Hamburg, FRG, pp 163–167

Schneider HJ (1987) Durchlässigkeit von geklüftetem Fels – eine experimentelle Studie unter besonderer Berücksichtigung des Wasserabpressversuches, Mitteilungen zur Ingenieurgeologie und Hydrogeologie, Heft 26, Dissertation, RWTH Aachen, FRG

Schneider P, Lippmann-Pipke J, Schoenherr J (2006) Quality assurance of water balance simulations at the landfill cover test fields Bautzen, Nadelwitz, Germany. In: Dimensioning landfill surface liner systems using water balance models. Wiss Ber Hochschule Zittau, Görlitz 86:91–106 (ISBN 3-9811021-0-X)

Schramm J (2006) http://www.friedensblitz.de/sterne/glanzzeiten/Graff.html (last accessed September 2006)

Schreiner M, Kreysing K (1998): Probennahme und Sondierungen. In: Handbuch zur Erkundung des Untergrundes von Deponien und Altlasten, Vol. 4-Geotechnik, Hydrogeologie; BGR (Bundesanstalt für Geowissenschaften und Rohstoffe), Springer, Heidelberg

Schulze O (1998) Auflockerung und Permeabilität im Steinsalz, Hauskolloquium Endlagerung an der BGR, 23.06.1998

Schultze E (1972) Frequency and correlations of soil properties. Statistics and probability in civil engineering, Hong Kong International Conference, Hong Kong University Press, distributed by Oxford University Press, London

Schüßler R (2003) Moral im Zweifel. Mentis Paderborn, Germany

Schwartz LM (1980) Multiparameter models and statistical uncertainties. Anal Chim Acta 122:291–301

Schwarz R, Croise J, Lavanchy JM, Schlickenrieder L (2000) Verfahren der in-situ Bestimmung hydraulischer Parameter zur Erstellung hydrogeologischer Modelle. Internet publication

Serkiz SM, Allison JD, Perdue EM, Allen HE, Brown DS (1996) Correcting errors in the thermodynamic equilibrium speciation model MINTEQA2. Water Res 30:1930–1933

SHEMAT (2006) http://www.geophysik.rwth-aachen.de/html/software.htm. (last accessed September 2006)

Sigg L, Stumm W (1996) Aquatische Chemie. Teubner, Stuttgart, FRG

Shakespeare W (ca. 1596) The Merchant of Venice. Stratford-on-Avon, UK

Simunek J, van Genuchten MT, Suarez DL (1995) Modelling multiple solute transport in variably saturated soils in: Groundwater quality: remediation and protection (Kovar, Krasny eds) IAHS Publ 225, 311–318

Sivia DS (1996) Data analysis – a bayesian tutorial. Clarendon Press, Oxford, 189 pp

Smith AFM, Gelfand AE (1992) Bayesian statistics without tears: A sampling-resampling perspective. Am Stat 46:84–88

Smith RM, Martell AE (1989) Critical stability constants, vol. 6. Plenum Press, New York

Smith AFM, Roberts GO (1993) Bayesian computation via the Gibbs Sampler and related Markov chain Monte Carlo methods. J R Stat Soc 55:3–23

Spendley W (1969) Non-linear least squares fitting using a modified Simplex minimization method. In: Fletcher R (ed) Optimization. Symposium of the Institute of Mathematics and its Application. Academic Press, New York, pp 259–270

Spiers CJ, Peach CJ, Brzesowsky RH, Schutjens PMTM, Lietzenberg JL, Zwart HJ (1989) Long-term rheological and transport properties of dry and wet salt rocks, REPORT EUR 11848 EN Utrecht

Spitzer P (1996) Summary report: comparison of definitive methods for pH measurement. Metrologia 33:95–96

Spitzer P (1997) Summary report: comparison of primary standard measurement devices for pH. Metrologia 34:375–376

Spitzer P (2001) Traceable measurements of pH. Accred Qual Assur 6:55–60

Spitzer P, Meinrath G (2002) Importance of traceable pH measurements. Anal Bioanal Chem 374:765–766

Spitzer P, Werner B (2002) Improved reliability of pH measurements. Anal Bioanal Chem 372:787–795

Stanislaw H (1986) Tests of computer simulation validity: what do they measure? Simulation and Games 17:173–191

Steefel CI, Yabusaki SB (1996) OS3D, GIMRT software for modeling multi-component-multidimensional reactive transport. User manual and programmer's guide. REPORT PNL-11166 Pacific Northwest National Laboratory Richmond, USA

Stine R (1990) An introduction to the bootstrap methods. In: Modern methods of data analysis. Sage Publications Newbury Park, pp 325–373

Stoer J (1976) Einfürung in die Numerische Mathmatik Bd. 1. Springer, Heidelberg

Stormont JC, Howard CH, Daemen JJK (1991) In situ measurements of rock salt permeability changes due to nearby excavation. Sandia Report SAND90-3134, Albuquerque, USA

Student (W Gosset) (1908) The probable error of mean. Biometrika 6:1–25

Stumm W, Morgan JJ (1996) Aquatic chemistry, 3rd edn. Wiley, New York, 1022 p

Stumm W, Hohl H, Dalang F (1976) Interaction of metal ions with hydrous oxide surfaces Croat Chem Acta 48:491–504

Suckow A, Sonntag C (1993) The influence of salt on the noble gas thermometer. IAEA-SM-329, 94, IAEA Wien, A, pp 307–318

Swartzendruber D (1962) Non-Darcy flow behaviour in liquid-saturated porous media. J Geophys. Res 67:5205–5213

Tauzin E (1997) Re-analysis of gas tests performed in boreholes RB522, RB524, RB539 and RB540. BGR Peine, FRG

Tauzin E, Johns RT (1997) A new borehole simulator for well test analysis in low-permeability formations. IAMG'97 Proc Ann Conf Int Assoc Math Geol, Barcelona, E 819–824

Thompson M (1994) Statistics – the curse of the analytical classes. Analyst 119:127N

Thompson M, Fearn T (1996) What exactly is fitness for purpose in analytical measurement. Analyst 121:275–278

Thomson W (1891) Electrical units of measurement. Nature Series, Macmillan, Co. London, 1:80–134

Trappe H, Kraft T, Schweitzer C (1995) Neuronale Netzwerke zur Permeabilitätsbestimmung in Rotliegendsandsteinen. Erdöl–Erdgas–Kohle 111:159–162

Truesdell AH, Jones BF (1974) WATEQ, a computer program for calculating chemical equilibria of natural waters. US Geol Survey J Res 2:233–248

USGS (2006) http://wwwbrr.cr.usgs.gov/projects/GWC_coupled/phreeqc, (last accessed November 2006)

Van der Heijde P, Bachmat Y, Bredehoeft J, Andrews B, Holtz D, Sebastians S (1985) Groundwater management: the use of numerical models. AGU Water Res Monogr Vol. 5 Washington D. C., USA, 178 pp

van Genuchten MT (1980) A closed form equation for predicting the hydraulic conductivity of unsaturated soils. Soil Sci Soc Am J 44:892–898

Van Genuchten MT, Gray WG (1978) Analysis of some dispersion-corrected schemes for solution of the trnsport equation. Int J Num Meth Eng 12:387–404

van Nevel L, Taylor PDP, Örnemark U, Moody JR, Heumann KG, de Bièvre P (1998) The international measurement evaluation Programme (IMEP) IMEP-6: "Trace elements in water". Accred Qual Assur 3:56–68

VIM (1994) International vocabulary of basic and general terms in metrology, 2nd edn. DIN Deutsches Institut für Normung. Beuth Verlag, Berlin

Visser R (2004) Measurement uncertainty: practical problems encountered by accredited testing laboratories. Accred Qual Assur 9:717–723

Vitorge P (2005) documents under http//www.vitorge.name/pierre/insultes

Vogelsang J, Hädrich J (1998) Limits of detection, identification and determination: a statistical approach for practioners. Accred Qual Assur 3:242–255

von Weizsäcker CF (1981) Ein Blick auf Platon. Reclam, Stuttgart

von Weizsäcker CF (1990) Die Tragweite der Wissenschaft 5th ed. S. Hirzel, Stuttgart, 481 pp

References   321

Voss CI (1993) A perspective on groundwater modeling for performance assessment. In:
Proc US DOE Low-Level Radioactive Waste Management Conference. Phoenix, USA Dec
1-3, 1993
Voss CI (1996) A perspective on groundwater modeling for performance assessment. In:
Joint USGS NUREG workshop on research related to low-level radioactive waste disposal.
US Geological Survey Water Resources Investigations Report WRI 95-4015, 46-52
Voss CI (1998) Editor's message-groundwater modeling: Simply powerful. Hydrol J 6:A4-A6
Wagner S, Drees W, Hohenthanner CR (1998) H2-Diffusionskoeffizienten im Festgestein,
Freiberger Forschungshefte A849:285-313 TU Bergakademie, Freiberg
Wallick GC, Aronofsky JS (1954) Effect of gas slip on unsteady flow of gas through porous
media, experimental verification. Trans ATME, 201:322-324
Walsh MC (2000) Proving beyond all reasonable doubt - analytical aspects. Fresenius J Anal
Chem 368:553-556
Walcher W (1988) Measurement and the progress of knowledge in physics. In: Kramer (ed)
The art of measurement. VCH, Weinheim, FRG, pp 1-29
Walter AF, Friend EO, Blowes DW, Ptacek CJ, Molson JW (1994) Modelling of multicom-
ponent reactive transport in groundwater. 1. Model development and evaluation. Water
Resources Res 30:3137-3148
Walter F (1995) Untersuchungen zum Kompaktionsverhalten von Salzgrus als Versatzma-
terial für Endlagerbergwerke, GSF, Braunschweig, FRG
Weber JR (1994) Untersuchungen zur Permeabilitätsdilatanz kristalliner Gesteine unter
deviatorischer Belastung, Dissertation TU Clausthal, ISBN 3-930697-18-1
Walter F, Wallmüller R (1994) Beiträge zur Mechanik des Deckgebirges der Asse Südflanke,
REPORT GSF 2, 94 Braunschweig, FRG
Wehrens R, Putter H, Buydens LMC (2000) The bootstrap: a tutorial. Chemom Intell Lab
Syst 54:35-52
Wells HG (1898) http://www.fourmilab.ch/etexts, www.warworlds/warw.html (last accessed
September 2006)
Westall JC (1982) FITEQL: a computer program for detemination of equilibrium constants
from experimental data. Report 82-01. Oregon State University Corvallis, USA
Westall J, Hohl H (1980) A comparison of electrostatic models for the oxide, solution
interface. Adv Coll Interface Sci 12:265-294
Wielgosz RI (2002) International comparability of chemical measurement results. Anal
Bioanal Chem 374:767-771
Wieczorek K (1996) Neue Messdaten zur Permeabilität in der Auflockerungszone. Gemein-
samer GRS, PTE-Workshop: Erzeugung und Verbleib von Gasen in Endlagern für ra-
dioaktive Abfälle, Braunschweig, FRG
Wilkinson JH (1965) The algebraic eigenvalue problem. Clarendon Press, Oxford, 662 pp
Williams, WH (1978) A sampler on sampling. Wiley, Chichester
Wittgenstein L (1921) Tractatus logico-philosphicus. Reprint (1963) Edition Suhrkamp,
Frankfurt, FRG
Wittke B (1999) Permeabilität von Steinsalz - Theorie und Experiment, Text und Anlagen,
WBI-PRINT 3, Geotechnik in Forschung und Praxis, Verlag Glückauf GmbH, Essen, FRG
Wold S (1993) Discussion of a paper by Frank and Friedman. Technometrics 35:136-139
Woolery TJ (1992) EQ3, 6 a software package for geochemical modeling of aqueous systems.
UCRL-MA-110662. Lawrence Livermore National Laboratory Berkeley, USA
Yaramanci Y, Lange G, Knödel K (2000) Ermittlung von Wassergehalt, Porosität, und Per-
meabilität mit dem neuen geophysikalischen Verfahren der Oberflächen NMR, Internet-
Veröffentlichung der TU Berlin und der BGR, Berlin

Yaramanci U, Lange G, Knödel K (1998a) Effects of regularisation in the inversion of surface NMR measurements. Proceedings of 60th Conference of European Association of Geoscientists, EAGE, Zeist, NL

Yaramanci U, Lange G, Knödel K (1998b) Effects of regularisation in the inversion of surface NMR measurements. Proc 60th Conf European Assoc Geosci, EAGE, Zeist, NL

Yucca (2002) Yucca Mountain science and engineering report. Technical information supporting site recommendation consideration. http://www.ocrwm.doe.gov /documents/ser_b/index.htm (last accessed September 2006)

Zenner MA (1998) Zur Analyse von Slug Tests in geklüfteten Gesteinen: Indikationen auf pseudoplastische Kluftströmungsprozesse, In: Eigenschaften von Gesteinen und Böden, Freiberger Forschungsheft A 849, Technische Universität Bergakademie Freiberg, FRG

Zschunke A (1998) Global comparability of analytical results. Accred Qual Assur 3:393–397

Subject Index

Printing: Krips bv, Meppel
Binding: Stürtz, Würzburg